Tourism in Emerging Economies

Wei-Ta Fang

Tourism in Emerging Economies

The Way We Green, Sustainable, and Healthy

Springer

Wei-Ta Fang
Graduate Institute of Environmental Education
National Taiwan Normal University
Taipei, Taiwan

ISBN 978-981-15-2462-2 ISBN 978-981-15-2463-9 (eBook)
https://doi.org/10.1007/978-981-15-2463-9

This Springer imprint is published by the registered company Springer Nature Singapore Pte Ltd.
The registered company address is: 152 Beach Road, #21-01/04 Gateway East, Singapore 189721, Singapore

Preface

Tourism is a booming international trade, a vibrant economic service sector that provides employments in travel, tourism and hospitality sectors, which are one of the most thrilling and pleasing encounters of meeting people from diverse cultural background. Tourism in Emerging Economies is a book that is here to bridge the gap on the vital issues of tourism businesses and practices and destinations. It's here to provide a comprehensive guide for students, professors, entrepreneurs, and travel and tourism organizations to tourism and its related disciplines on the trends, practices, and case studies covering from historical tourism to contemporary tourism with forecast on future tourism.

Today, tourism is a global business that supports many major economy service sectors providing various developments, social awareness and cross-cultural encounters in destinations. This book is written to assist the tourism students, industry staff, academicians, and researchers in the travel and tourism fraternity. It will serve as a comprehensive guide for all English speakers, who already read and write English. This book takes readers through process of comprehending, acquiring, analyzing the tourism developments and new thinking about trends of global tourism industry evolutions and growing complexity. Henceforth, this book will provide strategic ideas on various tourism developments that are applicable by government, private agencies, voluntary sectors, ancillary service providers, and entrepreneurs by being a treasured databank for those businesses, policy makers and entrepreneurs concern with this fragile industry, thus delivers an all-inclusive analysis of existing tourism development.

Tourism in Emerging Economies is a book that covers various topics and case studies from the Asian and African perspectives. Each chapter of this book includes intensive researches on tourism development and operations across Africa and Asia written by young pioneers in tourism education, people with both working experiences, travel experiences, and broader philosophies about the global tourism industry's practicing and operating in destinations. The content of this book shall give students, industry staff and tourism entrepreneurs' reward of being able to apprehend numerous facts of diverse tourism matters which reflects the ironic practice from rural communities of Africa to well-suited pristine environment of

Asia where your travel discoveries touch your heart. It is hoped that this book, Tourism in Emerging Economies, will be able to be used in colleges and universities for academic purposes, as well as a source of reference to graduate students. The book is expected to be an easy reading guide and understandable by carefully following the background information in each chapter. The book is diverse in content, format, and topic. It addresses different forms of global tourism. This is one of the cooperations and touch things by colleagues involving great stakeholders in tourism. Therefore, the presented perspectives are beyond regular textbooks we have read from Western cultures in the libraries.

Tourism in Emerging Economies by far is one of the best, original and authentic conscious written books of the last decade. It produces steady reflection treasures of the world tourism industry. Works cover conscious historical tourism operation with perspectives about present tourism has given us an insight into future predictions of the industry.

Tourism industry provides job opportunities, promotions of cultural authenticity, and environmental education and protections. This book, Tourism in Emerging Economies, is here to guide you along your road to a successful business. It covers everything from the history of tourism to the present well-structured operations in emerging and developing economies. The content of chapters in this book discusses crucial issues confronting the travel and tourism industry, with achievable outlines, strategies plan, evaluations from some general theories, practices and application of three dimensions socially, economically and environmentally with management structure of maximizing the cultural, social and ecological diversity of destinations and enhance the tourism experience. According to Sustainable Development Goals (SDGs) 8, 12 and 14, tourism highlights our potentials to contribute to economic growth, social inclusion, and environmental preservation (World Tourism Organization and Organization of American States 2018). I hope that this book can be an answer to the growing concepts and doubt about SDGs concepts, criticism and debates on tourism for the third generation.

Taipei, Taiwan Wei-Ta Fang, Ph.D.

Reference

World Tourism Organization and Organization of American States (2018) Tourism and the sustainable development goals—Good practices in the Americas. UNWTO, Madrid, Spain

Acknowledgements

The contributors of this book, *Tourism in Emerging Economies*, would like to acknowledge their sincere appreciation, especially to the following people for their wonderful contributions and enthusiastic supports in making this edition a successful project: Prof. and Dean Hsin-Wen Chang, College of Tourism at Chung Hua University; Prof. and Chair Lin-li Tai, Department of Tourism and Leisure, Chung Hua University; Prof. Chien-Shing Lee, Department of Athletic Performance, National Taiwan Normal University; Prof. Arba'at Hassan, Open University Malaysia, Kota Marudu, Sabah; Mr. Omar A. J. Saho, Mr. Raymond Mendy, and the publisher (Springer-Verlag GmbH) in helping this edition published. Special thanks for the precious and historical photos courtesy provided by Prof. Arba'at Hassan, Mr. Wei-Kuang Fang, and Mr. Yi-Te Chiang, my close brother. Because of their generosities, this book is appropriate for a new guide, as well as very helpful to read with beautiful photos and illustrations. Their great efforts are for coordinating and putting the inputs on this entire publication, thus making this book in your hands. All of these elites continued to be pioneers of endorsing the development of the global travel and tourism industry. It is sincerely instrumental in having the supports of these people and appreciative of their fantastic support in making this edition successful. I would like to acknowledge that this book was partially supported by Executive Detector, Mr. Jay Lee, National Geographic Society—Asia, the members of the Graduate Institute of Environmental Education, National Taiwan Normal University (NTNU), and the staffs of National Geographic Foundation for Science and Exploration Asia for their contributions to the manuscript. This book was also subsidized by the National Taiwan Normal University (NTNU) and was funded by the Ministry of Science and Technology under project grants. This book was also partly supported by citizen science in training and outreach of the National Geographic Foundation for Science and Exploration Asia, Transaction Reference PYMTE-142998 to Wei-Ta Fang, General Secretary, Taiwan Wetland Society.

Contents

About the Author

Wei-Ta Fang was born in Kaohsiung, Taiwan on February 14, 1966. He received B.A. in Land Economics and Administration from National Taipei University, Taipei, Taiwan in 1989. He received his first master degree in Environmental Planning (MEP) from Arizona State University, USA in 1994, and the second master degree in Landscape Architecture in Design studies (MDes.S.) from Graduate School of Design, Harvard University, USA in 2001. He obtained a Ph.D. from the Department of Ecosystem Science and Management, Texas A&M University in 2005. He served as a specialist in Taipei Land Management Bureau from 1991 to 1992 and a senior specialist in-charge on Environmental Impact Assessment (EIA) and Environmental Education in the Taiwan EPA Headquarters from 1994 to 2006. He is currently serving as a Distinguished Professor as the Director of the Graduate Institute of Environmental Education, National Taiwan Normal University and the President of Society of Wetland Scientists Asia Chapter. He lives in Taipei City with his wife, Chia-Ying Ho and two sons, Cheng-Jun (June) and Cheng-Shun (Sam). His contact e-mail is wawaf@hotmail.com.

Abbreviations

ABTA	Association of the British Travel Agency
ACCA	The Association for Cultural Centers in the Amboseli Ecosystem
AITO	The Associate of Independent Tour Operators
APPTDC	African Pro-Poor Tourism Development Centre
ASSET	The Association of Small Scale Enterprises in Tourism
ASTA	American Society of Travel Agents
ATA	Africa Travel Association
BBC	British Broadcasting Cooperation
CBT	Community Based Tourism
CNN	Cable News Network
CNTA	China National Tourism Administration
CRS	Central Reservation System
CSR	Corporate Social Responsibility
CVBs	Convention and Visitors Bureaus
DFID	Department for International Development
ESE	Economy, Social life, and Environment
ESEE	Environmental, Socio-Cultural, Economically and Environmentally
ETC	European Travel Commission
FHB	Farm Holiday Bureau
FTO	Federation of Tour Operators
GCET	Global Code of Ethics for Tourism
GDP	Gross Domestic Products
GDS	Global Distribution Systems
GRTP	Gambian Responsible Tourism Partnership
GTB	Gambia Tourism Board
ICCA	International Congress and Convention Association
ICOMOS	International Day on Monuments and Sites
ICRT	International Centre for Responsible Tourism
IDA	International Development Association
IDAF	Ireland Department of Agriculture and Food

IFC	International Finance Corporation
INTECOL	The International Association for Ecology
KART	Kartong Association of Responsible Tourism
LCD	Liquid Crystal Display
LDC	Least Developed Countries
LDCs	Low Developing Countries
MIC	Middle Income Countries
MICE	Meeting, Incentive Travel, Convention, and Exhibition
MRT	Mass Rapid Transit
NCRC	Nature Conservation Research Centre
NDT	National Department of Tourism
NEA	National Environment Agency
NGOs	Non-government Organizations
NTO	National Tourism Offices
ONITS	Organisation Nationale des Professionnels Intégrés du Tourisme du Sénégal
RBT	Recreation-based tourism
SDGs	Sustainable Development Goals
SMEs	Small-to-medium sized enterprises
SWOT	Strengths, Weaknesses, Opportunities, and Threats
T&T	Travel and Tourism
TDA	Tourism Development Authority
TPA	Travel Promotion Act
TTB	Tanzania Tourist Board
UFI	The Union of International Association
UK VSO	United Kingdom Volunteer Services
UK	United Kingdom
UN	United Nations
UNESCO	United Nations Educational, Scientific and Cultural Organization
UNWTO	United Nations World Tourism Organization
USAID	United States Agency for International Development
VAS	Value added services
VDC	Village Development Committee
VFR	Visiting Friends and Relatives
VSO	Voluntary Service Overseas
WIA	Women in Agriculture
WTO	World Tourism Organization
WTTC	World Travel and Tourism Council

Chapter 1
An Orientation

1.1 Introduction

Tourism is a diversity industry that helps to improve the gross domestic products (GDP) of so many countries (Pyo et al. 1991), especially that of the developing countries. Based on the following factors, like economical social and environmental, the trend of tourism today in most developing countries, are coming up with good innovative products. These can be service as a pull factor to motivate in more tourists to travel to their destination (Nella and Christou 2016).

In addition, tourism is one of the strongest drivers of world trade and prosperity (Dwyer 2015). Poverty alleviation is one of the greatest global challenges. Despite turbulent times for the world's economy, these basic facts are unlikely to change. Focusing the wealth creating power of tourism on people, most needs remain an immense task and opportunity.

Secondly, considering further factors of tourism and challenges, these will face at the beginning of the 21st century. The prognosis is very optimistic as suggested in a publicized rapport. The rapport talked about quick development and stated that during the next twenty years, tourism will be one of the fastest growing departments in the world's economy (Lanza and Pigliaru 2000). However, tourism is far from the end of its development, as for now, it consists of a small percentage of the world's citizens. The main hazards in development are problems with political situation in the world, especially conflicts within the Balkans, instability of the Arab world, Libya, and disorganization between authorities and conflict in the former territory of Soviet Empire. Slightly, less hazardous factors deal with economics like recession and the increase of gas prices. Nevertheless, the outlook on tourist development remains optimistic. The quantitative development of tourism is accompanied by multi aspects including qualitative and structural transformations.

W.-T. Fang, *Tourism in Emerging Economies*,
https://doi.org/10.1007/978-981-15-2463-9_1

1.1.1 What Is the Definition of Tourism

Tourism is an activity done by an individual or a group of individuals. It leads to a movement of people from one place to another. They move from a country to another to perform a specific task. Sometimes, they move to visit a place or several places in the purpose of getting an entertainment. This will lead to an awareness of other civilizations and cultures, as well as increasing the knowledge from the countries, cultures, and history they experience. Tourism has direct impacts on the national revenue for all touristic destinations. It creates work opportunities, industries, and several investments to serve and raise nations performance and cultures, also distributes their history, civilization, and traditions.

According to sources, tourism is important, and in some cases, vital for many countries. It was recognized in the Manila Declaration on World Tourism of 1980 as "*an activity essential to the life of nations because of its direct effects on the social, cultural, educational and economic sectors of national societies and on their international relations.*" Tourism brings in large amounts of income in payment for services available. It accounts for 30% of the world's exports of services, and 6% of overall exports of goods and services. It also creates opportunities for employment in the service sector of the economy, associated with tourisms. These service industries include transportation services like airlines, cruise ships, and taxicabs. Hospitality services and accommodations include hotels and many other facilities within the tourism industry. They are also supporting in the tourism development economic.

1.1.2 United Nation World Tourism Organization Definition About Tourism

The World Tourism Organization (UNWTO) is the United Nations agency responsible for the promotion of responsible, sustainable and universally accessible tourism. As the leading international organization in the field of tourism, UNWTO promotes tourism as a driver of economic growth, inclusive development and environmental sustainability. It also offers leadership and support to the sector in advancing knowledge and tourism policies worldwide.

The UNWTO encourages the implementation of Global Code of Ethics for Tourism, to maximize tourism's socio-economic contribution while minimizing its possible negative impacts. It is committed to promoting tourism as an instrument in achieving the United Nations Millennium Development Goals (MDGs) and Sustainable Development Goals (SDGs), geared towards reducing poverty and fostering sustainable development (Gasper et al. 2019).

UNWTO generates market knowledge, promotes competitive, sustainable tourism policies and instruments, fosters tourism education and training, and works to make tourism an effective tool for development through technical assistance projects in over 100 countries around the globe. UNWTO's membership includes 158 countries, six

(6) associate members and over 500 affiliate members representing private sectors, educational institutions, tourism associations and local tourism authorities in 2019.

The United Nations have taken the concepts of tourism as an image of holiday making. According to definition, tourism comprises activities of a person traveling and staying in a place outside his/her normal environment, moves to other place of his/her choices, not more than one year, either on business, leisure visiting friends and relatives (VFR) and other purposes.

1.1.3 Four Different Type of Travels

- **International tourism**
  - **Inbound tourism** is a visit to country by non residents.
  - **Outbound tourism** is a visit by the resident of a country to another country.
- **Domestic tourism** is a visitation by a resident of a country take holidays with their own country.
- **Internal tourism** refers to the visit by residents of a country and non residents of the country.
- **National tourism** refers to internal tourism and out bound tourism the resident market for airlines travel agents and other suppliers.

1.2 The Relationship Between Tourism, Culture, and Local Economics

Since 1956, International Finance Corporation (IFC) has invested over US$2.5 billion (including syndications) in 250 hotel projects in more than 83 countries; more than half of these investments (52%) are in International Development Association (IDA) countries, which are some of the poorest countries in the world (Abdo n.d.).

Hotels play a critical role in developments as they catalyze tourism and business infrastructure. In addition, hotels generate jobs, grow tax revenues, increase foreign exchange earnings, and provide better opportunities for small businesses. This ability will facilitate local, regional, and national economic growth, thereby will help reduce poverty. With this reason, the IFC is deeply committed to the hotel and tourism industry. Their commitment is to expand to environmental protection and cultural preservation with investments in hotels that complement unique natural habitats and enhance the attractiveness of historically significant sites. At the same time, it will bring great income to the local economic to the development of local communities.

Furthermore, tourism has assumed a vital role in the development of destinations around the world. In most cases, culture is a major asset for tourism development, as well as one of the major beneficiaries of this development. Culture is a major factor in the attractiveness of most destinations not only in terms of tourism but also in attracting residents and inward investment.

In this section of the book, the growing relationship between tourism and culture, and the way in which they have together become major drivers of regional attractiveness and competitiveness, will be examined.

During most of the 20th Century, tourism and culture were viewed as separate aspects of a destination. Cultural resources were seen as part of the cultural heritage of destinations, largely related to education of local population, and the underpinning of local or national cultural identities. Tourism, on the other hand, was largely viewed as a leisure-related activity separated from everyday life and culture of the local population. This gradually changed towards the end of the century, as the role of cultural assets in attracting tourists and distinguishing detestations from one another, become more obvious. In particular, from the 1980s onwards "cultural tourism" became viewed as a major source of economic development for many destinations (Wood 1984).

Cultural tourism is particularly attractive because of the benefit; it can deliver to local communities. According to the National Trust for Historic Preservation in the U.S., these benefits include several scopes as follows.

1.2.1 The Benefits of Culture Within Local Communities

- Building community pride in heritage.
- Increasing awareness of the site or area's significance.
- Preserving local traditions and culture.
- Creating opportunities for partnerships.
- Creating jobs and businesses.
- Attracting visitors interested in history and preservation.
- Generating local investment in historic resources.
- Increasing tax revenues.

Example, in Kenya, shows the relationship between tourism and cultural has played an essential role in the development of the local economy. The Maasai people earn money from these special homesteads where tourists come to visit, hear and experience Maasai culture. Each *manyatta* has a central market where people sell their craft products.

The cultural *manyatta* is a co-operative activity. On the arrival, each person of the tourists, pays an entrance fee. During entering, the tourists are welcomed with song and dance by Maasai women and children. A resident guide will then show them around the manyatta, there are demonstrations of Maasai life and towards the end of the visit the warriors stage a dance, which the visitors are permitted to join. The Association for Cultural Centers in the Amboseli Ecosystem (ACCA) was established to generate ideas and implement decisions that were favorable to the local community with a view to improving their incomes accrued from tourism activities (Richards

2009). While enhancing the dignity of Maasai aboriginal people and protecting the ecological integrity of the Amboseli ecosystem, most of these activities are taking place within the Amboseli National Park to help in promoting the Maasai community and its cultures.

1.2.2 Tourism and Poverty Reduction

Tourism has been one of the vital tools in reducing poverty around the world today. The latest research from the World Travel and Tourism Council (WTTC) suggests that one (1) in 10 jobs of the travel and tourism industry will generate 324 million direct and indirect jobs worldwide by 2021. In 2010, international visitors spent more than $900 billion on goods and services annually, of which 37% were spent in emerging market countries. Tourism-related inflows are often the primary source of foreign exchange in many developing countries (and the main source of foreign exchange for 47 of the world's 50 Least Developed Countries (LDC). No other sector spreads wealth and jobs across poor countries in the same way as tourism.

In International Development Association (IDA) eligible countries, the development of high-end hotels sends a positive signal to foreign investors. Hotels create critical business infrastructure and venues for international conferences. In fact, in many IDA countries, tourism is one of the first sectors to be developed and is a catalyst for the development of other sectors. In some destinations, for every US$4 spent by a tourist, US$1 reaches the poor. High-end hotels in particular create an average of 1.5 to 3 jobs per hotel room and can generate substantial tax revenue. However, in Middle Income Countries (MIC), there is often a need for two and 3-star hotels, as an emerging middle class begins to travel. IFC also supports integrated mixed-used developments, which include hotels along with commercial and retail complexes that enhance prime business infrastructure both for international and local business activities, and spur economic growth.

Private financial markets are often unwilling to fund hotel investments in developing countries on account of risk factors, such as, political or economic instability, seasonality, unstable cash flows, capital intensive nature of developments, etc. The industry provides some of the most flexible employment and working practices found across the entire economy. Hotels and catering, for example, typically employ more women, ethnic minorities, older people and part-time workers than almost any other sector, a diversity reflecting the highly flexible nature of many jobs. Jobs can often be carried out in many different locations away from the conventional workplace, such as at home. For women, this can play an important part in promoting gender equality empowering women at all levels of society.

For example, Travel & Tourism provides a vital source of income and employment in many developing countries around the world. As an example, Travel & Tourism is directly or indirectly supports 76% of all jobs in Aruba, 31% in Belize and 27% in Namibia. Without this economic lifeline, millions more people would live in poverty in these countries, or be forced to leave their home countries in search of work. In

Gambia, it was expected that in 2007, the Gambian tourism industry will create nearly $134 million US dollars of total economic activity in demand. It is predicted that in the same year, it real terms will expand by 3.7% and to continue growing annually after that at a rate of about 3.8%. It is further predicted to account for 17.6% of total Gross Domestic Product (GDP) and 14.5% of the employment total i.e. 85,000 jobs or one in every seven (7) jobs.

The development of tourism is very important for strengthening the local entrepreneurship. This is because tourism related economic activities are formatting impacts for better living standards in the vicinities. Thus, it is determined that bigger intensity of tourism economic impact and its territorial distribution they are appropriate for rural and peripheral areas and many economic enterprises are satisfying the needs of locals and visitors in touristy vicinities.

1.2.3 Tourism Depended on Economic Growth

Tourism has been criticized for having negative impacts on many of the destinations that tourists visit. There are considered to be three principal impacts on economic, environmental, and socio-cultural.

Shift in Emerging and Developing Economies Impacts could help us finds out some negative aspects that most economic consequences of tourism are beneficial. These benefits include the effects of price and income elasticity, and also the economic consequences of tourism spending, including the generation of foreign exchange.

This process stimulates the local economy through multiplier effect, which can be direct, indirect, and induced (Fig. 1.1). Although the economic effects of tourism are usually held to outweigh tourisms economic benefits, the negatives can be significant. These negatives relate particularly to a likely increase in demand for imported goods once tourists begin to appear, revenue leakages out of the economy, over-dependence on tourism, and land value inflation (Harcombe 1999).

The study concludes that the economic benefits of tourism normally outweigh what negative features there may be. However, it also acknowledges that the situation is much less clear-cut when considering environmental and socio-cultural issues as opposed to economic ones.

Secondly, tourism being the number one industry in the world plays an essential role in development of the economic. It created a lot of benefits to people in the tourism industry both directly and indirect.

During direct and indirect round of expenditure, money will be paid to local residents in the form of wages, salaries, rent, interest, and dividends; and also to local businesses for routine services. Some of this expenditure (called the induced multiplier) generates yet more rounds of economic activity—by being spent on local goods and services (Harcombe 1999). It is only when all three levels of impact (i.e., direct, indirect, induced) are assessed that the full nature of this particular effect of tourism can be identified (Lemma 2014).

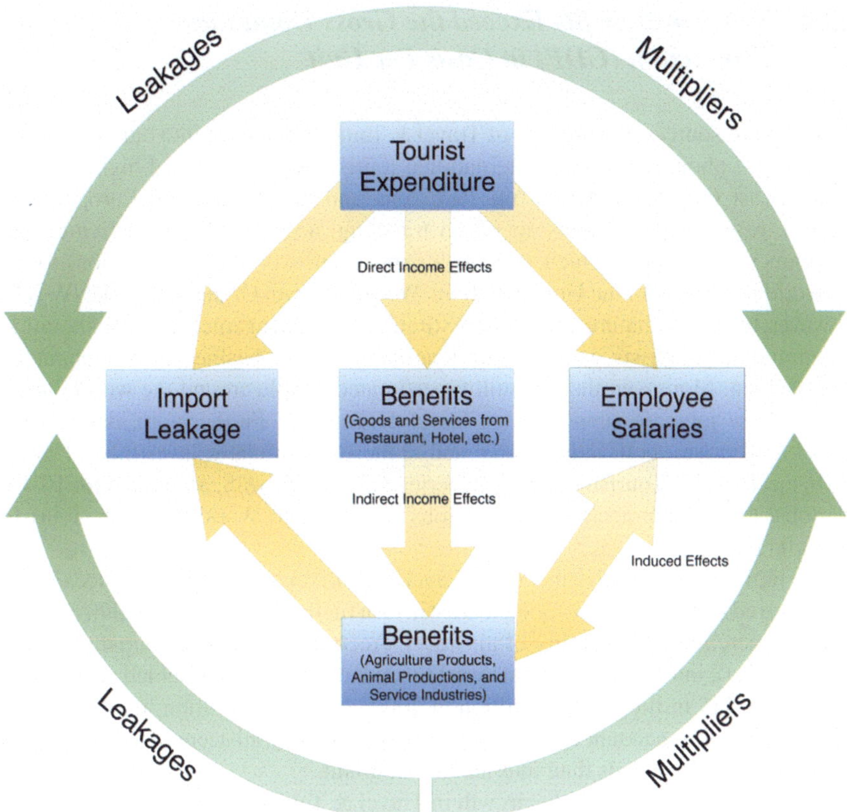

Fig. 1.1 Tourism has direct impacts on the national revenue for all touristic destinations. For example, service industries include transportation services like airlines, cruise ships, and taxi-cabs. All revenues create multiplier effects which mean an injection of extra income leads to more spending, which induces more income. Therefore, the increase in final income arises from new injection of spending on the rest of the economy to create more multiplier effects. (Illustrated by Wei-Ta Fang)

The economic effects of visitors' presence at destinations arise from the fact that travelers and tourists spend their money on a wide variety of goods and services. This expenditure can be seen as an injection of financial resources into the host economy, thereby creating new levels of consumer demand that as a result could help in developing the locals within the host communities.

1.2.4 *Tourism Benefits Exceed the Gross Domestic Production (GDP) in Costs Per Unit*

According to sources, the impacts of Travel & Tourism run deep into our economies. It is not just about the money. International tourists were accounted highest in the Middle East (up 10%), Africa (up 7%), Asia and the Pacific, and Europe (both up 6%), respectively. Visitors spend on traveling, accommodation, activities and souvenirs visitor exports which, in 2020, is expected to exceed US$1.4 trillion around the world according to the United Nations World Tourism Organization (UNWTO). However, by its very nature, the industry stimulates the engagement and collaboration of communities, tourists, governments, local suppliers and businesses throughout the supply chain. Moreover, the 319 million estimated people around the world during 2019, whose jobs are supported by Travel & Tourism. Furthermore, demand for Travel & Tourism, both international and domestic, stimulates investment. During 2018, the Travel & Tourism industry experienced 3.9% growth, accounted for 10.4% of global GDP, supporting one in ten jobs according to World Travel & Tourism Council (WTTC).

The rapid rise in global demand for Travel & Tourism over the past few decade has been spurred by the rise in living standards, in turn fuelled by growing wealth, coupled with the increased affordability of air travel. According to World Tourism Organization (WTO), international arrivals worldwide have more than doubled since 1990, rising from 435 million to 675 million in 2000, and to 940 million in 2010. These figures do not take account of domestic tourism, which by all accounts, generates up to ten times more arrivals than international tourism.

Between 2000 and 2010, the growth in Travel & Tourism's direct contribution to global GDP was volatile, with economic uncertainties, diseases, and terrorist attacks all having an impact. The decade between 2000 and 2010 recorded a growth of 9.7% overall, with Travel & Tourism's direct contribution to GDP ending the period at US$1,770 billion in 2010 (based on 2011 prices). During 2018, the Travel & Tourism industry still experienced 3.9% growth, compared to the global economy in slow growth rate (3.2%) according to World Travel & Tourism Council (WTTC). The direct contribution of Travel and Tourism (T&T) to Gross Domestic Product (GDP) will be expected to grow by 3.6% to USD4,065 billion by 2029 (Manzo 2019).

Regionally, the balance of growth was uneven according to 2019 Travel & Tourism Economic Impact (Manzo 2019). South Asia (7.2%), North East Asia (6.6%), South East Asia (6.2%), and Africa (5.6%) saw Travel & Tourism's direct contribution to Gross Domestic Product (GDP) rose by increased trends respectively, while other regions enjoyed increases slowly. Europe, meanwhile, suffered a slow increased percent (%) in Travel & Tourism GDP growth (3.1%) over the same period, and the North America, the world's largest market in monetary terms, was 2.1% (Travel & Tourism GDP growth) in regional performance (Manzo 2019).

In Africa, the direct contribution made by Travel & Tourism (T&T) to GDP was around US$194 billion in 2018 compared to US$77 billion in 2011. Adding the indirect and induced impacts would mean a total economic contribution of Travel &

Tourism in the region of some US 319 billion in 2018 compared to US$171 billion in 2011 in GDP and one-fifth jobs of all global jobs. Leading country contributors to these totals including Turkey (15%), China (7.3%), India (6.7%), and France (4.1%) in 2018 T&T contribution to GDP in growth rate (%), respectively (Manzo 2019).

Finally, different stakeholders see the impact of tourism in different perspectives. Economists generally see tourism as route to macro-economic growth, and particularly a means of generating foreign exchange. For the private sector, tourism is a commercial activity, so the main concerns are product development, competitiveness, and commercial returns. Many conservationists now see tourism as a form of sustainable used of wild resources, and hence as a way to enhance incentives for conservation. For rural people and the development Non-government Organizations (NGOs) that support them, tourism is one component of rural development.

According to sources, there is no doubt that if properly developed and managed, tourism can be a vital cog of the economies of many countries. It can also play a major role in poverty elevation, particularly in some countries, due to its large multiplier effect. This economic multiplier effect can be as large as 1.43 times from indirect impacts from local tourism promotion (Fleischer and Freeman 1997).

1.3 Recent Development in Tourism

The developments of the tourism industry have paved a great part in the tourisms industry. In contrast, tourism is becoming of increasing importance for economic growth and has led to wide research advances in several areas of the industry. The economic importance of tourism has caused governments and industries to en-gage in aggressive and intense promotional activities in order to develop and increase the dynamics and quality of the sector. It is also important that tourism be compatible with sustainable regional development and considers issues that involve information society.

Secondly, the recent development of tourism has really plays a vital role in the development of the following factors. These include:

- Air travel
- Road travel
- E-ticketing
- Sea travel
- Rail modernization
- Entertainment venues
- Accommodation etc.
- Meeting, Incentive, Convention, Exhibition/Event (MICE) industry.

All of these factors that have being stated play a unique role in the development of tourism in Emerging and Developing Economies. This also helps in promoting the development of the tourism industry economically, socially, and environmentally within the developed and developing countries at large. With the advanced development of tourism most of the developed and develop countries have come up with a new form for tourism to attract more tourism into their destinations. This form of tourism will be disused in the following chapter of the book. The recent development of tourism in emerging economy has played an essential role in the advancement of tourism economy. This leads into the following factors to help improve the living status of local people within the destinations. It has also help improve the GDP for developing destinations.

- Tourism destinations destination are becoming more creative to come up with new products to target in more customers into the destinations.
- The new development of tourism also has helped in creating jobs for people within their destinations both directly and directly.
- The recent development of tourism has also served as a pull factor to motivate more tourists to go on for holidays.

Inclusion, tourism in many developing and least developed countries is the most viable and sustainable economic development option, including in some countries, relating to the main source of foreign exchange earnings. Part of this income trickles down to different groups of the society and, if tourism is managed with a strong focus on poverty alleviation, it can directly benefit the poorer groups through employment of local people in tourism enterprises, goods and services provided to tourists, or the running of small and community-based enterprises, etc. This will have positive impacts on reducing their poverty levels.

Currently, many developing countries regard tourism as an important and integral part of their economic development strategies (Sinclair 1998). In many developing societies, tourism is perceived as a panacea for their fragile economies that is characterized by a scarcity of development resources, such as finance and expertise. These resources are needed to increase economic surplus, without which these countries would be forced to rely solely on international aid to support development efforts. Therefore, the well-recognized benefits of tourism are the usual reasons advanced for governments' support for tourism. The benefits are usually felt at two levels: macro or national level and micro or sub national or local level. At the first level, tourism is expected to foster economic growth through foreign exchange earnings and an increase in state revenue. The second level is an improvement in the people's well-being in the areas of job creation, revenue income distribution and balanced regional development. In this respect, tourism is described as an industry although it has no single production characteristics or defined operational parameters. Travel and Tourism (T&T) is also multi-faceted. Its economic dimension cannot occur without inputs of a social, cultural and environmental nature within a destination.

Tourism is more than an economic activity. It is, in essence, a massive interaction of people, demanding a wide range of services, facilities, and inputs that generate opportunities and challenges to host countries. For this reason, it is necessary to manage the growth of the tourism sector and to have clear guidelines to ensure that growth is compatible with national and sectorial objectives through emerging of tourism economics.

1.3.1 National Tourism

Tourism is one of the fastest growing sectors in global economy. Developing countries are attempting to cash in on this expanding industry in an attempt to boost foreign investment and financial reserves. While conceding that, the uncontrolled growth of this industry can result in serious environmental and social problems.

The growing competition between countries has fostered considerably over the recent decades that national tourism agencies have intensified promotional activity to adapt international promotional campaigns of the National Tourism Offices (NTO) to new demands of current market.

National tourism has played a significant role in the development of tourism in a destination. With the help of national tourism today, many tourism destinations have improved their products and services that offered to the people.

Government plays essential roles in national tourism by providing bodies that could play and active roles in the promotion of tourism in a destination. Government identified tourism as one of the key contributing sectors to the medium-term strategic priorities of growing the economy and creating decent work a good number of people within the tourism industry.

The role of a local government is to promote social, economic, environmental and cultural well–being of their communities and its involvements in tourism must be related to that. The government has the mandate to come up with their own tourism plans which sets out the priorities over the medium to longer term and how the local authority intends to contribute to community well–being. The plan must set out the community outcomes as a result of tourism development, how these have identified and how the local authority will contribute to these. The Local Government Tourism Strategy may contribute to economic development strategies and/or regional as well as national tourism strategies. They must serve as the basis of the role of government (Terpening 2015).

1.3.2 International Tourism

International tourism is when people travel globally outside of their region and home country. This means that they are in a temporary movement from their resident's place to another place for different reasons for a recreational, health, business or some other reason.

International tourism is spread out around the world quite unequally. First of all, it was explained by different rates of social and economic development of countries and regions. The greatest development of international tourism accounts for western European countries. The share of this region is above 28.8% of the world total tourist markets in global T&T spending (Manzo 2019). International tourist arrivals, up 6% in 2018, reach 1.4 billion. According to PR 19003, United Nations World Tourism Organization (UNWTO) in press release, the Middle East (+10%), Africa (+7%),

Asia and the Pacific and Europe (both at +6%) led growth in 2018. Arrivals to the Americas were below the world average (+3%) in 2018.

International tourism has also played a unique role in the development of the tourism industry. In that case, it helps developing countries so many incentives to the host countries that is servicing as a pull factor to motivate in more tourist into their destination.

- International tourism contributes to the diversification of economy, by creating new sectors that provide services to the sphere of tourism.
- A good numbers of new work places are created and the problem of unemployment is solved.
- International tourism the source of foreign currency inflows to the country and a tool to maintain work force employed.
- International tourism increases inputs to the cash flow balance sheets and GDP of the host country with the help of international tourism.

1.4 Newly Industrialized Countries

A term of Industrialized Countries is used by political scientists and economists to describe a country whose level of economic development ranks it somewhere between the developing and first world classifications (Majaski 2019). These countries have moved away from an agriculture-based economy to industrialized, countries.

Tourism has played a vital role in the development of industrialized countries. The least developing countries (LDC) have great benefits in tourism because it has serviced as a vital tool to boost their countries Gross Domestic Products (GDP).

Secondly, tourism is now a huge contributor to the economies of all countries. It is the biggest contributor to many of the African countries such as Gambia, Senegal, South Africa, Kenya, Tanzania, Ghana, etc. These countries have moved a step forward with the new developing trends in tourism. This will lead to the development of the following sectors, like economic factors, social factors, and environmental factors. For example, according to a research, Kenya has a long-established and highly successful tourist sector, catering to the conventional and ecotourism markets. Kenyan Tourism reported that their international tourism is a lucrative source of income for Kenya, accounting for 2.24% of the nation's GDP in 2006 because of the advanced development of their tourism sector. In 2015, Kenya's tourism GDP raised to 3.8%. According to sources from World Travel Tourism Council (WTTC), it is expected to rise 4.0% by the year 2025.

1.5 Conclusion

Tourism is one of the largest industries in the world. It is a highly productive service sector of most developing and developed countries. Tourism is being the most vibrant economic sector. It has the capacity to create jobs, and boost both local livelihoods, national and international economies. Tourism also be referred to as a double-edged sword, and unless managed properly, it can be a potentially negative force. This brings a lot of benefits to a destination that embraces it and manages it in a proper way by looking at it into various dimensions. Based on the following theme, socially, politically, and environmentally, tourism has a diverse face all of which make up the overall sundry industry in tourism destinations.

However, tourism is a thriving global industry with the power to shape developing countries in both positive and negative ways. The tourism sector has full of strength despite the transnational challenges posed by terrorism, health pandemics, and global financial crisis. In 2012, international tourist arrivals passed one billion. Since global tourism growth slowed down in 2018, but international tourist arrivals still hit 1.4 billion. Over last year, the developing nation tourisms were up to seize the economic opportunities that foreign visitors visited. Some countries had proved and more adapted than others in doing so. During 2018, Africa grew at 7.3 percent from international tourist arrivals, compared to global tourism arrivals grew by nearly 6 percent. Kenya has developed a lucrative tourism sector, powered in recent years by ecotourism. Some local communities have benefited directly through social and economic development.

Therefore, tourism can achieve above goals if tourists respect the environment and places host communities at the center of the development process. Responsibility lies with the governments of developing nations to ensure that tourism grows in a sustainable manner. As a driving force, tourism sector continues to grow based on the emerging economic, determine the role of tourism development in least developing countries (LDC) toward Sustainable Development Goals (SDGs). According to previous research, over the last decade, tourism has been experiencing continued growth and deepening diversification to become one of the fastest growing economic sectors in the world. Modern tourism is closely linked to the development and encompasses growing number of new destinations. These dynamics movements have turned tourism into a key driving factor for socio-economic progress.

Today, the business volume of tourism equals or even surpasses that of oil exports, food products, or automobiles. Tourism has become one of the major players in international commerce. At the same time, it represents one of the main income sources for many developing countries. This growth goes hand in hand with the increasing diversification and competition among destinations.

This global spread of tourism in industrialized and developed states has produced economic and employment benefits in many related sectors—from construction to agriculture or telecommunications, etc. The contribution of tourism to economic well-being depends on the quality and revenues of the tourism have offered. The UNWTO assists destinations in their sustainable positioning and even faces more complex on national and international markets. As the UN agency dedicated to tourism, UNWTO stated out that particularly developing countries stand to benefit from sustainable tourism and acts to make this into a reality according to Sustainable Development Goals (SDGs).

References

Abdo n.d. Poverty literature review–tourism sector. Global manufacturing, Ag ribusiness, and services (CMGSF), International Finance Corporation. Washington, DC, USA.

Dwyer (2015) Globalization of tourism: drivers and outcomes. Tour Recreat Res 40:326–339

Fleischer A, Freeman D (1997) Multi-regional input-output analysis: a tool for measuring the economic impact of tourism. Ann Tour Res 24:998–1001

Gasper D, Shah A, Tankha S (2019) The framing of sustainable consumption and production in SDG 12. Glob Policy 10:83–95

Harcombe DPT (1999) The Economic impacts of tourism. ABAC J 19(2):10–22. Assumption University, Bangkok, Thailand

Lanza A, Pigliaru F (2000) Why are tourism countries small and fast-growing?. Tourism and Sustainable Economic Development, Springer, Boston, MA, pp 57–69

Lemma AF (2014) Tourism impacts: evidence of impacts on employment, gender. Income. Overseas Development Institute, EPS-PEAKS, Brighton, UK, pp 1–22

Majaski C (2019) Newly industrialized country—NIC definition. Investopedia. https://www.investopedia.com/terms/n/newly-industrialized-country.asp

Manzo GG (2019) Travel & tourism economic impact 2019. World Travel & Tourism Council, London, UK

Nella A, Christou E (2016) Extending tourism marketing: Implications for targeting the senior tourists' segment. J Tour, Herit Serv Mark 2:36–42

Pyo SS, Uysal M, McLellan RW (1991) A linear expenditure model for tourism demand. Ann Tour Res 18:443–454

Richards G (2009) The Impact of Culture on Tourism. A report produced for the OECD on the synergies between culture and tourism. Organization for economic co-operation and development, Paris, France

Sinclair MT (1998) Tourism and economic development: a survey. J Dev Stud 34:1–51

Terpening J (2015) 2015 top markets report travel and tourism. Washington, DC, U.S.A, International Trade Administration

Wood RE (1984) Ethnic tourism, the state, and cultural change in Southeast Asia. Ann Tour Res 11:353–374

Chapter 2
Religious Tourism

2.1 Introduction

Religious Tourism is also known as faith based tourism stated by different authors. This is a form of tourism, which motivate many people from different parts of the world, to travel to destination of their choices, either individually or in groups. Basically, their visitation is based on religious reason, which can be either be on pilgrimage, missionary, or leisure (fellowship). This form of tourism is regarded as to be one of the oldest types of tourism (Rinschede 1992). It has been in existence since before the development of the age of steam. It has played a vital role to the improvement of the standard of the tourism industry and looking at it in the early days. Readers may also know that it is one of the oldest types of tourism and a worldwide phenomenon of religious history when people travelled to fulfill their faiths based on their beliefs (Fourie et al. 2015) (Fig. 2.1).

Looking at the trend of travel and tourism industry during the primitive days, religious tourism has always served as a push and pull factor in destinations. This type of tourism motivates good number of the masses to travel to so many countries on religious reason depending on their faith and beliefs. Below are some of the prominent destinations tourists did their traveling those days: Mecca in the Kingdom of Saudi Arabia, Fatima in Portugal, Vrindavan in India, El Camion in France, and Santiago de Compestilla in Spain.

In additions, tourism and religion can also have impacts on tourist behavior, for instance religious influence to choose a destination. Tourist product preferences and the offering of the religious related opportunities and facilities to tourists are similar to other tourism subgroups. Religious tourism should be seen more in the context of the current competitive environment, in which the tourism and the hospitality industry are constantly searching for new market segments. In this context the tourism industry often inhibits a competitive relationship where tourists feel they have no other alternative than to compromise, on their spiritual beliefs in favor of a tourist experience. Therefore, tourism and religion should pursue as a complementary, if not co-habitual relationship (Weidenfeld and Ron 2008). The way these relationships are

W.-T. Fang, *Tourism in Emerging Economies*,
https://doi.org/10.1007/978-981-15-2463-9_2

Fig. 2.1 *Ritual* in Hinduism, Bali, Indonesia. (Photos by Wei-Kuang Fang)

shaped that is also influenced by the way industry perceives and address the tourist religious need.

2.2 Different Type of Religions in the World

Before naming the different types of religions in the world, we will first of all, look into the definition of religion. This will at least give us a clear under-standing of what is the meaning of a religion and how it is importance in the development of the tourism industry. A religion is often defined as a set of beliefs that explains about the world and universe. The meaning of a religion is hard to define because so many people have many diverse definitions. There are many definitions on what to define a religion but not the one that could be said to be accurate about the definition of religion. Some of the widely used definitions about religions by authors are as follows (Sharply 2009; Timothy 2012):

1. An institution to express a belief in a divine power by exercising of a faith.
2. A belief concerning a supernatural, sacred or divine, and the practices of the institutions associated with such belief.
3. A strong belief in a supernatural power that control the human destiny and behaviors.

Basically, with the help of these definitions, readers will have a clear understanding about religions and their effects in the day to day activities of mankind (Fig. 2.2). Now, let us look at the most popular religions in the world in alphabetical order. A

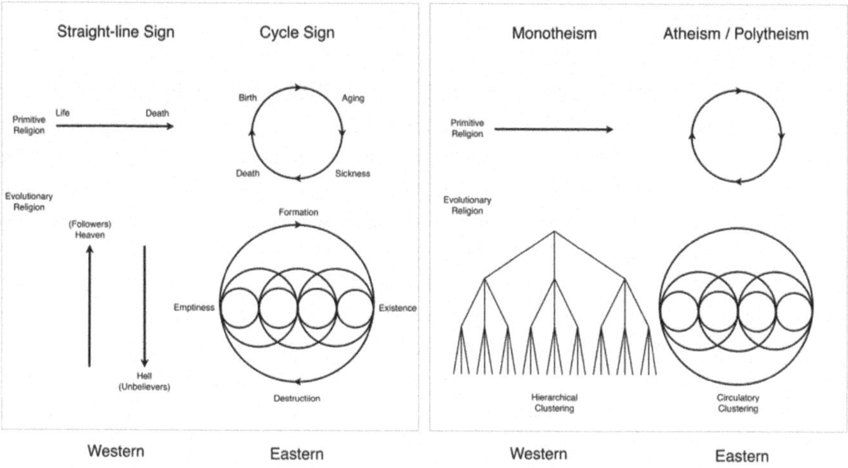

Fig. 2.2 The principles of Boolean logic are helpful to classify intensely current religions in Western and/or Islamic countries, i.e., monotheism. The hierarchical classification for denominational families is distinct and clear to western readers. However, the list of denominations in Oriental (South Asian and Far Eastern) countries derived from atheism/polytheism is very difficult to classify religious subgroups from their infused cultures of sharing broadly similar beliefs, practices, and historical ties. (Illustrated by Wei-Ta Fang)

brief explanation of each of them is elaborated, as well as the roles in the development of religious tourism in the world today.

- **Atheism**
- **Buddhism**
- **Candomble**
- **Christianity**
- **Hinduism**
- **Islam**
- **Jehovah's Witnesses**
- **Jainism**
- **Judaism**
- **Mormonism**
- **Rastafari**
- **Santeria**
- **Shinto**
- **Sikhism**
- **Taoism**
- **Unitarianism**
- **Zoroastrianism**

2.2.1 Atheism

Atheism is one of the religions and irreligion facets different in belief from the other religious set groups. They denied the concept that God, divinities, and spiritual are things of reality. Atheism trusts that, people do not need to rely on gods or holy writings to create moral codes to rule life as a mean of associating one-self with God. Numerous skeptics are enthusiastic secularists and are in contradiction with thought of the presence of belief in open concern of the society. They did not state that gods did not exist but simply said that gods were not involved in the life of humans. Most of them have personal reasons as to why they do not believe in the following religion. This religion is centered on God or gods because they believe that religions are irrational. There is not enough evidence to prove the existence of any religion. Most of the atheists based on their past, present, and future on the Greek and Roman philosophers' theories. Most of those theories rely on the thought of these philosophers, among them including the Darwin Theory of Evolution. However, some portions of them too believe in God. They are very pervious, do take more humanistic approach, and do not focus on the existence of God.

Historically, it is accepted as true that Clinton Richard Dawkins (1941–) is an extensively documented atheist (Dawkins 2016). He is well-thought-out to be the father of modern atheism. Some people who claim atheists are actually religious. Religions as Buddhist comply with some of the principles of atheism (Fig. 2.2), as do religions, grew out of Judaism and Christianity in Western cultures. On the other hand, many atheists think that religion does offer positive things like charities that help those in need, art, music, wisdom, and a sense of community. Since Atheism is not based on religious beliefs, each Atheist has a personal moral code which is based on secular factors. They believe that religious people are not self-reliant because they turn to God or gods. Atheism also appeals to people who have lost faith in religion or who are not interested in it (Bullivant 2012).

Atheism being a sort of traditional belief has a good social occasion time on the life platforms. They do lively celebrating on social events, like a newly delivery child, wedding of bride, and death. They are done with ceremonies that are generally important for them in gracing the occasion. The belief of such is an important venture for them in organizing and naming the ceremonies to welcome newly born babies into the world. The newly born baby is actually loved by friends and relatives. They attend the observance, usually features chosen mentors or supporting adults, who promise to support and protect the child. According to Atheism, people get married through a non-religious wedding, is very personal and devoid of any supernatural presence and religious rituals. Likewise, Atheist funerals are not religious. They are personal reflections of the person who has died.

2.2.2 Buddhism

Buddhism is one of the historical religions that based on spiritual traditional believe of deity. This religion was believed to be started in India and it has spread to south part of Asia to countries China, Indo-China, Japan, Korea, and Taiwan where it is mostly practices. According to statistics about three hundred and seventy six million people in the world today believe in the Buddhism.

Buddhism is a religion believes in philosophy based on the teachings of Siddhartha Gautama (c. 563 BC–483 BC). It is regarded as a spiritual awakened one called Buddha the only master to be worship, who they believe can end the suffering of humankind. According to Buddhism religion, all elements of life are interrelated, therefore, mercy is necessary to be shown to all elements. They also achieve enlightenment through leading lives of morality, meditation, and wisdom (Fig. 2.3).

There are many different sects in the Buddhist religion with two main sects, that is, the Theravada Buddhism and Mahayana Buddhism. The Mahayana Buddhism has started to spread widely in East Asia from Tang Dynasty (c. 618 AD–907 AD). The Mahayana Buddhism religion is one of the customs worshiping in the temples. During the year 2017, in Taiwan, for example, with a size of 36,000 square kilometers, there were registered 12,271 temples mixed with faiths in Polytheism of Temples of Taoism and Confucianism (Rujia) (78.3%) and Temples of Buddhism (19.6%). The worshippers enjoy seeing delicate decoration of temples, reading the

Fig. 2.3 The Reclining Buddha Statue, a giant 49 metres length, at Phan Thiet is the largest ones in Vietnam. (Photos by Wei-Kuang Fang)

Fig. 2.4 Temple of the Emerald Buddha, The Wat Phra Si Rattana Satsadaram, is famous for the sacred Buddhist temple in Bangkok, Thailand. (Photos by Wei-Kuang Fang)

sutra (the Buddhist Scripture), and/or pray with incense for their fortune and lucky from Buddha, Bodhisattva, Deity, and the guardian of Buddhism in the beginnings of the Chinese Lunar Year.

The pilgrimage is not forceful but suggested to be undertaken in one's life most which is done by visiting temples in Mainland China, Nepal, Sri Lanka, Cambodia, Thailand, Myanmar, Indonesia, Vietnam, Japan, Korea. Fraction followers in Taiwan are seeking for blessing by visiting the temples. These have boosted domestic and international tourism as people visit temples for religious purposes and also to grace important programs conducted at such venues (Figs. 2.4, 2.5).

2.2.3 *Candomble*

Candomble is an African-Brazilian faith consuming approximately two million supporters. It is a syncretic faith, meaning that it is a mixture of several beliefs. The origins of Candomble go back to era of the slave trades. There were African slaves strained to trail their spiritual beliefs and exercise their duties, but were not permissible by their masters. At the core of the Candomble convictions are the customary African philosophies of Yoruba and Bantu. They had basics of Christianity, mostly

Fig. 2.5 One of the famous tourist attractions with relics of temple complex of first Hindu, the Angkor Wat, is dedicated to the God Vishnu, then Buddhist at Angkor, Cambodia. (Photos by Wei-Kuang Fang)

of Catholicism. The word Candomble is defined as dance in uprightness of the gods. Accordingly, dance and music play important roles in the religion. Although many slaves were converted to Christianity, others still practiced their religions in secret or disguised their practices as those of Catholicism.

With regard to the development of tourism being the world global trade, various Africans nowadays visit the city to learn more about their ancestors and their religious beliefs. The Afro-American always visit Africa to trace back their roots. Among them is the famous Alex Healey for tracing back his roots to a village called Juffureh, the Gambia (Healey 2016). He is believed to have connection with Kunta Kinteh, not only roots but also cultural identity.

The aim of every person is to trace his or her origin, worshippers perform special dances to allow the *orisha (orixa)* to possess them. They are spirit of gods that link humans to the spiritual world. Each *orisha* is connected to a force in nature including certain food, animal, and more. A person's behavior is a consideration of their *orisha*. The devotion takes place in the system of dances and songs. The sacred places for followers of Candomble are the temples. These structures have indoor and outdoor sections and designated places for the gods. Worshippers who enter the temples must be in clean clothes. They also must splash water on themselves before entering the temple to clean from impurity from the outside world. Many people travel far and wide to visit temples at different locations of the world. It is gaining popularity boosting both domestic and international tourist market.

2.2.4 Christianity

Christianity is one of the most popular religions in the world. It has become exciting since in the first centuries. Christianity was believed to be founded by the followers of Jesus Christ of Nazareth, who was believed to be the Christ chosen from God, to come and save his people from sins. The founder of Christianity, where believed to be his followers. They are James, John, Peter, Mathew, Phillip, Bartholomew, and Judas (who betrayed Jesus), as well as other people. Some followers have never met him but still have been influenced by his teachings. The Gospel authors like Mark had mystical revelation of his divine natures, and Paul played a very vital role in the founding of Christian communities and churches after his conversion.

Christianity then started to spread from Jerusalem in the near east in early 4th centuries. During that period, it was successfully expected as a state religion by the Armenian in 301, Ethiopia in 380, and Georgia in 332. It has become common to be adopted in Europe during the middle ages. It was then spreading to other places of the world during Europeans age of exploration. As a result of these, it has become one of the world largest religions during its expansion around the world. Christianity has brought a lot of theological disputes that have eventually results to many distinct churches. The largest branches of these churches are the Roman Catholic, eastern orthodox, and Protestants, churches Methodist, churches Jehovah's witnesses, etc.

The religious event of Christians draws many peoples to visit monuments in Rome and Jerusalem, as well as attending major gathering stages at the cathedrals. Some of them participated in the events whilst other just travel to witness the events (Fig. 2.6).

2.2.5 Hinduism

Hinduism is also part of the accents' religions in the universe. A statistic revealed that Hinduism has more than nine millions disciples. Eighty percent of the Indian populations believe in this religion, the Hinduism. The Hindu people do not admit but recognize one God and external soul that they called Brahma.

Tracing back the history of the Hindus in the BCs, scholars stated that the Hindus religion was believed to be a mix of numerous traditions and faiths. Element of Hinduism can be recognized into three different deities, that is, the Shiva, Vishnu, and Narayana. All these gods have their roles and responsibilities in the Hindus religion. The *Shiva* is the destroyer and the mother goddess; *Vishnu* is the preserver and *Narayana* is the protector of the universe, plus the saver to people from disasters and oppressions.

Fig. 2.6 Completed in 1993, this Statue of Jesus Christ in Vung Tau, Vietnam, is lately famous in Asia among worldwide famous high replicas of Christ the Redeemer, such as the Christ the Redeemer in Rio de Janeiro, Brazil; Christ the Redeemer of Maratea, Italy; Cristo-Rei in Lisbon, Portugal; Cristo Rey del Cubilete in Guanajuato, Mexico; Christ of the Ozarks in Arkansas, United States; Christ the Redeemer of the Andes, Argentina; Cristo de la Concordia in Cochabamba, Bolivia; Cristo Blanco in Cuzco, Peru; and Cristo de las Noas in Torreon, Mexico, etc. (Photos by Wei-Kuang Fang)

By looking into the elements of Hinduism, it was also believed to trace from the recent years in Pakistan during the year of 3000 B.C. The nomadic Aryans had arrived in India spreading Hinduism to people. They were as far as to the north and south during that time. Being able to form regions around Russia through their expansion, they brought with them religious customer from India to worship God on their own believes during the rise of the Jainism and Buddhism in India. Scholar's states that it has affected the rising of the Hindus religion to spread fast within India around the 800 BC to 600 BC. However, none the least, they were able to come up with some poems and songs to make their religion and even motivate in more people to come and join the religion. Furthermore, with the coming of Islam in the region has affected them, too. But then, with the coming of Christianity in the region has created different effects to their religion. This made them to puss them harder to bring back the accent Hindus practices within these period. They had to distinguish themselves and separated into different schools to spread the religion more into the region.

Secondly, during this period of time, the Hindu followers' beliefs become popular and famous. One of their most popular holy day is the *Deepavali* or *Divalii*, meaning the festivals of light. They also referred to worship as Punjab they focused their worship either at their homes or in the temple by setting up shrine. These holy places could either be a room normally while they are worshiping and calling repeating the images and names of gods for several times. The Hindus customers throw some flowers, water, fruits and incense three times a day according to their beliefs. These temples also give the followers a significant meaning. These towers in the temple stand for the sprite accent of heaven. A pilgrimage plays significant roles in Hindu religion. It is believed that the followers are to worship their gods.

2.2.6 Islam

Islam is the second largest religion in the world, second after Christianity. It has more than one billion followers found around the globe. Islam means total submission to the will of God. There are many Islamic groups all united by their faith by believing that there is only one GOD, named or called as "Allah". The holy Qur'an is the scriptures in Arabic language. People read and proclaim that the Qur'an is massages from God. The two main groups of Islam's are the *Sunni* and *Shia or Shi'as (Shīʿah,* from *Shīʿatu ʿAlī).* The Muslims believe that the revelation of Qur'an is the central miracle of Islam. The Muslims people believe in Muhammad as a prophet delivered by "Allah" to come and preach the good news to lead his people to follow God commandments.

The Muslims conceal the fundamental faith that, there is no God but Allah. Mohammed is His messenger and servant. The Qur'an presents them as a framework for worship and a sign of commitment to the faith containing five principles or pillars. They are (1) *shahada* (testimony, evidence, creed)—The Muslim profession of faith ('there is no god but Allah, and Muhammad is the messenger of Allah'); (2) *solat* (five times daily prayers); (3) *sawm* (fasting from dawn until dusk during the month of Ramadan); (4) *zakāt* (almsgiving; payment made annually under Islamic law on certain kinds of property and used for charitable and religious purposes); and (5) *hajj* (the pilgrimage to Mecca; at least once in a lifetime for those who can afford). The Muslims also have good Islamic customs that every Muslims need to follow and practice every day during their life time. They must adhere to the five daily prayer of Islam. They must also have specific diets that are not supposed to be eaten, as it is an unlawful food according to the teaching of Islam. Some of the foods are like pork and other meats slaughtered by non-Muslims, drinking of alcohol, telling lies, adultery, fornication, and gambling.

The Sunni means "One of the Paths" is another branch of Islam. The Muslims in this path acknowledge the first four Khalifs to have been rightful successors of Muhammad. They accepted the "correct books" (Sihahu's Sittah), or six authentic Hadith, and belong to one of four schools of jurisprudence (Abu Hanifa, ash-Shafi'i, Malik, or Ahmad-ibn-Hanbal). This is by far the biggest group in Islam. Their spiritual center is the Ka'aba in Masjid al-Haram in Mecca, Saudi Arabia. Besides Mecca, Muslims are also encouraged to visit Prophet Muhammad SAW tomb located in

Fig. 2.7 The Ka'aba (in black), located in the centre of Masjid al-Haram in Mecca, Saudi Arabia. It is the spiritual center for Muslims around the world. Muslims are urged to perform hajj at least once during their life time here. (Photo by Arba'at Hassan)

Masjid Nabawi, Medina, in Saudi Arabia, too. Their number of worshippers makes up over 85% of all Muslims in the world today (Fig. 2.7).

The Shias meaning: "Followers", i.e., of Ali, they formed about 10 percent. They believe in heredity succession of Khalifs. To them, Ali, the nephew and son-in-law of Muhammad, was the first rightful Khalif. One of his son was assassinated, and the other abdicated the throne should have succeeded him. They are based the claim on Surah 2:124 of the Holy Qur'an that:

"I will make thee an Imam to the Nations". He pleaded: "and also (Imams) from my offspring!"

Many "Hadiths" are quoted by the Shias to support their case. According to tradition, the twelfth in heredity line, Imam-al-Mahdi (a kind of Messiah), is still alive and hiding to appear before judgment. Among the many splinter groups of the Shia, the Ismaili plays a very dominant role in East Africa. They are decidedly less fanatical than the Iranian Shi'as Muslims, but more devoted to philanthropic work. Their spiritual leader is the Agha Khan who has become a well-respected as an international figure.

The center of Shiite activity is not in Africa, but in Iran, where the Ayatollah (lit. 'Sign of Allah') Khomeini established a new base for a Shiite Islamic missionary thrust. It is felt in almost every corner of Africa today. Even in a small island like Lamu, known as "the Mecca of East Africa" Iranian Shiite influence and finance has gained control over one of the oldest and largest mosques on the island; a power struggle that is quite typical for many parts of Africa today.

2.2.7 *Jainism*

Jainism is an ancient religion like any other religions. The followers have a common philosophy and believe that happiness in life and hereafter can be achieved if one lives a harmless life, without causing hard and trouble to any form of living object on earth, including both humans and non-humans. It is a form of faith that doesn't subject itself to the worship of God but on its own way of doing things. Jainism like Sikhism is also among the religion originating and incepted in India. There are approximately 4.2 million supporters of Jainism in the earth majority of whom, are mostly, staying and existing in India.

It is historically reveals that Jainism is religion founded by Tirthankaras, a normal human being, involves in passionate thought, repentance, and progresses on how to improve soul to the state of purity. It is by getting close to the divine and supreme God. Jainism is a religion interested in the well-being of the universe and mankind. According to Jainism, creatures, animals, and plants have passions, ambience to be equal, thus should be esteemed. As for this reason, Jains are lacto-vegetarians and use natural possessions as competently and thoughtfully as imaginable for their daily meal as they support restoration. They don't eat any living object, specifically most creatures with no blood, simply they believe that should not use its intelligence over other living souls.

Jainism has three vocal points. These three vocal points serve as the beliefs of Janis that possess in their perception. They are (1) right belief; (2) right conduct; and (3) right knowledge. These beliefs serve as the guiding principles for the Janis have made them different from other worshippers. The uniqueness they possess is side-stepping violence, separating from belongings, communicating the reality, no theft and burglary, and performing sexual limitation. They also share some attribute with other religious facets among which includes the performing of pilgrimages and fasting during the holy month. That is almost similar to the Muslims. The common ration behind the reason of fasting is of cleansing of the body and mind. There are various types of fasting ranging from complete fast to cutting your favorite food only. Apart from temple, Janis also visits giant creature such as mountain in Shatrunjaya. Festivals of Jain are also amazing events that attract various participants around the world to witness like during the monsoon period.

2.2.8 *Jehovah's Witnesses*

The Jehovah's witnesses are another form of Christian religion that was formed in the late 1800 s in the United States by Charles Taze Russell (1852–1916). The movement was globally recognized in 1900s. Although some people abused and criticized them, but they still kept up the good move with their religion. The Jehovah's witnesses' belief from Christian based on religious denomination, who believes that most of the Christian churches have strayed. But for them, they consider themself has brotherhood joined by religious belief. According to statistics about 6.5 million followers in the world today are Jehovah's witnesses. Known for evangelism, the Jehovah's witnesses go from door to door into the neighborhood to preach the people and converts them to the religion. They used the new translation of the Holy Scriptures to their own translation. The Scriptures were translated to about 70 languages just to motivate people to come into their religion. Because they believe that missionary work is required of all Jehovah's Witnesses, as long as they are capable to spread the gospel.

Looking at the Jehovah's, witnesses believe and customs their belief in trinity, like Catholic and the Methodist do. They also believe that Christ Jesus died on a stake not on a cross. They don't also celebrate Christmas and Easter because they believe that it has paganism influence, and the Lord Jesus did not expect people to commemorate the memory of his death. They do commemorate the Memorial of Christ's Death which marks the anniversary of the Last Supper that Christ Jesus have with his disciples. They believe that their purpose in life is to serve God.

2.2.9 *Judaism*

Judaism is also one of the oldest religions in the world. It was believed to be found in the Middle East by Moses. According to statistics, about 12 million people in the world today believe in Judaism. Most of them are found in Israel and the United States. The holy book of Jews is called the Torah or Hebrew Holly Bible which is important book of the Jews people.

The Torah and Oral Law are ones of vital aspects of Judaism that must be studied by the Jews. Dietary Law is also required believes in the Jews customs that they should not eat certain foods as a Jews. They also have a way of life call the Halakha which is their way of life that every Jews must fellow those who believed in Judaism belief in One God. One of their customs is practicing prayers together. The Jews people also attend synagogues for praying likewise on their religious' events.

2.2.10 *Mormonism*

Mormonism is a Christian belief. However, it is virtually deviated from the supplementary groups of Christianity practice around the world. Christian cathedrals obligate absence, off course, and their church is the only one exhibited when the church envisioned by Jesus. They also used the holy book of Bible for the teaching and believing in guideline of Treaties, and Pearl of Great Price. According to Mormon faith, God created the world. He had a body, married, and was able to have children. Jesus is the central to Mormon faith. They believe that people become like divinities subsequently their lifetime on Earth, and that, people are like gods before birth. These are some of the concerned ideologies in Mormon. The Mormons follow an ancient custom baptized tithing and give the Church ten percent of their twelve-monthly income. Mormon rites include: marriage ceremony, inductions for the dead and, baptisms, funerals (burying is preferred), ordinances, and temple ordinances.

Mormonism is a religion, today, widely observed in America with so many books of revelation from Joseph Smith (1805–1844). He was believed to receive revelation from God, some of the texts that were written far back in 600 BC to 400 AD. The Mormons believe in God, Jesus, and the Holy Ghost, which together form the Godhead. Mormons believe in living prophets. Their prayer consists of a simple and honest talk with God. Sunday is the day of worship. Mormons fast on the first Sunday of each month. They refrain from two consecutive meals. Monday evenings are reserved for family times. The Mormons faith has also forbidden the malpractice of abortion, homosexuality, making a bet, pre-marital sex, pornography, consumption of coffee and tea, among other things. Major important days are marked by the Mormons like the Christians. They are the Christmas, Easter, and Pioneer Day. On July 27 is marking the anniversary of the Mormons transfer to Utah, United States. These days are grace by the Mormons, which is witnessed by lot of domestic and international participants around the world.

2.2.11 *Rastafari*

In Jamaican history, the Political frontrunners namely Marcus Garvey (1887–1940), completed a divination that a king would be capped in Africa. That king would convert the black people in a region of believe. Based on this philosophy, the Rastafarian concept was delivered. Rastafarianism is a relatively new religion based on African customs. They hassle the thought of the civilization of God and the religion of humans, with a significance that God can seem to be in the form of a human. They promoted the idea of humanity than the worldly materiality. Rastafarian was recognized as a belief in Haile Selassie (1892–1975), the king of Ethiopia. Today it is a widely practiced religion in Jamaica and other part of the world. The basis of this faith is to improve the situation of black people in Africa and other part of the world for them to live a decent and standard live. While some Rastafarians perceive on to the old-style belief that blacks will one day overpower the western world, the Rastafarians have faith in contemporary diverse society. However, all Rastafarians believe in re-embodiment and everlasting life. They defend humankind and natural life. They oppose Paganism, contraception, and killing—humans and animals.

Rasta is practiced in follower's homes or in a community center in the forms of dances, musical shows, and concerts. They do not have a specific venue for collective adoration of prayers, like other religions have: churches, temples, and mosques. Their gatherings comprise of dancing, drumming, singing, praying, and chanting in squares and halls.

Reggae and dancehall music is a significant component of the ceremonies of the dreads old-fashioned melody, namely Nyabingi. This is a mix of gospel sounds and African drum. Most Rastafarians are vegetarians. Rastafarians are encouraged the followers strict diets of clean food and drinks, like vegetables, fruits, and herbal tea. Marijuana is a smoking substance widely used by Rasta. They advocate that it is natural herb that needs to be legalized in the world. It is thought to enrich mystical awareness called wisdom weed. They must avoid alcohol, coffee, milk, meat (especially pork), shellfishes, and fishes that measure more than a foot long. Rastafarians are not allowed to use sharp instruments on human bodies. They are not allowed to cut hair, shave, tattoo, etc. As a result, Rastafarians grow their hair and have to twirl it into dreadlocks, symbolizing a lion mane (the Rastafarian symbol is the lion).

The tourism industry is boom with the Rastas. The events are happening and staged performances done internationally. Some of them are reggae musically concerts in Jamaica, Netherlands, and Germany, all of which host carnivals. These Rasta's concerts have invited lots of tourists to witness those events. They purchase the tourism products by staying in hotels, consuming foods and drinking beverages in the destinations. All of those activities have supported the local tourism industries and enhance the international economic growth.

2.2.12 Santeria

Santeria is an Afro-Caribbean religion that developed from the Trans-Atlantic slave trades. The religion is centered on African traditions and has some elements of Christianity. It is an oral tradition. The exact number of Santeria followers is unknown. It has being a clandestine faith for decades, but has since revealed itself in so many parts of the world. It is conceived to be started by the Nigerians, who have a mixed of Yoruba religious traditions with elements of the Catholic faith of the plantation owners. They try to convert the enslaved Africans his religion.

In the past, Santeria was viewed as a faith for the illiterate and underprivileged in the South America. However, now, it also has a middle class proficient following into countries like the United States. Santeria inspires constructing affiliations with state of mind that is appearances of God. Santeria advocate that people must worship God, so that their feelings would keep on existing. The Roman Catholic fundamental of Santeria is connected with the acceptance attitude in all embracing saints and symbols. They have certain powers like predicting the future by using coconut halves, nuts, or shells. Supporters of the confidence often dress in white as a symbol of cleanliness. It is also habitual for Santeria supporters to participate in numerous rituals in order to stay in touch with Gods.

Santeria priests also provide support to sick people. They believe that the priest can predict the future of the followers. They have a wealth of knowledge on the medicinal properties of herbs, holy water, and natural elements, as dictated by Catholic and African belief systems. Santeria rituals include engaging the *orisha* through dancing, eating, singing, drumming, and more. *Bemb* is an example of a ceremony that welcomes the *orisha* to join in the festivities.

The rituals of Santeria can take place in a house with altars or in a hall, which is properly decorated. If the *orisha* possesses a person, the individual starts dancing dramatically and shares messages from the *orisha* with others in the community. One of the major forms of worship to ensure the survival of *orisha* is animal sacrifice. Animals are sacrificed as food for the *orisha*, who only drink the blood, while humans eat the meat

2.2.13 Shinto

Shinto is one of the religions founded in Japan. This is a religion based on a spirit called *kami.* The Shintos believe to God, namely Kami, as spirits or powers. Shinto does not have a God, scriptures, commandments, or a founder. It does have a priesthood, which does not, however, follow a rigid structure, like priesthoods of other religions. Shinto religion is one of the most popular religions in Japan. Looking at the history, the Shinto religion in the eighteen century, this religion is considered as an official religion together with Buddhist.

The belief of Shinto followers is that if anyone believes in Kami and worship the Kami rituals in a proper way, he or she should worship Kami in full respond. The prayers will result the followers with success, good health, and happiness. Since it is based on ritual and not on belief, many people do not consider Shinto as a religion instead consider it as a way of life, whereby anyone who believe in it can also believe in other religion.

The Shinto people recognize natural features like the Springs Mountains and believe to have connected to Kami. This is because they believe that nature is a very vital part in their religion.

Participating in the festivals and going to shrines are parts of the Shinto customs every follower should adhere, too. The Shinto religion is also sharing some common relations with Buddhist. Some of them believe in Buddhism and at the same time believes in Shinto. Normally, their weddings are performed by the Shinto priest but during the funerals it is done by Buddhist priest. When Shinto followers celebrate a ceremony in the shrine, normally, they celebrate in the form of dancing and cleansing to worship Kami. They would wear beautiful and charming dresses, which they believe could help and protect them from bad spirits.

2.2.14 Sikhism

A religious state of an individual is of great significance to the religion. Sikhism is a monotheistic faith with so many supporters living in Asian continent. According to the history, there are over 23 million internationally constitute of people believed to be Sikhs. Majority of them lives in India. Sikhs believe they should lead good lives and meditate, rather than focusing on performing rituals and sacrifices. The duties of a Sikh are praying, giving charity, helping those who are poor and sick, working honestly, and having God in their mind all times.

Approximately, Sikhs make up 0.39% of the world population of which 83% lives in India. They form a majority of 59.9% of the population of India. Sikhism was originated in Punjab region of modern-day of India and Pakistan in about 1500 BC by Guru Nanak. Historically, the Sikhs are being subjected to some rivalry and fight by other religion, for example of which includes Islam and Hinduism. These factors deter them from expansion. Over the years, this has become a trouble for the Sikhs. In recent development, they get conflict with the Punjab province of India that led to the loss of life of thousands of Sikhs. The anti-Sikhs notably by Muslims and Hindus in India had resulted in massive killings in pursuit of faith as a holy war against the Sikhs.

Sikhs led by Guru Nanak mostly based there on his teaching. The Guru Nanak, taught them to promote a good life by treating others equally with a compassionate manner. Leading a truthful life is of prime importance through the five vices, like anger, greed, lust, pride, and attachment to material things. In order to be liberated, a follower must stay away from these vices. Liberation involves experiencing God through worship, love, and uniting with God. It is not earned but given by God.

The Sikhs have a central place of worship call *gurdwara*. The worships comprise of days. They are observed as most important holy days in Sikhism. Those days are (1) Diwali, which marks the prison release of the sixth Guru in 1619; (2) Gurpurbs, which are festivals associated with Gurus; (3) Vaisakhi or Baisakhi, which is the Sikh New Year festival. In 1699, Guru Gobind Singh created the 5Ks to distinguish Sikhs who have been initiated into the Khalsa, or Sikh community. Khalsa members wear the following 5Ks: a steel bracelet called Kara, cotton underwear called Kaccha, a wooden comb called Kanga, uncut hair called Kesh, and a steel sword called Kirpan. The 5Ks is common bonds between all Sikhs. Sikhs events have a trend in the tourism industry.

Religious tourism has becoming a popular activity marked by religious set groups around the world in their specific geographic locations. Sikhs are not an exception. Sikhs does series festivals, which are marked by several audiences around the world. These festivals are occasions for Sikhs to rededicate themselves to their faith, comprising of executions and death anniversaries of the Gurus to inspire the faithful and remind them of their history and the value of sacrifice for a good cause. People of various faiths are invited to these celebrations to give them a view of the Sikh faith and way of life. Tourists also travel to witness this amazing and aspiring event of festival of light, New Year festival, and other festivals steered by the Sikhs.

2.2.15 Taoism

Taoism or Daoism, is also an ancient religion based on various Chinese philosophies, religions, and customs. The Tao, so called as the "way" or "principle", is central to Taoism. It is connected to all things. Taoists may not worship the Tao, but they worship various deities that depend on the historical saints, sages, heroes, heroines, and legendary figures that followed the Tao. It is a religion without a specific founder, and its development was influenced by many religions and philosophies that existed in ancient China. It is believed that it first developed as a religion in the 4th or 3rd centuries BC. In the year 142 AD in Mainland China, the Tao revealed himself to Zhang Doling (34–156 AD). He formed the first organized school of the religion.

He also became the first Celestial Master, a tradition that survives today. During the Tang and Song dynasties, Taoism was a semi-official religion in China. On the other hand, Taoism was still well spread. However, Taoism underwent critics after communist acquired Mainland China and disqualified the religion. Taoism, then moved to its new home in Taiwan. After some time, Chinese executives commence pleasant somewhat accepting the beliefs of Taoism and taking place in mainland China, the religion underscores the amalgamation of the contradictory and harmonizing powers of the world. It pressures the synchronization of environment, immortality, generosity, self-improvement, as well as involvement in serving humanity and the world.

The Tao is the ultimate creative energy of the universe. It is also the organic universe or cosmos itself, as well as the way of nature and the world. Deities, along with everything in the universe rely on the Tao. It includes the guiding words on life and describes the nature of the universe in a poetic style. Rituals of Taoism attempt to bring harmony to the universe, the world, society, and individuals. Due to their complicated nature, priests perform Taoist rituals and their assistants, who chant, dance, and play percussion and wind instruments. The ceremony is also involved rituals asking for order for the universe and for the village. Through temple rituals, Taoists seek to stabilize the yin and yang for individuals and the community as a whole. Worship of the Tao includes praying to divine being, refining, thought on objects, and creation presents to divinities (Fig. 2.8).

Fig. 2.8 Today temples are marked as places of worship in Taiwan. Creatures like dragon, tiger, and other objects are placed in the temple, which is visited by both residents and non-residents to perform prayers. Right and upper left: Dragon Tiger Tunnel/Pagoda (built in 1976), Lianchihtan, Kaohsiung, Taiwan. (Photos by Wei-Ta Fang)

2.2.16 Unitarianism

Unitarianism is a personal approach to a religion that promotes religious freedom. It is open to various beliefs. With around 800,000 followers worldwide, Unitarianism stresses the oneness of God and the unity of all humans. Unitarianism includes people who believe in different religions like Christianity, Judaism, Buddhism, Paganism, and Atheism can follow Unitarianism. While Unitarianism does not follow a specific set of beliefs and is tolerant, it does have general guidelines. Unitarians support personal and rational thinking to reach one own conclusions in judging to things and coming to a final decision. Unitarianism promotes diversity and believes that is necessary in the world. Human differences are creative, not destructive. They believe that religion should play a proactive role in the world. Thus, Unitarians often engage in community work. Unitarian collective worship lacks a formality and ritual, but includes readings, prayers, sermons, hymns, songs, and periods of silence. Because of the gender- and religion-inclusive nature of Unitarianism, the service uses language that is not based on a particular gender or religion.

According to Unitarianism, God is one being (the Mother or the Father) and is present in all things; Jesus is a simple man. God is not the union of three beings (God the Father, Son, and the Holy Spirit) that Christianity promotes. God is loving and compassionate, and not harsh as in other religions. It is an open and tolerant tradition;

collective worship is designed in a way that all Unitarians with varying beliefs and differing relationships with God can feel welcome. Worship may sometimes focus on celebrating the life, so that all followers present can participate. Actions speak volumes in Unitarianism; therefore, followers are usually involved in social initiatives. People who want to make social changes are often interested in joining the church.

The Unitarians are open willing comfortable all interrogations and uncertainties, and are precarious of classifications with harsh religious rules, scriptures, and conclusions. Unitarians believe that humans have the potential be good. They are responsible for the evil in the world; therefore, humans are responsible for correcting their wrongdoings. Equality is practiced. Women and homosexuals can become Unitarian ministers. Some Unitarians do not believe in God but many do.

2.2.17 Zoroastrianism

Zoroastrianism is a belief to be among the eldest faiths observed in the world. The central concept surrounded by the Zoroastrianism is that there is only one God named Ahura Mazda. Ahura Mazda means Wise Lord. It needs to be praised and worshiped by the human beings. Records have shown that; today it is currently one of the smallest religions in the world, with one-third 1/3 of every million is a Zoroastrian. They are present all over the world.

Historically, Zoroastrianism is a religion believed to originate from prehistoric Iran about 3500 years ago by the prescient Zoroaster or Zarathustra. Zoroaster was born into multi-practice faiths like the early Hinduism in the Northeast Iran or Southwest Afghanistan. Bestowing to Zoroastrianism, it supporters can pray as frequently as they like by facing a basis of divine light like the sun or fire which represents energy or wisdom of God. This idea of praying is a tradition that started in the early life stage of children. They are inducted into the system when they are at the age of seven. They must drive over and done with a commencement formality. By giving them a terrified top and rope, they must wear the top and tie the rope round it three times. It means that they have faith in noble words, opinions, and actions, as a mean to better life.

Zoroastrianism encourages its followers to think good thoughts, speak good words, and perform good deeds. They are also obliged to wash down their fingers as a cleansing ritually before praying to God. Apart from individual prayers, congregational worship generally takes place round the time of anniversaries. These are marking a stages of life in which the light in a form of fire is always a part of the occasion. Zoroastrianism constructs their faith in the family welfare and community. They do not conceive the faith in celibacy. This religion also cheers captivating an energetic part in enlightening the social order of communities in which it is observed. By part of their generosities and philanthropies, they enthusiastically contribute to welfare of the society funding social and academic assistances. This is added with a trust that all societies are similar based on what we need to consider each other and offer assistance to those in need.

Today's life is empowered volunteerism among other religious travellers. They also travel for helping the locals by teaching children and public, and help developing sustainable projects in the communities. Zoroastrianism was the official religion of Persia from 600 BC to 650 AD.

At the principal faiths of other religions, for example Christianity, they raise that manhood is God's off springs. In Zoroastrianism, they believe that human beings are God assistants. The optimistic and intelligent adoptions of human beings will thrust evil away and God Paradise will then be set up on Earth, which includes commentaries, myths, ritual, and descriptions.

2.3 The Importance of Religious Tourism

Religious tourism is very elastic industries. It has a lot to offer to its customers and destinations. As a result, it is believed to be one of the most popular growing industries in the world comprises of the following twelve factors. These factors are the (1) pilgrimage, (2) conventions, (3) meetings and events, (4) cruising, (5) missionary travel, (6) safaris, (7) adventure travel, (8) volunteer vacations, (9) weekend gateways, (10) retreats, (11) visiting attractions and, (12) leisure vocations. All these factors mentioned have played very vital roles in the development of tourism industry, in both, developed and developing countries.

2.3.1 Pilgrimage

Pilgrimage plays a very vital role in the development of religious tourism. It services are pull and push factors in order to motivate people to travel to different parts of the world, to visit shrines, holy places, seek healing, prayers, and transformation monuments purposely on religious reasons, depending on their customs and beliefs.

Pilgrimage is defined as a journey undertaking for religious motive to sacred places or shrines, purposely for prayers. Looking at the world scenarios today, there are many people going on pilgrimage to different countries to cities like Mecca in Saudi Arabia, Jerusalem in Palestine, Palmyra in New York (site of the first version city where books of Mormon was firsts printed), Lopburi in Thailand, Vezelay Abbey in France, Saint Thomas Becket (1118–1170 AD) at Canterbury a shrine in England, Fatima in Portugal, Vrindavan in India, El Camino in France and Santiago de Compostela in Spain, Padre Pio of Pietrelcina in Italy, Mount Kailas in China, etc. In the early middle ages, the churches encouraged Christians to travel to make pilgrimages to special holy places called shrines. It was believed that if you pray at these shrines, you might be forgiven your sins and have the chances of going to heaven. It has motivated a lot of Christians from different parts of the world to travel to shrines hopping that their sins will be forgiven.

According to The Pew Forum on Religion & Public Life, about 84% of the world population is religious (Hackett et al. 2012). Such a large number of religious believers are definitely indeed an in-credible driving force for the development of religious tourism. Due to these factors, many countries around the world today have put in some great emphases on the development and utilization of their religious cultural resources to attract in more tourists to their destinations. Statistics shows that more than three billion people of the three major religions in the world have a deep seated needed for this kind of religious tourism as pilgrimage.

2.3.2 Convention

A convention is a gathering of individuals who meet at an arranged place and on a specific time frame. They come to convention to discuss or engage in some common interest. Convention is also an essential role to play in the development of the tourism industry. Many people today in the world travel to so many different destinations on religious purposes to attend conventions.

2.3.3 Meetings and Events

Meetings and events is also another growing sector in the development of religious tourism. These meetings and events have motivated a lot of people to travel based on religious reasons to different countries. Millions of Christians in the world today, travel to different part of the world on Christian meetings and events within the four corners of the world. They meet people to share the word of the gospel to its believers and also motivate in other people to come and join in the Christian family.

2.3.4 Cruising

Cruises have added benefit of having good attendance at all services in order to make travel easy for religious travels. Looking at the strength of religious tourism today, it has motivated so many religious travellers on cruising to holy land voyages in different parts of the universe. For example, in America today, the most popular types of religious cruises today are Christian cruises. These are two basic types of cruises: (1) a charter cruise where the group takes over the entire ship or (2) a group sailing with a smaller number of people cruising with regular passengers on an open sailing. With a group cruise, members book through the travel agencies organizing the cruise. In return, they will get special amenities on board. Generally, they pay a premium for these amenities with the profits going back into the religious organization's coffers. The group usually brings along some of their own entertainers and speakers, offering special activities on board, such as church services specifically for the group.

2.3.5 Missionary Travel

A missionary travel is a member of a religious group went to an area to act as evangelism or ministries. Based on religious purposes, they talk to people in that host community about God and convince them to follow God and worship him as their personal lords and saviors. Missionaries' travellers play a very vital role in the development of tourism in the last centuries. Many Christian missionaries' travel to all destinations around the world to proclaim good news of the lord Jesus Christ. They convince people to believe in him. Islam is another religion that has also missionaries across the world to preach the holy Qur'an and make people to believe in Muhammad as the last prophet.

2.3.6 Safaris

Safari tourism is another form traveling where religious people are adventuring. Since the expansions of religions in the world, youth groups travel to different destinations across the world to go on safaris tours. They hold land to see the beauty of the flora and fauna based on these travels. They also share the words of God within them, and people accompany with them they travel to fascinating places. The most exciting part of tours is the preaching of God words among themselves and place they have stopover.

2.3.7 Adventure Travel

Adventure travels play a very vital role in the growth of the religious tourism. Millions of people travel all around the world to different religious attractions sites on adventures either to trace the roots of their religion or explore their beautiful, land. For example, Christians travel to Bethlehem in Jerusalem to visit the land where their savior Jesus Christ was born. In India, too, they are richly endowed with ancient temples and religious festivals which motivate so many Hindus and Buddhists to go to India to trace back their roots by taking adventures through the land of India; whilst the Muslims travel to the Kingdom of Saudi Arabia to perform religious prayers most of who travel on land using camels.

2.3.8 Volunteer Vacations

Volunteer vacation is also another popular form of traveling. In the recent years, it has becoming an emerging trend in the travel and tourism industry. Millions of tourists travel to different parts of the world on volunteer missions, especially to the African content, to spread the Lord's Words and teaching of the religion to the people of the land. This form of travels started as being a global focus and a growing trend. Tourism researchers stated that one of the motivated factors of people to travel on volunteer holidays to use part of their trip participating in unpaid assistant works in a form of humanitarian activities considering that missions often has connotations of a religious purpose, doing services on the way of God. For example in the Gambia, Christian missionaries from Europe go into the country purposely to volunteers, helping the needy people by building schools, health, and sponsoring students in other to create a good living standard for the community.

2.3.9 Weekend Gateways

Weekend gateways are one of the popular sectors in the tourism industry typically on religious tourism. Tourists travel to different destination on weekends to go on visiting religious grounds and attractions. Looking at its trend, today it has motivated so many tour operators to organize packages to destinations, like China, India, and Spain in other to attract in religious travelers.

2.3.10 Retreats

In the world today, millions of religious believers travel to different parts of the world to go on retreats. Religious tourism offers interesting attractions like basilicas, monasteries, and sanctuaries and place of retreat situated both along the Adriatic and the hinterland of the Emilia Romagna region. In Rome today, there are many Catholics travel to the destination on religious retreats. In England, one of the center of spirituality at Westminster cathedral, is as also motivated a lot of tourists to travel on retreat programs.

Secondly, the faith movement is a pastoral and evangelical movement, which organizes catechetical events, groups, and retreats, producing theological publications and resources. The essential charisma of the faith movement is the promotion of a new synthesis of modern knowledge and divine revelation. The movement organizes summer and winter retreats, and holds regular forums on contemporary topics around the United Kingdom. This is to motivate many tourists to visit on either taking part in the retreat events.

2.3.11 *Visiting Attractions*

Attractions site are one of the key factors that motivate a lot of tourists to travel to different parts of world. They visit attractions sites which could either be manmade attractions or natural attractions that portray significant symbols of religious values. Millions of them travel to countries like Saudi Arabia, Spain, China, United States, France, Italy, and India. They travel there to visit religious attractions based on their faiths and believes. All of these great countries have a lot to offer in terms of attractions especially religious attractions ranging from cathedrals, temples, mosques, which could be visited by either believers or non-believers.

2.3.12 *Leisure Vacations*

Leisure vacations have motivated millions of people from different parts of world. They take part in leisure travel and go to destinations of their choices. Leisure trips embrace a lot of prospects. People can do well for lots of fun and recreations, provide them with unique opportunities to visit attractions and landmarks of the world. Leisure time is important for us to fulfill with, so we can experience and feel these benefits.

At the end of this section, the lists on Table 2.1 and Table 2.2 provide some data of statistics on religious tourism in the world, and some information for the infrastructure involving religious tourisms.

2.4 Impacts of Religious Tourism on Different Cultures

Religious tourism has a very great impact on the life style of the people today, depending on different cultural background. This can be explained as the three triple bottom lines as below: (1) Socio-culturally, (2) Environmentally and (3) Economically (Fig. 2.9). Based on these three triple bottom lines, it will surely help us to get

Table 2.1 Statistics on religious tourism in the world

Tourism types	People/year
Religious travelers	3000,000,000
Faith base trips taken	6000,000,000
Religious meeting attendee	15,000,000
Churches and religious organizations	400,000
Religious travel programs	50,000
Participate in volunteer vacations	1,500,000

Table 2.2 Infrastructures involved religious tourism to move the industry fast growing

Travel agents	Tourist boards	Lodging/accommodations
Tour operators	CVBs	Attractions
Ground operators	DMOs	Motor coach operators
Receptive operator	Travel wholesalers	Air services companies
Incoming operators	suppliers	Transportation of all kind
Destinations	Insurance companies	Convention centers
Missionary travel	Rail and train	Car hire/rental cars
Agencies	Consortiums	Retailers
Airlines	Trade associations	Consolidators
Museums	Cruise lines	Charter transportations
Travel guides	Nature reserves	Services providers
Technologies	Shrines	Online travel companies

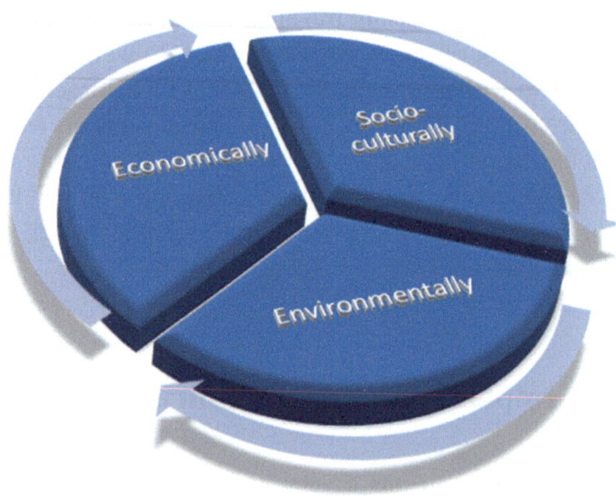

Fig. 2.9 The positive feedback loop illustrates whilst the quality of life is link-ages from the surrounding arrows connecting public goods of tourisms regarding to three major aspects in the (1) Socio-cultural, (2) Environmental, and (3) Economic conditions, respectively

a clear picture of the importance of religious tourism and its impacts on our different cultural background.

2.4.1 *Socio-Culturally*

Religious tourism has great effects on the cultural lives of the community. It is one of the most important issues debated by the tourism research and the academics today. It helps to improve the quality of life in different cultural backgrounds, and that makes people learn more about their world, exposing themselves to new perspectives, experiencing different cultural practices, broadening horizons, and increasing insight and appreciation for different approaches to live in an area. Religious tourism also offers the residents an opportunity to meet interesting people, making friends, learning about world, and exposing themselves to experience different cultural practice from different backgrounds. The possible impact that could cause in the destination for religious tourism:

- Improved the living standard of the people;
- Improved the understanding of different communities;
- Increased demand for history and cultural exchange;
- Preserved cultural identity of the host area; and
- Promoted cultural exchange.
- Positive change in values and customs, etc.

2.4.2 Environmentally

Areas with high-value natural resources such as oceans, natural and artificial lakes, spas, mountains, distinctive values of flora and fauna which possesses an authentic natural beauty, serves as a motivating factor, attracting good numbers of international tourists and natives. They like to try to find exciting and holy association with nature environment. People tend to value the pristine natural features of the destination that preserve, conserve, and restore the ecological values of the developing and industrialized nations. Land mass portrays unique features with lodging, camping, and other accommodation facilities. Travellers can stay overnight during attending major religious feasts. The places like parks offer facilities like volleyball courts, basketball lawns, football fields, and other recreation facilities. They are constructed for recreation to the visitors.

Religious tourism generates income for environmental protection and development projects. This will pave sound ecological initiates to beautifying the scenery. The various available funds and budgets support the preservation and conversation of historical and monuments that serve at traditional landmarks, of developing economies. These are of great religious significance visited by high number of domestic and international tourists.

2.4.3 Economically

Economically religious tourism increases job opportunities of a destination. These opportunities provide employment for residents of the destination. They can work in accommodation and catering industry, airport operation, cabin crews among other all of which help promote a better living condition for citizens of a destination. Furthermore, this industry diversifies work nature of the employees, varying from average low wage entry level to high paying proficient position. From the aspect of view in management, tourism administration for programs spawns the income and lifts standard of living of the local people in destination as faith travelers spending during their vacations in host communities. These proceedings help both urban and rural regions. The diversification shapes by religious tourism lend a hand to local communities that are probably reliant on travel and tourism industry.

2.5 Conclusion

Religious tourism grows and expands in recent years, coupling with chances that are created for both the local tourists and foreigners for investment of small medium and micro enterprise business enterprises. The developments of various projects are also implemented to be of standard with customer's requirements in meeting their need and demand, and infrastructure spending.

Religious tourism or faith travels frequently will encourage the development of public utilities. They are like tap burn water, sewage drain, sidewalks, illumination of light, car parking lots in counties and city centers, public restrooms, litter and waste disposal, and landscaping. Such upgrading for development is needed to benefits people from faith travelers. Likewise, tourism encourages development for transportation infrastructure, ensuring in advancement of good roads, expansion of airports facilities to meet the standard required by public transportation, and non-traditional transportation. Some examples are like train stations which are constructed for facilities of tourists' movement.

Religious tourism encourages new fundamentals to fasten together to put on the market ever-increasing opportunities for local development. Local people will have a better share of the profit generated by tourism business, adding healthy competitive value to the reputation of a particular tourist destination which is often regards as increases in community's revenues. Accommodation, lodging, camping, and sales taxes of bed and breakfast in the hospitality industry, increase excise revenues coming of air travel, other transportation taxes, business taxes, and fuel taxes are new jobs that generate more income and tax revenues.

When considering economic impacts of tourism, it is essential to understand that tourism businesses often include significant numbers of people with low-paying jobs, often at minimum wage or less. These jobs are often seasonal, causing under-employment or unemployment during off-seasons. Labor may be imported, rather than hired locally, especially if particular skills or expertise are required, or if local labor is unavailable. Some tourism-related businesses are volatile and high-risk ventures as they are unsustainable. Greater demand for goods, services, land, and housing may increase prices that in turn will increase the cost of living. Tourism businesses may claim land that could have higher value or other uses. Additionally, non-local owners, and corporations may export profits out of the community. The community may have to generate funds (possibly through increased taxes) to maintain roads and transportation systems that become more heavily used. Similarly, if additional infrastructure (water, sewer, power, fuel, medical, etc.) is required, additional taxes may also be needed to pay for the community.

References

Bullivant S (2012) Not so indifferent after all? self-conscious atheism and the secularisation thesis. Approaching Relig 2:100–106

Dawkins R (2016) The god delusion, 10th anniversary edition. Transworld Publishers Ltd, Ealing, London, UK

Fourie J, Rosselló J, Santana-Gallego M (2015) Religion, religious diversity and tourism. Krklos 68:51–64

Hackett C, Stonawski M, Skirbekk V, Grim BJ (2012) The global religious landscape. The pew forum on religion & public life. Pew Research center. Retrieved 31 March 2019

Healey A (2016) Roots: the saga of an American family. Da Capo Press, Cambridge, MA, USA

Rinschede G (1992) Forms of religious tourism. Ann Tour Res 19:51–67

Sharply R (2009) Tourism, religion and spirituality. In: Jamal T, Robinson M (eds) The SAGE handbook of tourism studies. Sage, Thousand Oaks, CA, U.S.A

Timothy D (2012) Religious views of the environment: sanctification of nature and implications for tourism. In: Fennell DA (ed) Holden A. The Routledge Handbook of Tourism and the Environment. Taylor & Francis, Abingdon Oxford, UK

Weidenfeld A, Ron A (2008) Religious needs in the tourism industry. Anatolia 19:357–361

Chapter 3
Mass Tourism

3.1 Introduction

Mass Tourism plays a very essential role in the development of the tourism industry after the World War II (Jayawardena 2002). Due to that, the trend and popularity has helped in the up grading the mass tourism cross to domestic products. The destinations have involved in motivating tourist to come in large numbers to various destinations. It has led to the expansions of the destinations in various aspects, based on the following factors: (1) economically, (2) environmentally, and (3) social-culturally. Mass tourism has been recognized as one of the major sources of foreign revenues, employment opportunities for the local peoples, and economic growth for many developing and least developed countries.

Today, tourism is one of the largest and most dynamic developing sectors of external economic activities. Its growth and development rates are considered as high volumes of foreign currency inflows, infrastructure development, and introduction of new management and educational experience. They are all actively affected various sectors of economy. It is positively contributed to the social and economic development. The highly developed western countries like Switzerland, Austria, and France, have accumulated big chunks of their social and economic welfare on profits from tourism. According to recent statistics, tourism provides about 10% of the world's GDP and employs almost one tenth of the world's workforce (Manzo 2019). All of these factors considered tourism as actual and potential economic impacts to economy. Many people emphasize positive aspects of tourism as a source of foreign exchange, a way to balance foreign trades. Looking at it from an academic point of view, tourism has numerous definitions by different scholars and schools of thought, few of them are discussed below.

Before discussing about mass tourism, the writer would like to first look to the meaning of mass tourism. This will provide us clear understanding about what mass tourism is all about relating to the concept, and operation. These are some of few definitions for mass tourism stated by different type of writers (Singh et al. 2003; Vainikka 2013).

W.-T. Fang, *Tourism in Emerging Economies*,
https://doi.org/10.1007/978-981-15-2463-9_3

- Mass tourism is an accumulated appearance of tourist in a special destination.
- It is the act of visiting a destination with large number of people at one time to places of leisure and interest.

3.2 The World's Top Tourism Destinations on International Tourist Arrivals

Annually, the United Nations World Tourism Organization (UNWTO) collects data every year to the most visited countries based on the number of international tourist arrivals for every country around the world. Dating back to 2012, to 2015 list has not showed many changes. The European market seems to still leading with tourism arrivals every year. For example, the European destinations, like France, Spain, and Italy are still leading with the big number of arrivals. On the other hand, China, Turkey, and Russia do still hold the same positions. The country holds number ten is Mexico replacing Thailand. The top 10 of most visited countries as compiled by WTO in 2015 are as follows.

3.2.1 France

France received 83.7 million visitors. The City of Lights, Paris, is easily become the most sought destination city in the travel world. It is really no surprise that France has always at top of the chart over and over again. The services provided France has become a pull factor in motivating many visitors from around the world. Whether you are traveling to visit the famous Musée du Louvre or Eiffel Tower in Paris, or going wine tasting in Bordeaux, and castle hopping in the Loire Valley, the destination has always provide visitors plenty to offer. As an European country, France is also relatively compact. It is convenience to travel from one region to another within a short time.

3.2.2 The U.S.A

The average visitors of 74.8 million come to the United States of America (U.S.A) annually. Stretching from California in the west to New York in the east, the United States of America covers an area of 3.79 million square miles—that is a whole lot of country to explore. The U.S.A is as diverse as it is huge; with many national parks and vibrant cities visited by foreign tourists, local visitors and business travelers. According to records, New York City, Los Angeles, Orlando, Las Vegas, and Chicago were the top five most popular visited cities in the U.S.A. (Schmalbruch

Fig. 3.1 There are many national parks in the USA. One of the favorite tourists' park is the Yellowstone National Park. (Photos by Wei-Ta Fang)

2017). We rather prefer the national parks—some of our favorite parks are the Yellow Stone National Park, Petrified National Park, Mount Rushmore National Park, Grand Canyon National Park, Zion National Park, Arches National Park, and Sequoia National Park, to name few (Fig. 3.1).

3.2.3 Spain

Every year around 65.0 million visitors come to Spain. Since 1970s, Spain has always attracted a lot of European tourists, mostly to the shores. In summer, and spring vacations, Costa del Sol and many of the Balearic Islands, are often flooded with German, British and Dutch tourists. Spain has a lot more than pretty beaches. The country also has rich vibrant cultures to showcase to its visitors. They also have lots of historical sites to see and rich culture to be experienced the vibrant hip metropolis of Barcelona, traditional towns in Andalusia and Castilla-La Mancha. Spain makes for an excellent travel destination with its world-class cuisine, centuries-old history and vibrant, and the characteristic cities.

3.2.4 *China*

About 55.6 million visitors from all over the world come here annually. Since it opened its doors to tourism, the number of visitors to China has skyrocketed, largely due to an increase in global business dealings and cultural exchange. It doesn't look like that or is it going to stop anytime soon? A large number of companies have moved their production chains to China, so have banks and big name franchise stores. At the same time, China is becoming more and more popular as for travel destinations. China can offer ancient cities, historical sites and good food with wine for tourists (Fig. 3.2).

They can dine under limelight night markets in most of the city centers, where visitors can eat different type of local foods, even in the crowded capital city of Beijing, Shanghai, and so on. There are lots of secret corners with local flair and cultural heritage. China also provides a lot to offer in rural areas. Many parts of China are beautiful in landscape, fauna and flora. Some places are hidden but tourists can experience things like farming in Yangshuo, climbing up the world's tallest Buddha statue, and hiking the Mount Emei, a unique place for tourists to experience.

3.2.5 *Italy*

Travelers come to Italy reaching 48.6 million every year. From romantic canals of Venice to the history rich piazzas of Rome and pebbled beaches in the Amalfi Coast, Italy's attractions are so many. The country provides a diverse travel experience which services have a pull factor in motivating visitors to come. It is obvious that Italy is one of the most favorite countries to be visited. As far as tourism trends go, it might increase its number of visitors in the feature before 2020.

3.2.6 *Turkey*

About 39.8 million visitors come to Turkey annually. Spanning across Europe and Asia, Turkey is famous for its distinctive cultures, age-old traditions and varied landscapes—ranging from the snow-peaked Mountains of Nemrut to the unique valleys of Cappadocia. Historically, Turkey was the home to distinctive civilizations through the centuries, with impressive Roman ruins and Ottoman architecture to bear witness. Most importantly, Istanbul is the crowd-puller, attracting business travelers and tourists alike with its extreme characteristics: with futuristic skyscrapers alongside traditional mosques and boutique hotels next to street food stands (Figure 3.3).

Fig. 3.2 China has beautiful historical buildings and culture for mass visitors to discover. (Photos by Wei-Ta Fang)

Fig. 3.3 Istanbul, Turkey has a lot to offer to tourists. Besides Blue Mosque (left), Galata Tower (center), the Grand Bazaar (right), all are extravagant places tourists should discover. (Photos by Arba'at Hassan)

3.2.7 Germany

About 33.0 million visitors come here every year. In UNWTO's ranking of world tourism expenditure, Germany is placed at number one. It looks like that all these money spent on boosting its tourism is well worth it. Many of Germany's major cities are tourist magnets, especially in festive season—drawing in visitors with their Christmas markets, beer festivals, and summer events. Munich's Oktoberfest it counts for a big percentage of the annual visitors, while Berlin and Frankfurt attract mainly for European city lovers.

3.2.8 The United Kingdom

Although it numbers are a far cry to that of Italy but UK receives about 32.6 million visitors every year. It also has quite a lot to offer on its shores, like the old-world charm of England, dramatic landscapes of Scotland, and the lesser known coastline of Wales. Let us not forget that London is a miniature world where global cultures converge, and headquarters of almost every international company is located.

3.2.9 Russia

The Russia Federation counting for about 29.8 million visitors yearly. In recent year, Russia has joined the top tourism contenders, just under the United Kingdom in the listing. Tourism to the country has grown up tremendously. The country has boasted spectacular natural attractions, like Lake Baikal, as well as marvelous historic cities: Moscow and St. Petersburg. Moreover, Russia's interesting culture and language make it appealing to foreign tourists looking for a unique experience.

3.2.10 Mexico

Mexico receives 29.1 million visitors every year. Mexico has climbed up the chart to replace Thailand in our last report, claiming the tenth spot in this year's ranking. In America's ranking, Mexico comes in the second right after the USA, beating Canada and every other country in the continent. The number of visitors to Mexico alone is higher than the total number of visitors in entire Caribbean (22.8 million). It is imagined that most of these numbers come from the Yucatan Peninsula, an extremely popular holiday destination, lined with gorgeous beaches, ancient Mayan ruins, lush rainforests, Mexican hats (sombrero) (Fig. 3.4) and vibrant nightlife.

3.3 The Beginning of Mass Tourism

The history of mass tourism can be traced back during the ancient package travel by Thomas Cook. On this package, tours for temperance meetings and aristocratic within Europe, was found to be a trend until this day. Remaining details showed evidence in the evolution of tourism development was a variety of enabling factors. They were influences underpinning the growth of tourism in particular factors adding

Fig. 3.4 In Nogales, Mexico, tourists are attracted to Mexican people who are friendly and nice, and their handcrafts, like sombreros (left), sold in the bazaar with affordable price. (Photos by Arba'at Hassan)

the success of travel and tourism namely; technological innovation, increase in time and wealth and, the emergency of the tourism industry.

Mass tourism was being a phenomenon of large scale packaging of standardized leisure activities and experiences of a destination at fixed cost price for sale to mass clients. This was brought sharply into international focus at the end of the World War II, within 20 years of industry took-off. This international tourism displayed nearly all characteristics of its manufacturing counterparts. It was mass, standardized, and rigidly packaged tours offered on a large scale.

In the opening of the 20th century, travel continued to expand encouraged by the increasing wealth and improvement in transportation. Tourism development became the fashion or trend, which was set by the upper class or elite, only to be emulated by the middle classes. It has been an evident since the grand tours and travelers continues to this day. Mass tourism is being practiced and still remains a program for exclusive destinations. Again a previously up-market activity has also been in a packaged type.

3.3.1 Package Holidays

Package holidays are also known as package tours. They used to be one of the most popular holidays that motivate good numbers of the masses traditionally. The package consists of a scheduled flight known as charted flights, accommodations, excursions, foods and so on. These types of packages are usually organized by tour operators which are sold to the travel agents or directly to the customers, all are in inclusive packages. Thomas Cook was one of the first tour operators, chartered his first trip by train on the 5th July, 1941 from Leicester to a rally in Loughborough. The distance was about twenty miles away on the mission of a temperance meeting, coupled with the commencement of noble travel among the royal British family. They used to travel around the area to gain skills and education, and social to fit for positions in the government departments or agencies. After Thomas Cook started to gain popularity from the first tour, he started to undertake wider tours worldwide around 1872. In the year 1948, the Thomas Cook Tour operating company was nationalized. After World War II, when the touring declined, he visited the British seaside resorts then began to do some promotional works. This had brought back the company to its feet by promoting holidays to destinations like Switzerland, Spain, Italy, and many more. As a result of his efforts, many tourists were motivated to book the package holidays with the company again.

In addition, Vladimir Raitz, the co-founder of the Horizon Holiday Group, pioneered the first mass package holidays abroad with chartered flights between Gatwick Airport and Corsica in 1950. He also organized the first package holiday to Palma in 1952, Lourdes in 1953, and the Costa Brava and Sardinia in 1954. Moreover, with the amendment made in Montreal to the Convention on International Civil Aviation on June 14, 1954, there was very liberal to Spain, allowing impetus for mass tourism using chartered planes.

Box 3.1 The package holidays in UK Principles Services

The package holidays are divided into three groups. These groups are the (1) transport carriers like, airlines, road transportation, ferries, and cruise lines. The second group is (2) accommodation providers. These include the campsite hotels, hotels, villas, apartments, holiday centers, motels, and guest houses. The final group three consists of (3) ancillary services like, travel insurance, car hire transfer agents, and excursions operators. These will consist of operators like in the UK Principles Services. Some major UK Principles Coach Transport National Express, Shearing Rail Transport EuroStar, Lei Shuttle Air Transport BA, Virgin Airways, Sea Transport, Sea Link Hotels Campsites & Caravan sites Haven Warner, Ladbrokes Holiday Centers and Parks Centers, Butlins Car Hire Hertz, Eurodollar Guiding Service Guide on Friday, Blue Badge Guide. The chain of distribution massive range of products and services required to meet travel need of leisure and business tourists. The relationship between producers, wholesalers, and retailers are from the supply structure, or chain of distribution. This is the chain of distribution with business logo customers, and family's chain of distribution.

Component parts are package holidays, transportation, accommodation, transfers, and other services. Wholesaler tour operators are either direct or indirect retailer travel agent, then show to the customers. Holiday packages are usually sold to clients through travel agents indirectly. These travel agents receive a commission for acting as a link between the customers and suppliers. The standard travel agency commission on a package holiday sale is around 10%. Association of the British Travel Agency (ABTA) estimates that 70% of all-inclusive tours are sold by the travel agencies. The example for indirect holiday sales is the First Choice, while some tour operators are direct holidays sales like Portland Holiday.

3.3.2 Arrival of Holiday Camps

At the beginning of the 20th century as travel continued to flourish globally. There has been numerous grow in tourism industry. Among major tourism development of 1930s was the creation of holiday camps. The holiday camps aimed at the growing low income market for holidays. These holiday camps offered 24 h entertainment services, all at inclusive price in order to motivate more tourists. They can travel to different destination of their choices and stay in holiday camps. The holiday camps were efficient in operation, with overnight facilities. After the War, there was a great rush to the coast. Many people had not had a holiday for years and could not wait to get away. The holiday camps provided what they were looking for. Prices were reasonable, food was plentiful for the time, and were many activities they could engage on to enjoy their holidays. The holiday camp sectors expanded many holiday camps were then taken over for military uses. They were then opened to holiday makers.

In Britain, the first holiday by Butlins during 1936 provided cheap rates. The purposes were to build accommodation with mass catering facilities, wide range of entertainment, and amusements on board, mostly on the seaside resorts. The accommodations were typically consisted of chalets, like small flats apartments. The infrastructures were arranged blocks of three or four storey building houses, and terrace of ten to twenty long. In the UK, large numbers (some in many hundreds) of static caravans are termed holiday camps.

Holidaymakers would pay a fee for their accommodation and decide whether to go full board (all meals would be included in the price), half board (only the main meal would be included) or self catering (no meals provided).

3.3.3 Development of the Spas

Spas were already well established during the time of the Roman Empire, but their popularity was mainly based for medical benefits of the water, lapsed in subsequent centuries. They were never entirely out of favor, however, with the development of spa and seaside resorts. The heyday for the domestic market was when the middle and lower classes started to travel on railways. Spas at Bath and Buxton became famous for their health-giving waters. Spa bathing was considered very therapeutic. With rapid expansion of railways opening up, a mass market for travel, places back in many areas like in Brighton.

Baths were an important part of the daily life for both Roman men and women of all social classes. The ancient Romans managed to transform it into an art, while members of the Roman upper class, frequently built private baths in their homes or villas. They still favored the public bathhouses, anyway. These large public baths were known as *hermae* and frequently spanned several city blocks. There was a fee for using the *hermae*, but it was minimal and could be afforded by freest Romans. A visit to the bath would traditionally last several hours. They included exercise, bathing, and socializing. Roman baths were often included gardens, a library or reading room, restaurant, bar, marketplace, museums or theatres. Bathers moved from a room to other room at a leisurely pace, enjoying the company of fellow Romans.

In the 11th century, the King Bath was built over the ruins of the temple of Minerva at Bath. By the Elizabethan Era, the popularity of hot springs at Bath had increased greatly. At this time, the use of spas was becoming more widely accepted attracting many visitors who were searching for cures to various illnesses and ailments. The growth and development continued into the 17th and 18th century, when spas were frequently built in secluded mountain towns providing visitors with majestic views.

It also became a practice for spas to be staffed by medical professionals. They were prescribed and carefully monitored the treatments provided for each visitor. The treatments of this time consisted primarily of either soaking in or drinking water. These spas were tremendously successful and they grew rapidly, eventually expanded to restaurants and casinos.

In North America, native Indians were also enjoying the benefits of hot spring therapy as well. Native Americans bathed in mineral springs to enhance their physical and spiritual health. In New York State at present day, the Mohawk Indians use hot springs for their healing properties. The oldest spring known to have been used by the Mohawks is the Saratoga Hot Springs. The meaning of Saratoga Hot Springs is the place of medicine water of Great Spirit. In 1790, the Saratoga Hot Springs began offering both spa treatments and accommodations for visitors. As America expanded westward, new mineral springs were discovered. The discovery prompted to the development of an elaborate new classification system for developing spas, based on the study of geography, geology, mineralogy, and climatology. By European standards, American spas are still in their infancy and the differences between American and European spas are numerous.

Traditionally, Europeans viewed spas as a venue for the treatment of present illnesses and the prevention of future ailments. In addition, Europeans also placed special emphasis on the importance of spas in helping visitors relax by combating the stresses of everyday life. In contrast, traditional American spas have maintained their focus on wellness, attracting individuals who are already healthy with programs in nutrition, exercise, and beauty. More recently, American spas have begun to adopt numerous new programs, including meditation and spirituality. While the origins of spas are rooted in the healing waters of natural hot springs, the focus of spas around the world, including in the United States, has shifted. Advances in technology and medicine have developed and replaced treatments to more traditional.

Box 3.2 The Grand Tours

This parallel development from renaissance in the 17th century had positive outcomes from freedom of tourism. Between 17th and 19th centuries, wealthy travelers from noble families in Europe were influential in shaping the oversea travel market. Young English elites of the 17th and 18th centuries often spent two to four years traveling around Europe in an effort to broaden their horizons and scope of understanding academically. They learned about language, architecture, geography, and culture in order to experience the Grand Tour. The Grand Tour gained popularity during the 17th century during development for their social, artistic, and diplomatic skills. They saw them while visiting Italy, Switzerland and Germany. As more of the middle classes started to travel, destination in Greece, Portugal and the Middle East were added to their itineraries. In the mean times, it became recognized that education of gentlemen should be completed by the Grand Tour.

3.4 Mass Tourism in Developed and Less Developed Countries

Looking at the trend of tourism globally, it has started with the mass tourism business where the people travel in their large numbers to different place of interest in destinations of their choices. This is just for the purpose of pursuing in leisure activities due to these activities to many destinations around the world. Then, there has been come up with the infrastructural and super-structural development in order to meet the demand, need and wants of the tourists coming to the destination. One of the push and pull factors that motivated tourists to go in large numbers to destination of their choices are the triple S factor, means the sun, sand and sea. Therefore, many travelers in the world started coming in huge numbers to many destinations. This was the result from marketing the triple S where they are motivated to experience in their destination and enjoyed from the sun, sand, and sea.

Secondly, the developed countries, like America, Spain, England, Italy, France, Portugal, started from this form of tourism. They started marketing their destination through mass tourism destination by using the **three-S (3S)**—the **sun, sand** and the **sea.** The tourists were attracted to come to their destination (Claver-Cortés et al. 2007). As a result of this, it has motivated a lot of tourists from different parts of the world to develop on building of seaside resorts, amusements parks, recreations centers, and parks, etc. After that, when the tourism industry had fully developed and started to gain in popularity, they started to look into other form of tourism. That could help them attracted more tourist to their destinations. As a result of this, it has made most of these countries to be into top ten tourist destination. This can be monitored by looking at the statistics of tourist arrivals going into this destination every year. These developing countries have also started to market their destination through lower cost of package to destination. Therefore, tourists are motivated to come to their destinations. They did marketing similar products and concepts to developed countries, that is, the sun, sand, and sea.

The third one is the grand tour has also played an important role in the development of mass tourism. For example, in Italy, the Grand Tour is another motivating factor of traveling in the 17th century. This was mainly only for academics, priests and important people. However, by the 18th century, wealthy families also commonly took part in the Grand Tour. In around 1840, when rail transport was introduced and the Grand Tour began to slightly fall out of vogue, visiting Italy was no longer considered something for the elites, and the first form of mass-tourism was introduced. The 1840s saw the period in which the Victorian middle-classes began to tour the country too, unlike the Georgian upper classes in the 18th century. Several Americans were also able to visit Italy, and several more tourists came to the country. Places like Venice, Florence, Rome, Naples and Sicily were still remained at the top attractions. They were attracting in masses of tourists from different parts of Europe traveling to Italy. During this period there was a rapid development of seaside resorts to accommodate tourists in their mass number.

The developing countries like the Gambia started venturing into tourism since the 1965. During this period, they came up with the Gambia Tourism Authority, then, was renamed to Gambia Tourism Board. In the tourism market, Gambia offered cheap costing with numerous destination points using the **triple-S (3S)** concepts, i.e., the sun, sand, and sea, as explained earlier. With that new concept, it has pull factor to attract tourists from European countries to Gambia. Due to this reason, they pulled tourists coming from different parts of Europe to the Gambia, similar to destinations like Spain, France, Turkey and Cape Verde Island, when their winter season came.

Table 3.1 The advantages of mass tourism in three aspects

Social	Economic	Environmental
Leisure facilities are put in place where the local people could benefit from	Mass tourism creates many jobs in the industry for local people	Great awareness of the landscape, wildlife preservation, and monuments preservation become tourist attractions
New infrastructure are put in place, examples like airport hotels, electricity, water scan telecommunications. Local community gets benefits	Great earning of foreign exchange	Income from tourists may be paid for management conservations and repairs
Income from tourism industry can be invested in the community	It also help to improve the gross domestic products of a destination	
Mass tourism also helps the local and the tourist to learn from each other culture. This can help in building a mutual understanding between the locals and the tourists		

Good number of tourists from Europe and Scandinavia came in large numbers for vacations, and leisure.

During winter season in their countries, they can have good time under the sun, on the sand, and in the sea. Within these periods of time, the number of hotels, motels, beach bars, restaurants clubs, casinos, and etc. started to be occupied in full percentage with the Tourism Development Authority's (TDA) help. As a result, this also attracted many investors to come and invest in Gambia. This is really helping Gambia in up lifting the standard of tourism industry and economy. When tourists started coming in large number, especially during winter at their home countries in Europe, they look into other forms of tourism, thus developing tourism industry to widen the other products and services for money. Therefore, every year, Gambian Government under the Gambia Tourism Board will have positive income. Table 3.1 indicated the advantages of mass tourism in three aspects in developed and less developed countries. However, some disadvantages of mass tourism have been detected in three aspects in Table 3.2.

3.5 Factors Involved in Making a Mass Tourism Successful

From previous introduction, we can try more and take some advantages of mass tourism in three aspects, but we should avoid the disadvantages of mass tourism. Various organization and enterprises are involved in the packaging, planning and

Table 3.2 The disadvantages of mass tourism in three aspects

Social	Economic	Environmental
Loss of peace in local communities because the desire to make money supersedes health and safety regulations	Many jobs in tourism industries are seasonal. Good numbers of jobs are low pay job because of its demand	Destruction of the environment and loss of habitats occur in order to build hotels and roads
Locals want to copy the life style of the tourist coming into county and due to this factor they lost the norm and values of their cultures which could eventually fled away	The industry is dominated by large travel companies which sell package holidays by brochures or internet	Pollution problems liters untreated waste going into rivers, sea, and air
Traffic congestion will delay the transactions of the local communities	High cost of living for the locals. Most products and services could be very expensive for the locals to afford due to high demand of tourist in the destination	Gully erosion is likely to happen due to high demand in constructing more hotels in the destination
Drugs trafficking could easily be in the country due to the high demand of mass tourism which could eventually lead to the loss of life	The tourist may prefer familiar goods, which has to be imported, rather than the local products	Locals could be displace due to the development of costal resorts
Mass tourism can easy tarnished the image of the country if tourism industry is not properly managed	Mass tourist resorts often owned by foreign investors and at the end of the day all the profit will go out of the country and most of this normally happen in the less develop countries	Damage to natural flora and fauna
Locals people can't rarely effort the used of the services provided		Extreme pressure put on the delicate ecosystems
It could also increase a rift between the rich and the poor people		Conflict between the environment and the economy

administrating of package tours. By working on the logistic of the tour from the originating destination to the host destinations, we do all recommendations to the health and safety checks, as well as analyze the deployment of human resources and contracting of hotel accommodations in the following aspects:

- **Tour operators**
- **Travel agents**
- **Accommodation providers**
- **Transportation**

3.5.1 Tour Operators

The role of tour operators is unlike the travel agencies. They sell holiday and range of other travel products to meet the need of their customers. Tour operators actually come up with different verity of holiday's packages that include accommodation, facilities, transfers, excursion, and other services. If we want to consider, travel agents are the retailer arms of travel business, then the tour operators can be linked to wholesalers, since they buy in 'bulk' from the providers of travel services. The travel service providers are like the airliners and hoteliers. They will break the 'bulk' into manageable packages and offer the finished product on whole packages like inclusive tour. Here they will either be sold to the travel agencies or direct to the consumers. These are some of the products tour operators come up with to make up the package holiday. There are different type of tour operators as follows.

- ***Specialist Tour Operators***: A specialist tour operator is different from mass-market tour operator. Here, they put together packages with a combination of accommodation, travels, and sometimes transfers. A specialist operator is more flexible to customers to be able to do what they want in order to meet the need and wants of their customers without difficulty. They only specialize in cater to the tourists interests of choice of activities during holidays, like, bird watching, cycling, trekking, hiking etc. They also sometimes have few clients than mass operators.
- ***Incoming Tour Operators***: Incoming tour operators purchase tourism services in their own countries and package them into products that are subsequently marketed abroad by partners. Some of the incoming tour operators also play an essential complementary role to the outgoing tour operator by providing a range of optional services at destination. They are like passenger transfers, tours, sports activities, special visits, etc.
- ***Mass Tour Operators***: A mass-market tour operator is a tour operator, like Thomas Cook or Thomson's who deal with about any type of holiday. They put together packages to suit every ones needs whether it is wild two weeks in the party central Ibiza or a romantic weekend getaway in the city of love, Paris. They cater for just about everything. A mass market tour operator supplies their customers with accommodations, transfers, transport and other extra services that their customers may need in order to meet the demand and want of their customers.
- ***Domestic Tour Operators***: A domestic tour operators play a very vital role in the development of the tourism industry. They are responsible of selling packages holidays with the country they are operating in. For example, a domestic tour operators in the USA would sell packages for exploring the United States, like trips to national parks (e.g. Zion National Park, Badlands National Park, Gardens of Gods National Park), canyons (e.g.: Grand Canyon National Park, Black Canyon; Nine Mile Canyon), or something like that, as opposed to packages to foreign clients.

Box 3.3 The Federation of Tour Operators

The Federation of Tour Operators is known as FTO. It plays a significance role to make sure that there is long-term success of the air inclusive holiday. This holiday is through the influence of governments and opinion from formers on the benefits to consumers and other stakeholders as compared to other forms of holiday arrangements.

FTO and its members have been proactive in the field of holidays' health, safety and hygiene for many years. By setting up a committed and proactive Health and Safety Committee attended by industry professionals, are supported by the FTO Head of Health, Safety and Operational Support and independent technical experts in such areas as fire safety, security, food hygiene, health and transportation. Through partnership, education, information and support, they also continue to exercise a powerful influence for change on issues such as, health, safety and hygiene, and regulatory improvements overseas. These are done to make sure that any tour operator that registers with them has to fulfill the rules and regulation, basically to promote safety and security of the customers at any particular destination.

Secondly, one of the most essential roles of the federation of tour operators (FTO) is to make sure that if any of their members are in trouble in bringing back their customers due to war, strikes, earthquakes etc. in a country during holiday, the federation always come in and take action to fly their customers back to their countries safely. This is done because the federation has a contingency play in case of emergency happens to any tour operators registered with the organization.

How tour operators plan and sell their package holiday? These are organized trips of predetermined activities that involve several tourism services. Tour operators design these holidays and sell them at a fixed price. This could also be a package as all-inclusive package. Tour operators typically follow the pattern in terms of planning and creation of package holidays and later sell them to customers, through:

- **Research**
- **Contracting**
- **Pricing**
- **Target Groups**
- **Integration**

3.5.1.1 Research

It is essential that tour operators undertake a market research to find out good products and services. This will help them to have a clear understanding to see if they package of their products to a particular destination is achieved or not. The research will also help them to know their target market. As result, this will enable them to have a decision as weather to go head to lunch a package or not with an established programmed. The role of a market research will vary between established programmers and the introduction of a new programmer based on the need and want of the customers they are targeting to go in for the product and packaging of a particular destination.

Research has to take place on a continuous basis. Tour operators are constantly monitoring their sales, competitors and the market as a whole. With this, they can make decision about capacity to be offered to customers in an existing destination or to decide which destination shall be dropped. They use many sources of data and trends to information of the process.

- ***Sales Figures***: When a figure goes down, this is the time to leave the destination and if it the figure goes up at some point is the time worth investing in more capacity. This is the way operators figure out their sales.
- ***Communication***: This is where sales are make to people in reservations and with travel agents make a report on customers demand, in the internal communication there should be efficient marketing staff, whose should work in conjunction with operations staff for smoothness of the research.
- ***Research Findings***: Research finding also plays a very essential role for the tour operators to have a good knowledge. They should know what products and services are available in a particular season. Therefore, before they are able to design specific products and services that will meet the need and wants of their customers and as a result of these it will help them to meet their target market by putting up good packages programs for the customers.
- ***Pest Analysis***: In research, there could be pest analysis. With this, it will enable the tour operators to have a clear understanding of a particular destination. They can detect by looking at the following factors: political, economic, social and technological. Without the help of these factors, they will not ne bale to have a clear knowledge of a destination in other for them to see if they could package in their products and services to that particular destination

An organization which is aware of these factors and used the information or the tour operators will come to a decision making about which destination to include in their package holiday programmed. In spite of this, there can be occasions when an operator has to pull out of a destination because of factors which could not be foreseen. An example can be like political crises or strikes, with the help of these factors, they will be able to set up a contingency play in cases of and problem.

3.5.1.2 Contracting

Tour operators fixed typically contract about twelve months ahead of the holiday season. This means that tour operators need to make an estimation of capacity in order to agree with bookings of the hotels and airlines. There can be some adjustment done later but it will be difficult to make major changes due to prices have been agreed and brochures have gone for printing. Is also difficult to contract too far in advance as tour operators have to make decision, for accommodation providers whenever they are packaging their products.

- ***A Fixed Contract***: These are not advantageous to the suppliers as it means, that if the accommodation is not sold they will still pay the accommodation provided. This can be a disadvantage to the tour operator.
- ***An Allocation Contract***: This is one of the most popular contract operators signed with hoteliers, to ensure that all their hotel beds are sold. Sometimes they contract with tour operators for more than the number of beds they have. However, they have to confirm by an agreed date, how many beds will actually be taken. Allocation contract is not only applied to the hoteliers but also applied to airline companies. Smaller tour operators cannot fill a whole chartered flight (aircraft), so they buy allocation of seats from them. For example, in Gambia tour operators had dictated what priced to pay to the hoteliers for the beds. This was because they are the ones in bringing in good number of tourists in the tourism industry. If the hoteliers did not take the price they offered to them they would just recommended their tourist to other hotels. The next hotels would probably be ready to take their offer. For in this case, they would form a vertical integration with the hotel.
- ***A Sale-Only Contract***: This means that the tour operators estimate the amount of accommodations needed and then pay for what is actually sold. This can leave hotelier with unsold rooms, which they can resell in less than minutes to other agents. The other agents need their services or to other tour operating company.

3.5.1.3 Pricing

There are number of pricing strategies which are commonly used by tour operators. These can include the following pricing strategies: fluid pricing, discount pricing, promotional pricing, competitive pricing, and seasonal pricing. These can be explained in the following context.

- ***Fluid Pricing***: This is where the brochures are issued by tour operator long time before the season started. They are printed with prices. The prices are put in at these stage may be true reflection of the price that the customer will ultimately pay to the tour operator when booking their holidays with that particular tour operating company.
- ***Discount Pricing***: Pricing may be discounted if a holiday remains unsold and the customers are not going for what tour operators are offering to them. By fixing contracts and therefore, prices a long time ahead of the season, tour operators purchase their cheap prices for flight and accommodation to ensure that all holiday packages are not left unsold. This will help others not to lose on their business.
- ***Promotional Pricing***: Tour operators tried various promotions linked to prices over recent years, including two for one offer. This was not very successful as many customers believed that the original prices had been raised to cover the offer of the second one in the package. With the help of these pricing systems, it will always help in tour operators to motivate more tourists to book with their tour operating company in their masses.

- ***Competitive Pricing***: Mass-market tour operators are keen to remain competitive on the price as they often match prices offered by competition. Some times more up-market operators will do this as well, so it is worth customers asking.
- ***Seasonal Pricing***: Tour operators divided the year into different seasons. First, there is a broad division into winter and summer seasons. In the summer, usually they offered more beach holidays while in the winter, they offered indoor activities like, indoor games and ice ski. A summer brochure will typically cover from May to November and winter brochure from November to April. As a result, it helps them to put in different pricing on each and every one of them depend on the time the customers are booking with them.

3.5.1.4 Target Groups

Target groups are exceptionally valuable to tour operators as they are effectively deals with group of students, academic, volunteers, and educators, as well as families and youths. They handle group booking, offer special discount, and arrange them closer to each other, and so on. These are done for seat in the flight and room in the hotels or other type accommodation. A room in the hotel sometimes can accommodate 3 persons with the facilities of conference hall, swimming pool depending on how the customers' preference. They will want their holidays to be like that because the tour operating companies are always there to set up their itineraries to suite customers need and expectations.

The identification of different types of customers, they categorized by customers coming from various European and domestic travelers. The tourism industry has a wide range of customers to provide excellent services from catering different customer types in meeting to their needs and expectations. As part of their daily business, the tour operators will always come across the following types of customers whenever they sell their products to a particular destination.

- ***Socio-Economic Group***: People have been traditionally divided into socio-economic groups depending on their incomes and employment status. Tour operators offer products and services that appeal to all types of customers, regardless whether they are retired, unskilled, skilled, junior staff, middle managers, as well as senior managers. This is to suit different spending and purchasing pockets, offering products and services that both low income earners and higher earners can afford to have similar vacations with similar programs. These range from business class and economic class for the air flight. For hotel accommodations they can have VIP rooms, executive rooms, and standard room. These different type of offers will meet the demand and wants of their customers.
- ***Age***: Age is the first thing considered in segmenting target market, offering products and services that appeal to customers. The ages ranges are from 18 to 70, as well as children travel with their parents. They are given much attention and care. Tour operators target customers of different ages. For example, twenty persons are aimed at people in their 18 s. They are mostly students and Saga for the over

50 s. They are over special excursion for families. They are offered nannies for parents travelling with their kids, as well as accommodation. Travellers going with parents are also given special attention especially children with special menu and playground for them to play with their friends. These are all done to meet their customer's expectation in order to have a good repeating business with them in the future.

- ***Customers with Special Needs***: This is potentially an important area that tour operators consider much. It is through the confining its operations to customer with specific needs. Any aspects of their life are values and respected, offering them quality and standard services at all time, giving them priority boarding on flights in order to make them comfortable and feel at home plus extra services their customers want from them.
- ***Families***: People travelling with their children, both European and domestic, tour operators provide value packages alongside reliability and safety, give them products and services are as stated in the brochure. Facilities and services are also allocated for them, they are provided separate swimming pool for kids, rooms for parents and their kids can stay together. Sometimes, they are also provided room with facilities with TV play stations.
- ***Disabilities***: European and domestic tour operators receive disable or handicap customers for their travel business. They reserve rooms full with facilities suit to customers with disabilities. The rooms have disable access toilets and showers, ramps for wheelchair, and many more. The customers can go round in the premises of their accommodation. The tour operators have to ensure that there are ramps accessible by wheelchair users.
- ***Independent Travelers***: Tour operators cater for independence travelers, i.e., people travelling without group. They try to get away from normal hassle and burden of being in the group due to the congestion. These are people who travel alone and with feeling to travel without all hassles and compromises attendant on travelling with others, being your own master. Accommodation and foods are made available for the tourists travelling independently. They are only provided either bed and breakfast or bed only. They are always in need of special excursions and want to attend special event like parties and buffets.
- ***Special Interest***: Both European and domestic tour operators offer tailor made products and services, specifically for tourists seeking a particular features of destinations. Different activities are organized for them with different interest includes ornithological tours, cultural orientation, drumming, fishing and dancing (for those with interest in cultural experience). Packages like these are also put in place by tour operators to meet the need of customers based on which activities they would love to do during their holidays.

3.5.1.5 Integration

Firstly, integration in general sense means when two or more companies doing the same type of business. They come together to make the most profit in ordinary tariff as against the product and services coming from non-member organization. At the same time, they can trade among members free to some points.

Since many of the rising organizations lack necessary resource of pursing economic development single handily, integration is the best opinion for them. On the other hand, over the previous in the tourism industry the process of integration has taken place between sectors in the tourism industry in other to make the tourism industry to be flexible so that companies within the industry will be able to come up with more products and services to meet the needs and expectations of their customers.

Today, travel and tourism organization are merging, collaborating, and working together for a common interest. They have common interest into amalgamation of organization for various reasons like:

- It helps them to gather and take care their package holiday within themselves by organizing accommodation in the package.
- It is an advantage to a large company of integrating to a smaller so as to generate more profit.
- They want to keep everything within themselves; they do not want other smaller business to survive.
- It is also disadvantage to smaller companies due to that larger companies control most of the operation in the industry.

3.5.2 *Travel Agents*

Travel agent industry also is another factor in motivating a good number of the mass travel around the world. Travellers go to different destination depending on their choice. They look at the Travel Agencies to find out of different shapes and sizes. Some customers are specialized in leisure travel, some in-group tours, business travel but few do very well with all of them. Most of the services provided by travel agencies are provided at no additional cost to the travelers. They earn commission from vendor who provides actual mode of transportation or accommodation.

The travel agencies deduct of cost of the airline ticket for almost 9 percent. The rest goes to the airline, government, and credit-card companies. The same price is charged by the airline, whether the air tickets are distributed through a particular travel agency or not. More than 80 percent of all airline tickets are now distributed through travel agents. For example ASTA, short name for the American Society of Travel Agents, is the world's largest association of travel professionals. These members include travel agents and companies whose products sell in the form of tours, cruises, hotels, car rentals, etc. They are one of the leading advocates for travel agents, travel industry and traveling public.

ASTA offers many benefits to its members and the traveling public. The majority of ASTA's members are travel agencies. However, travel suppliers, such as airlines, hotels, car rental firms, cruise lines, and tour operators, are also join the ASTA. They have membership categories for students, travel schools, retail travel sellers, and others. ASTA is the organization for anyone related to travel industry.

In addition, the Association of British Travel Agency (ABTA) has really play an importance role in helping their members to grow up. Their successful business should be able to flourish in their future. The Association of the British Travel Agent has been in the heart of travel more than sixty years. The ABTA has been helping the customers and public to have confidence in their travel experience, thus customers do not face difficulty in making a choice to the destination for holidays. They also help in their members by providing a scheme of financial protection and a course of redress in case something goes wrong. The organization also makes sure that there is a voice between its members and the government in order to ensure that the industry and the public get a fair deal. Promoting health safety and security is also one of their number priorities. The roles of a travel agent should be taken the following affairs.

- The travel agents design itineraries to fit the need and wants of their customers.
- Travel agents provide travel information and event schedules for customers pertaining to destination they are selling their products to.
- Travel agents serve as a link between the customers and tour operators companies by selling their flight tickets for them.
- Travel agents serve as a marketing tool to destinations they are selling their tickets too.
- Travel agents help in booking accommodations for their customers.

3.5.2.1 Travel Agents

The computer reservation system is called the Central Reservation System (CRS) for travel agent. It is a computer system used to store information and retrieve the information. The development of the CRS has played a very essential role in the development of tourism industry. With the use of the computer reservation system, it has helped to improve the work of the travel agent by making their works easy and simple. This can help the travel agents to make reservations direct from their based to any airline anywhere without going through a coordination process or contract negotiations with the tour operating companies they are partnering with. Furthermore, with the help of the computer reservation system in travel and tourism industry, it helps travel agents in processing reservation. Such products will cover presentation and reservations fare, quote ticketing, and other additional services to make their work more efficient.

The primary reason for using computers in handling reservations is to increase yield metrics, but this entirely depends on the level of systems integration. This is because computer reservation systems not only helps tremendously in the reservation processes, but also support decision-making in marketing and sales (e.g. yield management, discount policies, the creation of guest records, etc.

Box 3.4 Strengths and weaknesses of a travel when dealing with customers
The Strengths

- Travel agents save time of their customers when they make reservations with the agent.
- Travel agents' calls also provide special deals, offers, and their customers will not know what happen on the actual ground.
- The travel agent has a good peace of mind when reservations are put into their control.

The Weaknesses

- Travel agents sometimes find very difficult to retain travel products for their customers. They can only sell their products direct to the customers.
- Travel agent sometimes find it very hard to give full information to their customers because most of the time they only consternate on the products they can have profit on from their contract company.

3.5.2.2 E-Ticketing

An electronic ticket or *e-ticket* is a means by which a passenger or a group of passengers can ensure their seat in a commercial airline through booking over the Internet. A confirmation number is assigned to the passenger, along with the flight number, date, departure location, and destination location. The Internet has provided these new economic environments to conduct business. E-commerce is also a growing sector and many tourism businesses are involved in developing their Internet services, including traditional travel agents, tour operators, national tourist offices, airlines, hotels and/or other accommodation providers, and car rental agencies. The key differences from traditional markets are the speed at which information can be communicated, global accessibility, and the minimal costs of establishing a business online.

In addition, an e-ticket allows authorized travel agents to transmit ticketing information direct to the airline's database, enabling passengers to check-in and board the flight without holding a paper ticket. E-tickets substitute the paper-based flight coupons by an electronic ticket image that is stored in the airline's database. Very convenient, E-ticketing is a paperless and hassle-free alternative to the conventional paper ticket. With the introduction of e-tickets, all a passengers need to do is to show up at the airport without printed paper ticket. For travel agents, e-ticketing will lead to productivity enhancement by cutting down ticketing requirements and consequently printing of flight coupons. Travel agents can now focus on their core competence, which is travel consultancy and research, rather than on ticket issue and delivery. Access to critical passenger data gives flexibility in modifying the ticket at any point and anytime. The system automatically reports the electronic ticket sales thus lightening the work load for travel agents.

3.5.3 Accommodations Providers

The accommodation sector is one of the most essential sectors in travel and tourism industry. As we all know, without accommodation there will be no tourism. Due to this reason, it has helped us to figure out the importance in the travel and tourism industry, thus holds a central role in the tourism business. The movements of people in their large number today count on the availability of accommodations on a particular destination. As a result of this, it has also play a vital role in the development of mass tourism in many parts of the world, especially in the developed countries like France, Italy, America, Spain, England, etc. They have built many resorts in order to invite people to come to their destination in their large numbers. As a result of this factor, it has really improved the structure of their tourism industry.

In addition to the development in the tourism industry, accommodation providers are able to equip different type of accommodation systems to meet the need of these masse travels. They come from various destinations. These accommodations, process for their customers bed and breakfast, half board, and full board to meet the needs of the customers:

- Bed and breakfast (where the price is only included for the provision of bed and breakfast. It may be cooked in buffet style or continental.
- Full board (where accommodation and all meals are all provided)
- Half board (with bed, breakfast and either lunch or evening meal included in the price)
- All inclusive (also include accommodations and all meals snacks and beverages for the period of stay)

Box 3.5 The Different Types of Accommodations

- Hotels
- Guest houses
- Self catering apartments
- Tents
- Lodges
- Caravans mobile homes
- Villages
- Boarding houses
- Holidays camps etc.

3.5.4 Transportation

Development of transport has taken place since the beginning of industrial revolution which linked the growing of economic opportunities. After World War II, each stage of human societal development, particularly transportation mode has been developed or adapted. This includes air travel, cruise ship, railways and roads. They have been observed that throughout history, these transportation modes are solely responsible for the economic growth. They are linking the early stages of European expansion from the 16th to 19th centuries. They support the development of international trade through colonial empires but were constrained by limited inland access.

The first stage of industrial revolution in the late 18th and early 19th centuries was linked to the development of canal systems in Europe and America. They were mainly to transport heavy goods. This was permitted the development of rudimentary and constrained inland distribution systems. The railway stage of industrial revolution in the 19th century was intimately linked to the development and implementation of rail systems, some transcontinental, enabling a more flexible inland transportation system. This allows many people to travel for either leisure, business, or educational trips, especially after World War II.

The second stage was when the air travel began after World War II. Then came a surplus of aeronautical technology and ex-military pilots were more than ready to fly. It had contributed to the development of tourism but by then, it was only the rich who could afford to go for holidays. The airfare was very important because it helped people who always wanted to travel for holiday. It should be an easy and a faster way to get to their destinations. When air travel is affordable, it contributes to international mass tourism, pretty much as we know it today. Over the years different developments in tourism, like technology, safety and security, have changed as the way people travels. That was also the number one priority for every organization. The operations during those days were just to make travel convenience for travellers. With the development of carrier technology, wider bodies of aircrafts were built to carry people in big numbers. Boeing 747 and Airbus A380, known as jumbo jets, were also introduced to make travel convenience and accessible to customers to travel. Many travellers can be flown at one time to a destination at one time. These wider jets were designed for a combination of efficiency and passengers' comfort.

Thirdly, the development of transport, specifically cruise ships also played another importance role in motivating tourists to travel in large numbers. They travel or cruise to destinations of their choices around the world by sea and ocean. Cruising has becoming a major part of tourism industry, accounting for U.S. $29.4 billion with over 19 million passengers worldwide in 2011. The industry's rapid growth has seen nine or more newly built ships catering to a North America clientele added every year since 2001, as well as other servicing European clientele. Smaller market like the Asia Pacific region is generally serviced by older ships. They are then displaced by new ships in the high growth areas. The cruise industry is one of the rapid growing sectors in tourism industry. Its growth is expected to continue because more new ships are coming up with modern facilities. This can target better number of the masses customers cruising with them to different choice of their destinations.

3.5.4.1 Long Haul Travel

The advancement of transportation system in tourism good number of masses is motivated by travel long haul destinations. They travel for holidays without difficulties. Long haul journeys are typically done by wide body aircraft involving long distance, typically beyond six and half hours, and often are non-stop flight. On some long-haul flights, jet airliners do have to have one or more stops to refuel the airplane to reach the destination. Long-haul flights often involve intercontinental travel, like from Americas to Africa. These are always long-haul flights due to the distance, which must be covered by the aircraft.

For travelers, a long-haul flight can be demanding. People do enjoy being fascinated in the plane for an extended period of time. They may start to feel bored or restless In-flight entertainment, like music and movies, is often provided. In this case, many long distance airlines provide private entertainment stations at each seat which allow passengers to choose entertainment according to their choice. This can be either a movie, television show, games, or/and music. Food is also served on a long-haul flight, along with beverages. Sometimes, travellers can have food or beverage at their preference.

3.5.4.2 Short Haul Travel

Mass package traveling is available by means of short haul destinations. Millions of tourists today go on short haul traveling. Short haul traveling is considered to be destinations that tourists fly within five hours and below. For example, tourists travel from the UK to countries within Europe like France, Spain, Italy, and Portugal are considered in this category.

In Asia, traveling within the Asia Pacific is considered as short haul traveling. Short haul traveling is more popular than ever, with 90% of all outbound departures from Vietnam, Cambodia, Laos and Macau, going to destinations within the region, according to research from Euro Monitor International. Not only does the short-haul market dominate departures in Asia, it is also growing rapidly, with departures up by 20% between the year 2000 and 2005 as compared to long-haul's growth rate of 12% for the same period of years.

Box 3.6 Case Study on Barbados Tourism

Barbados is an island in the country named Lesser Antilles. It is 34 kilometers (21 mi) in length and 23 kilometers (14 mi) in width, totaling to 431 square kilometers (166 sq mi). It is situated in the western area of North Atlantic and 100 kilometers (62 mi) east of the Windward Islands and Caribbean Sea (Fig. 3.5). It is about 168 kilometers (104 mi) east of the islands of Saint Vincent and the Grenadines, and 400 kilometers (250 mi) northeast of Trinidad and Tobago. Barbados is outside of the principal Atlantic hurricane belt.

Fig. 3.5 Location of Barbados. (Illustrated by Yi-Te Chiang)

The beautiful Caribbean island of Barbados is one of the most sought after holiday and tourism destinations in the Caribbean. In recent years, the island has been voted the leading destination in the area. One of the most attractive features of Barbados in its unique of charming comes from the harmonious mixture of the old world and the new. The island boasts beautiful white sandy beaches, hotels, fine dining restaurants and an exciting nightlife, Barbados has proven itself to be a great place to live, work, play and invest which is one of the pull factors that has motivate in tourist and investors to go to the destination.

Over the years, the island has really adapted to accommodate the tourism trade with lots of hotels, villas, and beach resorts springing up all over Barbados. Tourists can choose from a range of different transportation method to ensure that they are able to get the most out of the island. Barbados is also a very suitable area for business growth and investment in other to expand the test and fashion of their tourism products to pull in tourist in their masses to come to the island.

Secondly, Barbados really does make the most of everything it has got. Visiting Barbados is such a special experience and tourists are always guaranteed to take away some amazing memories to treasure for life. Half million people visit the Island every year and got the most repeat visitors of the any of the Caribbean Island, according to the tourism records.

Looking at its strengths in the development of tourism, it has made the island to be one of the most visited destinations in the Caribbean. Moreover, it has also contributed 15% of the island GDP. The island has also continued to attract many visitors' and crew ships passengers in 2009 to 2010. Barbados tourism broke a 25-year record in the first quarter of 2015 (DECSBF 2015), according to the Barbados Tourism Marketing, Inc.

The island received 171,471 stay over arrivals in the first quarter, a 15.1 percent increase over the first quarter last year; the highest performance of any comparative period in 25 years. With sport and health tourism being viable economic options for the island, tourists are seeking sports, fun entertainment, and health related treatment. Each year, approximately 500 cruise ships make Barbados a stop on their travel itinerary.

The major market of the Barbados tourist industry comes mainly from Canada, continental Europe, and the United States. Majority of them come from this destinations because of the following reasons:

- Beautiful sandy beach
- Friendly people
- Solid historical and cultural legacy
- Political stability
- The beautiful natural environment, etc.

In conclusion, Barbados has been the number one in most receiving tourist destinations in the Caribbean. It continues to motivate more tourists from different parts of the globe to Barbados. By setting up type of attractions can be appealing to the tourist to meet their expectations. The customers contribute such big amount of monetary to Barbados and have repeatedly improved the business to the island.

3.6 Conclusion

Mass tourism has played a very important role in the development of the tourism industry. Because it is one of the main highly demanding sectors in the tourism industry, it has helped to improve in the development of to many countries around world today. As a result, the mass tourism came with higher influx of travelers. We should also look at mass tourism as an attractive tool for economic development, specifically in the developing world. Tourism is one of the major services in contributing immense development and economic growth of nations. This can be viewed as an export industry of three Gs, means: "get them in, get their money, and get them out." This is said to the philosophy in destinations. Tourism has assisted in the progress of many developing countries to move away from a dependency as mass tourism becomes a pull factor to motivate in tourists to come into their countries for holidays and vacations. Mass tourism is a form of an activity involving tens of thousands of people going to the same resorts, often at the same year. It is the most popular form of tourism as it is considered as the cheapest way for holidaying. This has resulting the mass tourism and what often is the cheapest way of holidays tourism sold in package with competitiveness deal with the promotion of sun, sand and sea (3S) as the motivation (Claver-Cortés et al. 2007).

References

Claver-Cortés E, Molina-Azorı´n JF, Pereira-Moliner J (2007) Competitiveness in mass tourism. Ann Tour Res 34:727–745
DECSBF (2015) Barbados breaks a 25 year record for tourist arrivals. Business focus Retrieved 31 March 2019
Jayawardena C (2002) Mastering caribbean tourism. Int J Contemp Hosp Manag 14:88–93
Manzo GG (2019) Travel & tourism economic impact 2019. World Travel & Tourism Council, London, UK
Schmalbruch S (2017) The 10 most-visited cities in the US this year. Insider Retrieved 31 March 2019
Singh S, Timothy DJ, Dowling RK (2003) Tourism in destination communities. Oxford University Press, Oxford, UK
Vainikka V (2013) Rethinking mass tourism. Tour Stud 13:268–286

Chapter 4
Cultural Tourism

4.1 Introduction

Cultural tourism has been identified as one of the most important areas for global tourism demand. The importance of this market has created a need for information on the characteristics, behaviors, and motivations of cultural tourists. These include experiencing the local culture, tradition and lifestyle, participating in arts-related activities, and also visiting museums, monuments and heritage sites (Richards 2001; Barton 2005). Cultural tourism has being the world most emerging trend of the overall travel and tourism in gaining reputation in recent years after ecotourism (Chap. 8).

Recently, some specialists suggest that the concept of cultural tourism should be placed under the lens of scrutiny because of two reasons. First and foremost, the culture seems to be a concept applied only upon aborigine communities as representative's stereotypes. Secondly, cultural tourism is being usufruct as a new way of continuing the colonization between center and its peripheries. Cultural tourism leads to an ethnocentric construction of uniqueness of cultural diversity of local people in destination, mainly focusing on artistic skills and ethnicity of locals for tourism experience, through encounters with the local cultures.

4.1.1 Definition of Cultural Tourism

The definition of cultural tourism will give us a better understanding of what cultural tourism is. At the place, one needs to know about the culture of a given society. These types of tourism can be defined as a form of tourism that is centered on the following key concepts (Wood 1984; Richards 2001; Mckercher and du Cros 2002):

- Travel to experience the arts or history of a location or travel to immerse oneself in the language, society, or culture of a region of destinations.
- Movement of people to more specific cultural attractions like heritage sites, artistic, cultural manifestation, arts and drama outside their normal place of resident.

W.-T. Fang, *Tourism in Emerging Economies*,
https://doi.org/10.1007/978-981-15-2463-9_4

- The movement of persons to cultural attractions away from their normal place of resident with the intension of gathering new information and experience to satisfy their needs pertaining to the cultural features of the host community.
- Visit by persons from outside the host community motivated wholly or in part by interest in the historical, artistic or lifestyle heritage offering of a community religion groups or intuitions.

4.1.2 Origin of Cultural Tourism

Cultural tourism is not new phenomenon. Since the ancient times, it has been a motivation for travel, although only among a tiny minority of the general public. The Grand Tour was once considered as an essential part of a gentleman's education added on which an acquisition was for artistic and cultural skills (Towner 1985; Cohen 2001). Today, cultural tourism is a worldwide mass trend that people travel to experience other cultures. However, it is only since the early 1980s that cultural tourism has been recognized as distinct from recreational and conventional mass package tourism.

In addition to the above description, the writer would like the readers to have a clear understanding what culture is. To be specific, culture is natural development during birth process what we call is behavior, that changes from individual to individual broadened when lives in a family. The family is part of a society (group or community), behavior of the group is culture; and its evolution is based on intelligence reaction to experience and need. For example, initial stages of humans, behaviors are similar to other animals living in groups. Animals too, live in groups (society or community), and their group behaviors and practices is their culture. Human intelligence (Guilford 1967) is reacted stronger to experience and needs that widened the cultural status. The culture of every given society is the social civilization set up of the communities backed by the traditional norms and values.

Culture and tourism have a mutually beneficial relationship. They can strengthen the attractiveness and competitiveness of destinations, regions and countries. Culture is an increasingly important element of tourism product as it creates distinctiveness in a crowded global market place (Barton 2005). At the same time, tourism provides an important means of enhancing culture and creating income to support and strengthen cultural heritage, cultural production, and creativity. Culture and tourism are linked because of their obvious synergies and their growth potential.

Cultural tourism is one of the largest and fastest growing global tourism markets. Culture and creative industries are increasingly being used to promote destinations. The increasing use of culture and creativity to market destinations is also adding to the pressure of differentiating regional identities and images. A growing range of cultural elements are being employed to brand and market regions in order to motivate more tourists. As a result of these, it has made organization to have partnership. Partnership is one of the most essential factors in tourism industry.

In addition, the complexity of both tourism and cultural sectors imply that, platforms must be created to support collaboration. Mechanisms must be found to ensure that these two sectors can communicate effectively. Local communities are beginning to come together to develop cultural products for tourism rather than competing directly with one another. New policies are likely to feature new structures and projects, involving public-private partnership and bringing together a wider range of stakeholders. The use of culture will not only to make destinations attractive for visitors but also to promote regions as destinations to live, work, and invest. Culture in all forms is likely to figure out strongly in the tourism product and promotion in most regions. The culture has traditionally relied on the natural assets, such as beach, sun or mountains, for their attractiveness that will serve as a pull factor to the tourists.

In conclusion, destinations are also trying to increase their comparative advantage by adding their stock of cultural attractions to make it more popular to the tourists. These will put them to develop their intangible culture and creativity.

4.1.2.1 Cultural Tourism in Europe

Cultural tourism has made a major contribution to the expansion of European tourism industry in recent decades. This contribution has made the industry to grow and gain popularity. In Europe, cultural tourism has been important in the expansion of cultural facilities, in the growth of tourism employment, development of European tourism, and cultural policies (Richards 1996, 2001). As a result of these, it has helped the tourism industry to grow and well preserved the traditional cultural heritage formation. At the same time, the motivation attracts tourists to visit museums and heritages sites to learn about historical cultures of aboriginals and learn about how their old ancestors were living during the older days.

As an example, according to sources, the Council of Europe was found in 1949. It was aimed to achieve greater unities to its members to safeguard European heritage and facilitate economic, social, cultural, educational and scientific legal and administrative matter. In maintaining human right, the Council of European is an international body, relies on specific power beyond those it national members. It has no independence sources of finding.

The Council of Europe has also had a role in promoting conservation of cultural landscape and historical heritage sites of various landmarks. The Council of Europe initially concentrated on individuals' buildings designated under the convention of the architecture heritage of Europe singing in Granada in 1985. In recent years, however, attention has been paid to wider issue of whole cultural landscape, including both natural built heritage features and archeological monuments. The council of Europe has, therefore, in the process of recommendation on the conservation and the management of cultural landscape areas in order to make sure that they are protected and well managed these heritages sites from vanish.

Secondly, the council of Europe has been particularly active in the areas of cultural heritage tourism as a mean of spreading awareness. The European heritage promotes cultural tourism to increase the coming tourists. The aim of the council of Europe in relation to cultural tourism is primarily related to culture, a feature that declares its policies different from European Union. According to sources, in early 1964, the council of European began to consider the development of cultural roots as a means of increasing cultural tourism to serve as a pull factor to tourists from different parts of Europe to visit historical sites and museums. It is expected that tourists can learn about their own culture and be able to know what was happening during the colonel period. During those days cultural tourism has always been seen as a major object of travel as the development for grand tours from the 16 century. Around Twenty Century, cultural tourism had become a case to be the objective of tourism. Sites like the British Museum in London, Louvre in Paris are points of attraction for millions of tourists. They could have a chance to see the displays of their own cultures.

Cultural tourism in broad spectrum is simply an emerging trend in almost all destinations in all continents. In this regard, we take a look at cultural tourism case studies and initiation, planning, development and administrations in diverse areas of various continents.

4.1.2.2 Cultural Tourism in Asia

Tourism is one of the most vital industries in Asian countries. It is considered the fastest growing industry. According to sources, in the year 1997, there were more than 90 million international visitors coming to East and Southeast in Asian continent (UNWTO 1998). Current projections by PR 19003 UNWTO indicated that by 2018, total of global visitor arrivals grew 6%, totaling 1.4 billion, more than 343 million international visitors arrived at Asian and the Pacific destinations. These figures do not take into account domestic tourists. For example, the China National Tourism Administration reported that China received 47 million overseas visitors in 1995, there were estimated 290 million were domestic tourists. In Indonesia, prior to economic 'melt-down' in 1997, there were found almost 4 million overseas visitors but more than 50 million were domestic tourists.

This trend has generated concern on socio-cultural, economic and environmental impacts to Asian societies and communities. Until recently, most commentary was made from a western orientation. It was assumed that tourism as Caucasian will impact western value. There is a similar misconception about host communities. They tend not to be passive unitary receptacles of tourism: they are also highly differentiated and often pro-active in adapting different categories of tourists. What once may have been interpreted as adverse impacts caused by secularism of modern western values introduced by Caucasian, tourists may have a multi-faceted basis. There is a need to re-conceptualize socio-cultural change in tourism. It may have to be re-examined many of the prevailing stereotypes about tourism's impacts that affected the tourism industry.

It is also imperative to examine tourism in the context of ethnicity, definitions of culture, and the state. There is a natural affinity between tourism and the nation-state in the sense that both have a profound interest in presenting the space as differentiate and unique, with boundaries around both geographical and socio-cultural place (Leong 1989). This is particularly true of multi-ethnic states and newly independent states (Hall and Page 2000; Teo et al. 2001). During postcolonial period, it might face problem of trying to create a sense of and commitment to national unity. Unfortunately, such sentiment did not exist. Ethnicity has become 'more than a neutral social scientific term; it has become part of the way people factually and prescriptively see themselves and others' (Picard and Wood 1997). International tourism plays a major role in this discourse and nation states will utilize the opportunities provided by ethnic tourism to objectify symbols and markers to meet political and policy objectives.

This chapter will help us to explore some implications of the tourism development and its socio-cultural impacts in Asian countries. Asian continent plays a vital role in the development of cultural tourism. It is motivated millions of tourists from different parts of the world to visit countries of their choices. In Asia, it is learned that indigenous cultures of local people live in the communities. For example, visitors watch the changing of the guards at the main gate of Gwanghwamun as one of their favorite activities in Seoul, Republic of Korea (see Fig. 4.1). Another case seeing in north east of Thailand in Koo Ka-Saing Village, this area is generally known as the most arid high land areas. Most native people live according to the history sources of civilization. Many things, like pottery utensils portraying late prehistory times in Thailand, were discovered. In addition, Mukdahan and Kalasin Provinces, including other provinces in the same southeast Thailand, play a vital role in maintaining their local traditions. They do create homestay accommodations for the tourists to come, stay, and experience their life style and learn their cultures and traditions in the local communities.

Fig. 4.1 Visitors watch the changing of the guards at the main gate of Gwanghwamun as one of their favorite activities in Seoul, Republic of Korea. (Photos by Wei-Ta Fang)

India is another country in Asia. It is also rich in cultural tourism. Cultural tourism in India is one of the vital tourism sectors that contribute a lot to its GDP. The predominant factor behind India's meteoric rises in the tourism segment in recent years. Looking back to the history, India is considered as the land of ancient history, heritage, and culture. These are some of the popular states in India which service as a pull factor to motivate tourists to visit places in India, like:

- Uttaranchal
- Rajasthan
- Tamil Nadu
- Uttar Pradesh

Among these states, Rajasthan is one of the most popular destinations because of its richness in cultural heritage.

Furthermore, cultural tourism in China is experiencing a golden opportunity for development because of its favorable policies and growing demand. Cultural tourism is one of the most essential areas in the tourisms industry that motivate millions of tourists to come to China. According to the Minister of Culture in China, "*The cultural element is essential to transform and sustain development of tourism. It can bring significance taste, diversity, fame, and grand effect to the tourism industry*" (Xinhua 2010). The Chinese folk's cultural villages also play a role in the promotion of cultural tourism (Fig. 4.2). They demonstrate traditional skills such as the manufacture of artifacts. According to traditional methods, the making of the artifacts utilize traditional materials, sing traditional songs in their own language, play a wide range of traditional musical instruments, dance and present other aspects of folklore in diverse cultures (i.e., camel riding, acrobatics, cooking local dishes in form of promoting cultural tourism in their destination).

Finally, Taiwan is one of the island nations in Asia that also promote cultural tourism. This has resulting many visitors from different parts of the world come to Taiwan and experience a good. According to Trends in Taiwan's domestic tourism and international tourism industry, they emphasize a growing preference for the destination to provide substantive access to outdoor cultural and ecological attractions.

Fig. 4.2 Most yurts stay with sheep herds are operated out of Hohhot in Inner Mongolia, People's Republic of China (PRC). (Photos by Wei-Ta Fang)

Fig. 4.3 All of these aborigines are of different cultures and stay in different areas in Taiwan. They are rich, lively culture, and lifestyles of each tribe are all precious cultural heritage. (Photos by Wei-Ta Fang)

These pulling and motivating factors to tourists provide visitors to experience the Formosan Aboriginals, for example. Tourists continue to develop a rich variety of activities and programs ensuring that they visit other indigenous culture. Below is an itemized list of the sixteen different ethnic of Taiwan indigenous people (Fig. 4.3).

•Amis •Atayal •Paiwan •Bunun •Tsao •Rukai •Pinuyumayan •Saisiyat •Yami •Thao •Kavalan •Truku •Sakizaya •Sediq •Hla'alua •Kanakanavu

4.1.2.3 Cultural Tourism in Africa

The culture of every society serves as the most important factor for tourism experience. This persuades tourists on their decision making processes to travel to destinations for sightseeing, leisure, academic, business or any other business. According to majority of nations, predominantly in Africa, some in North America and Europe, it is observed that cultural appeals of the destination have befall significantly in the development and growth of tourisms. Comprehensively, cultural attractions are usually perceived as being icons of important streams of global culture (Richards 2001; 2006). Culture as being part of human life since inception of human civilizations, the worldwide formation of ethnicity has guided us to the designation of UNESCO (United Nations Educational Scientific and Cultural Organization) World Heritage sites. The World Heritage sites today create a center of attention of higher magnitude of travelers annually. Both locally and internationally, culture is seen performing a significant role in establishing and reinforcing people's unique identities and sense of belonging to a particular locality and nurtured a sense of pride among local people. The recognition role of culture in creating and reinforcing people's identity has, in recent years, played a significant role in growing interest in diverse aspects of heritage tourism, especially in the developed world. As Richards (1996) in this volume *The Development of Cultural Tourism in Europe* stated, "it seems that the combination of nostalgia for the past, the need to reassert national and local identities have had a dramatic effect on the supply of cultural tourism." Thus, it can be argued that cultural attractions are critical for the development of tourism in the local, regional and international level.

Box 4.1 A Case Study Cultural Tourism in Kenya

Republic of Kenya is a country in East Africa which lies on the equator. With the Indian Ocean to its southeast, it is bordered by Tanzania to the south, Uganda to the west, South Sudan to the northwest, Ethiopia to the north, and Somalia to the northeast (Fig. 4.4).

Tourism can be referred to as traveling for discovery or for pleasure to destinations that are of great interest. Tourists have interests in people of the places they visit. They want to familiarize with their culture, their lifestyle, and their practices. For example, looking at Kenya, as a cultural tourism, its practiced with an aim of understanding the origins of a people's culture, their attachment to some practices, their norms and taboos, as well as their outlook on the meaning of life.

Cultural experiences provide tourists with an experience they would otherwise not getting elsewhere as a result thousands of tourists are motivated to travel to Kenya just to have a good experience about the Kenyan culture. Cultural tourism in Kenya mostly is based on the indigenous communities. They are still holding on to their cultures and traditional way of life style. Most of the communities targeted by tourists are the Maasai, Samburu, and Turkana. These tribes are known to be rich in traditional songs and dances, not forgetting that, they form the population of cattle herders and pastoralist. Their food culture is also another aspect that seems to promote authenticity of cultural tourism in destination. Bear in mind that Kenya most favorite drink is milk mixed with fresh animal blood.

Samburu women are in-charge in gathering roots and vegetables, attending children and collecting water within their society. They are also in-charge of maintaining their homes. Samburu girls generally help their mothers with their domestic chores. Entry into womanhood is also marked with a circumcision ceremony.

In Kenya, they have about forty two (42) different tribes. Each of these tribes has some form different communities with different cultural practices. As the result motive, there are many tourists travel to Kenya. They are purposely coming on cultural tourism to learn about different cultures. These cultures are spread throughout Kenya. While the aforementioned communities of the Maasai and Samburu mostly occupy the Great Rift Valley. Others like the Swahili and Digo are found in the coastal region of the country. The cultural tourism around the area goes hand in hand with coastal tourism that is mainly an attraction due to the Indian Ocean, coral reefs and ferries.

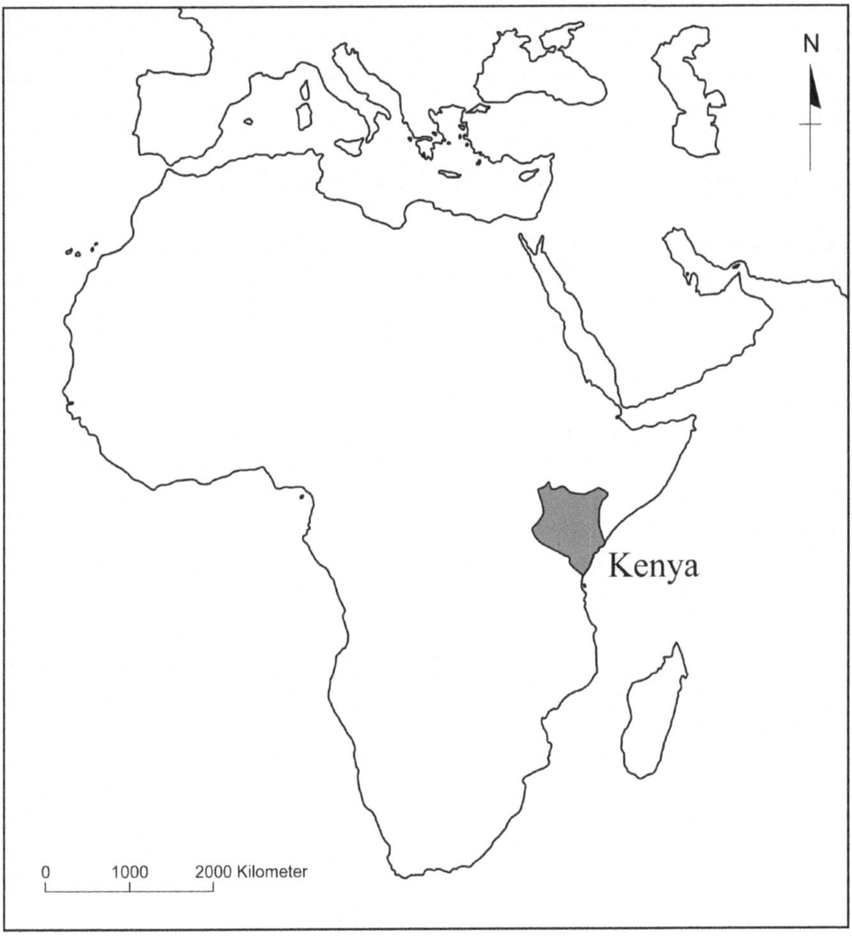

Fig. 4.4 Location of Kenya. (Illustrated by Yi-Te Chiang)

The coastal people have numerous festivals and ceremonies that are marked with traditional dances that attract many tourists. To the west of the country are the Luo and Luhya communities. These are known to be the occupants of the Great Lake like Lake Victoria. Their culture is marked with fishing, and tourists enjoy engaging in several activities as well. Cultural tourism in Kenya has helped the Kenyan communities to hold on to their cultural practices and pass it on from generation to generation. It gives much joy and pride when people travel from all over the world just to come and interact with the people of Kenya and to have a feel of what it is like to have some traditional practices to identify with. This culture gives a sense of identity and belonging to the whole Kenyan community. Although their culture is being passed on from generation to generation, tourism has its effects on the cultural rituals performed in public. Some rituals have been adjusted to suit the western values when performed for tourists; this is mainly because shocking and traumatizing happenings are not valued and accepted by families with small children.

4.1.2.4 Cultural Creativity in Developing Countries

The growth of tourism in Africa in broad-spectrum and the development of cultural tourism in particular, are at its developing phase in most of developing destinations. The point of cultural creativity is just at premature stage. However, there is a great variation in the level of tourism development in the 53 African countries. As Dieke (2000) contended, the difference in the imaginary tourism development variety ranges from the dominant (i.e., most developed) to late entrants (i.e., least developed). Few examples of this field of cultural tourism development, countries like South Africa and Zimbabwe in the South, Kenya in the east, Mauritius and Seychelles in the Indian Ocean, Morocco and Tunisia in the North, and Cote d'ivoire and Senegal in the west, (the so-called African success stories), have a sound reputable cultural tourism industry. Whereas, other countries like Nigeria, Cameroon, Eritrea, Sierra Leone, Angola, and Zambia, for one motive or another, have inadequate tourism development. By evaluation, these countries have considerable potential for future development, mostly due to their over-relying on mineral recourses at their disposal.

Cultural tourism has to be given details within broader perspective of ethnicity and ways of locals' life. In this view, culture can be scrutiny in two inter-related viewpoints: the emotional outlook mainly on what people think (i.e., attitudes, beliefs, ideas and values), and what people do (i.e., ways of life, artworks, artifacts and cultural products). From psychological perspective, culture is seen as the organized systems of knowledge and beliefs where people structure their experiences, formulate perceptions acts, and make choices among different alternatives (Roth and Moorman 1988). Thus, it can be argued that culture is in cooperation between emotional and substantial experience. As read in this dimensions clarifies, that culture is a creation of individual mental feeling which has expression in the physical world, typically fashioned as a determined skill which improve the well-being of the inhabitants, including ornamentation of scientific substance or talents of the young generation. In a good illustration, cultural tourism is the physical characteristic of discovering the cultural features of the destinations. Therefore, visiting cultural sites, museums, local fairs, exhibitions and monuments, has given a propensity to dictate the progress of cultural tourism in destinations.

4.1.3 Niche Tourist Satisfaction

In an ideal world, cultural tourism must engross mutually the emotional and substantial appearance of inhabitants in a specified location. Thus as Richards (2001) states, "cultural tourism … covers not just the consumption of the cultural products of the past, but also of contemporary culture or the 'way of life' of people or a region. Cultural tourism can for that reason be seen as casing both 'heritage tourism' (related to artifacts of the past) and 'art tourism' (related to contemporary cultural products)." The development of cultural tourism in Africa should take into consideration the two perspectives of culture. This is due to the fact that when we talk of African culture as it relates to tourism, we are talking about a 'living culture' of the African people. It is usually based on art performances and dance, and the contemporary ways of life of indigenous African communities (Hottola 2009). It is also more unlike Western cultural tourism initiatives that are primarily based on monuments and other forms of western material heritage. Therefore, as further understand in this regards, the culture of every given society is diverse from the other. In so doing, it has diverse patterns from one destination to another. Africa in the imagination of outcasts is implicit to be the structure of origin of human civilization starting from the ancient Egypt with lot artistic and historical heritages.

Most tourists may request to comprehend a fragment of themselves during an encounter with African civilization. Obtainable fashion and culture in accumulation to or perhaps in spite of, the artifacts and trinkets, are purchased to take home as souvenirs and gifts. The following critical issues on cultural tourism development in Africa run through most of the papers. First, there is clear indication that starting from initial stages of cultural tourism product development, tourism planners and developers from private and public sector should look at both the supply and demand side.

At the supply side, the most critical question that has to be answered is who is specifically entitled to decide and choose the form of cultural manifestations that should be developed and presented to the tourists. In many African countries as is the case with most countries in developing world, the government is seen as the main caretaker of the countries' cultural manifestations. The represented in the diverse indigenous cultures is mostly done in consultation with the local communities and peer groups. In the interest of different kinds of role that should other interest groups, mainly members of local cultural communities and concealed segment legislative body, engage in recreation in the commoditization process of developing and marketing process of cultural products and services in destinations.

Product research and development teams have shown that; there is generally a danger of cultural tourism studies creature product-based relatively than market-based. As examples, with increasing supply of cultural attractions, there is stiff competition among attractions at the local, regional and international level. In the content signify, rising cultural attractions for tourism does not signify predictably those tourists will come knocking at the door. As Richards (2001, 2006) comments that tourists will not come unless their needs are met. This means that it is significant to appreciate the organization and growth of the cultural tourism market. Cultural tourists are generally well educated as compared to mass package market segments which heavily replied on the sun, sand and sea. However, the cultural tourist may have the same in common but the main motivation for visits to cultural attractions is desire to learn 'new things' that local people have in their disposal. This is the authenticity of local cultures, music, dance, food and artistic values. Mainly cultural tourists are generally enthusiastic to discover the past and local culture of the places they visit. In this facet hence Africa has a relative plus than most destinations (continents) because most of the African culture is the living heritage of modern African people—with their diverse art performances, dance, and other forms of contemporary cultural expressions. These are attractions that postmodern cultural tourists are in search of. Fourth, it is argued that after a given cultural attraction, say ethnic art or indigenous dance performance has been developed for tourism; one of the main challenges that have to be tackled is development of appropriate interpretation services to meet visitor needs and expectations.

The local African communities are not well signified in design, planning, development and management of their relevant cultural possessions for tourism motivation in drawing their attentions for experience. Thus, constituent of particular local communities are typically not suitably characterized in the commercialization course of action. Therefore, the local literary possessions are altered into tourism yields to be accessible and sold to clients (tourists). However, by virtue of different academic researchers shown, those cultural appeals are frequently obtainable to sightseers in an approach that people themselves may not appreciate to be offered to the outside world.

The lack of proper planning in the product development of local cultural features of the destinations has led to questions, predominantly with considering the genuineness of civilizing, going to places of interest products and tenure of the tourism assets. This unavoidably results to the subjects of impartial allocation of tourism proceedings which gain from showcasing, and commercialization of cultural lures to tourists. The outlines of images regarding the confined inhabitants are presented to tourists. The product development and marketing of culture assets by tourism business has supplementary innovative significance to the sun, sea and surf repute. It has acquired a good image over the past decades, flourishing most regions of the world as a vibrant economic venture for under-developed, developing and developed economics. Cultural tourism in most destinations has initiates suppression, menial position, and useful as a means or an instrument; serving to promote an end.

In order to embark on touring cultural heritage sites of destinations, there is a need to products of cultural legacy to instigate cultural tourism development. This shall profit local communities sold as potential tourism products, abide by fundamental guiding principle and institutional methods. The methods should give confidence to local contribution and involvement in the project plan, accomplishment and administration of the artistic cultural possessions. Tourism development of every destination relies on the involvement of the local counterparts of the local people and communities with rich cultural authenticity.

4.2 The Importance of Cultural Tourism in a Destination

Before delving into the importance of culture in destinations, it is important to understand what culture is. Culture is one of those nebulous and complicated terms. Culture is referred to, again and again, in various contexts, and different people perceive it in different ways. Thus, to paraphrase the famous saying, "when you hear the word *culture*, you reach for your dictionary". However, if you reach for your dictionary and start to look at mass of multidisciplinary literature, the dictionary will explain the meaning of culture (Kulich 2011). It did not take long before you start drowning in different theories and get confused. For example, the famous work *Culture: A Critical Review of Concepts and Definitions* by Kroeber and Kluckhohn (1952) discovered more than 150 definitions concept on culture. Since then, the term culture has been used more widely. This reflects the complex history of the term culture and the range of usage. It has provided a focus of inquiry for several academic disciplines including anthropology, history, literary studies, human geography and sociology. Although there are no easy answers to understand the meaning to what is culture, one needs to have an overview of the theory to help you take an informed view on how important culture is to destinations.

Raymond Williams (1921–1988), one of the founders of cultural studies in Britain, had traced the development of the concept of culture and provided useful ordering of its modern uses in a series of influential books—*The Long Revolution* (1961), *Marxism and Literature* (1978) and *Culture and Society* (1983). He concluded that, outside of natural sciences, the term culture is chiefly used in three relatively distinct senses to refer to:

- Cultures as Arts and Artistic Activity
- Culture as a Learned, Primarily Symbolic Feature of Particular Way of Life
- Culture as a Process of Development.

4.2.1 Cultures as Arts and Artistic Activity

This is referred to culture with capital C, typically part of everyday talk. It is understood culture to consist of the works and practices of intellectual and especially artistic activity. Thus, culture is a word that describes music, literature, painting and sculpture, theatre and film (Williams 1983). Culture in this sense is widely believed to concern refined pursuits in which the cultured person engages (Longhurst et al. 2016).

4.2.2 *Culture as a Learned, Primarily Symbolic Feature of Particular Way of Life*

In human sciences the word culture is often used to refer to creation and use of symbols to distinguish a particular way of life, whether a person, a period or a group, or humanity in general (Williams 1983). Only humans argue but capable of creating and transmitting culture. We are able to do this because we create and use symbols. To study culture from this point of view is thus to ask, for example, what is the meaning of a style of dress (see Fig. 4.5), a code of manners, a system of belief etc. In light of this explanation, culture might be thought of as being everything everywhere and many anthropologists have adopted such a broad position, as illustrated by one of the most famous nineteenth century anthropological definitions of culture by Edward Tylor, who defined culture as "that complex whole which includes knowledge, belief, art, morals, law, custom and any other capabilities and habits acquired by man as a member of society" (Tylor 1871).

This definition emphasizes that culture is the product of humans living together and that it is learned (i.e., passed on socially not biologically). It also infers that culture is in some sense a complex whole. Interestingly, it also talks of culture in the singular, making the assumption that all societies possess a more or less advanced version of the same heritage. This explanation of culture by Tylor was successfully challenged

Fig. 4.5 A cultural presentation and analysis should engage more than the barter of simple information on cultural diversity, cultural showcasing. This should inspire and even incite the tourist to be able to cross cultural encounter of understanding and in all probability to revive cultural familiarity with local host communities from the meaning of a style of dress. (Photos by Wei-Ta Fang)

by Franz Boas (1911) who disagreed with the concept of culture in the singular and the assumption that all societies shared a more or less advanced version of the same heritage. Instead, Boas offered plural cultures that were different and could not be measured against some supposed single standard of advancement. Boas also disputed the late-nineteenth century theories that attributed most human differences to race, and his ideas were elaborated on by his students who focused on the diversity of culture.

4.2.3 Culture as a Process of Development

The earliest use of the word culture in the late Middle Ages refers to the tending or cultivation of crops and animals, A little later the same sense was used to describe the cultivation of people's minds. More recently the word culture has been used to describe the development of the individual capacities and it has been extended to embrace the idea that cultivation is itself a general, social and historical process (Williams 1983). Longhurst et al. (2016) provide an excellent introduction to these three senses on which the concept of culture is used which I strongly suggest you to read. These three senses are illustrated in the following example a play by Stan Lai (1954–) (unarguably Taiwan's best-known playwright) might be said to be a distinct piece of cultural to be a product of a particular way of life and to represent a certain stage of cultural development.

In view of the above, tourism is important because for almost all destinations, cultural facilities and cultural displays represent an important element of their infrastructure and makes destination more appealing. For many destinations, the integration of culture and economic activities has been at the forefront of growth and development initiatives. If a destination develops a thriving cultural industry, it is likely to attract more visitors or even investors.

4.3 The Advantages and Disadvantages of Cultural Tourism

A community heritage is the substance opportunity offers visitors. It protects that heritage which is essential to a major challenge in the culture tourism. The programs ensure that increased tourism does not destroy the quality that attracts visitors in the first place. When tourism is highly sophisticated to fast changing industry, it presents its own challenges, like advantages and disadvantages. Tourism is generally a clean industry without smokestacks or dangerous chemicals. However, it does put damage on the infrastructure on roads, airports, water supplies, and public services like police and the fire protection.

When travelers' problems increase in numbers, these will add stress and strain the infrastructure at heritage sites. They are safer only at beginning. The travel industry is already addressing them. However, a challenge results not only from visitor's impacts but also from visitor expectations for quality products and services. Tourism is essentially a service industry, means it depends on the competence of the people in many different jobs and locations. Tourism, while not a panacea, is an attractive form of economic development induced the advantages and disadvantages of cultural tourism as follows.

4.3.1 Advantages of Cultural Tourism

Cultural tourism has other positive proceedings and negative impacts both of which are stated here as advantage and disadvantages in previous discussion. The outline below is first and foremost stating the **advantages**, leading to the function of various types of cultural tourism. This can result to negative impacts otherwise disadvantage.

- Cultural tourism paves a sound way for the provision of job opportunities for the local people in destinations (World Tourism Organization and Organization of American States 2018). It also helps in generating economic activity and local or regional development. Most of works in the management and administration of tourism development projects happen within their territory. Cultural tourism plays an important role in ensuring the local people in destinations are employed. Local residents' works in accommodations and catering outlets, entertainment, tour operations and guides, and tourism industry are regarded as the vibrant service sectors for cultural tourism.
- Generally, tourism development in destinations constitutes most of the infrastructure development. It also helps in revitalize traditional buildings and craft industries. They commence with the construction and rehabilitation for good road networks, drinkable running water, full time electricity supply, all of which local people benefit from these facilities at their disposal.
- Cultural tourism promotes preservation and conservation of the cultural assets. It generates financial motivations to safeguard the integrity of the remaining cultural features local people. Products and services of the destinations such as groceries, life-style are the events and substantial records of the past.
- Cultural tourism in most instance calls for the protection and development of ecological features of plants and animals. They serve as demand serving to motivate tourist visits. Today, it is widely perceived as the pull factor for tourist to destination, mainly for the interest for wildlife spotting and encounters with the fauna and flora existed in rural settings.
- Financially, the economic benefit of cultural tourism cannot be over estimate as it fetches foreign exchange to the destination. Tourists from all over the world come and bring in their currency. They spend them upon arrival at their place

of dismemberment. This, in turn benefits the local tourism business for foreign commercial transactions.

- Cultural tourism in its authentic form operates to maintain edifying range of the destination. It is cooperated with different peoples from diverse cultural background to share many things in common building. It provides a vehicle for cultural exchange and understanding between the community and the tourist. This can promote sense of pride in their ethnicity, which is portrayed to tourist experience.
- Cultural tourism being a growing trend in the tourism business, connect shows and deeds, serves as an supplement worth to the existing historical and cultural legacy of the destinations.
- Advantages of cultural tourism in most destinations are highlighted. Cultural tourism helps in stimulate conservation and heritage preservations efforts. They start from the sustainability of traditional cultural legacy, helping in capability building programs through the teaching and educating of confined traditions, to the history, natural science, etc.
- Cultural tourism advances in the top quality growth of deeds that draw from local traditions and add value to the community. It enhances community cohesiveness and pride in the cultural identity.

4.3.2 Disadvantages of Cultural Tourism

As contrast, the above mentioned statements have stated the advantages of cultural tourism. The outlines below underline the demerits of cultural tourism in destinations. Most of them are rapidly increasing and altering the sustainability use of the tourism resources available at our disposal.

- Cultural tourism in destinations led to the destruction, culture of local destinations, local language of the people, interactions of the local with tourists' influence. This influences not only the dress codes but also the use of foreign languages. They are not necessarily deploy in the right order during conversations, religious values, but also depreciate tourists while tolerating less religious practice in the destinations—which are most likely violate locals during events.
- Cultural tourism development depreciates the ecological features alongside the cultural experience in destinations. It is added with couple by the ancillary services and products offered in conducting excursions. This will eventually lead to extinction of endangered species and depreciation of the forest cover.
- Job opportunities among the locals are trivial and imperial. They are mostly based on nepotism and favors. Statistically, it is observed that nine out of every ten jobs are offered to locals. The jobs are low proficiency with small earnings, and tedious services. On the other hand, prostitution is sky-rocketing, youths are venturing to drug and substantial abuses, experts predict that cultural tourism has caused the rapid increment of violence and theft cases in destination. This is intimately associated to transition of development in rural communities, youths and becoming

awareness of the worldly material world resulting them in engaging in robbery and malpractices.

- Cultural tourism has a negative impact on the culture of destinations. It has often been said that locals of most destination are getting westernized. They have gradually lost their identities as they try to become like the tourists initially adventuring into skills. Such performers, choral groups, rappers etc. are influencing locals especially the manner local speaks and pronounce words. The expressions and sentence structure has over the years become ever so more rich to meet the interest of the tourists.
- Cultural tourism is some aspects are moderate. The desire in changing of cultural dress is following the way the tourist wear their dress. Local people in destination imitate the western or tourist types of dress with the perceptions they sound more great and superior as compared to their traditional dress. This is eventually the result to a shift in dress pattern among locals to a more luxurious ways of dressing.
- Cultural tourism influences the increase in misdeed as more persons become compressed into rural settings. These are compactly inhabited by locals with high interest of tapping tourism benefits through their interactions with tourists coming in their communities. The misdemeanor charge rise as individuals in these compactly occupied rural settlements want to increase their standard of living and attain to the standard of living of the tourists. They perceive the tourists look prosperous and local youths turn to business illegal drugs. The prostitution has rapidly increased which at the same time increasing the disease of sexual transmitted infections. With higher rates in rape cases in police stations and courts have alarming locals for having a safe health.
- Facilities such as good water supply and electricity supply are decreasing in meeting the demand of local populations, higher percentage of which are used to supply tourist facilities.

4.4 The Essential New Trends that Could Have an Impact on Cultural Tourism

Cultural heritage tourism can have a tremendous economic impact on the local economy. To economic benefits like news business, jobs and higher property values, tourism adds less tangible but equally important payoffs. A well-managed tourism programs improve the quality of life as a resident takes advantage of the services and tourism attractions. It promotes community pride which grows as people work together to develop a thriving tourist industry.

An area that develops potential for cultural heritage tourism creates new opportunities for tourists to gain opportunities of an unfamiliar place, people or time. The arrivals of visitors, in turn, come new opportunities for preservation. Well-interpreted sites teach visitors their importance and extension and the importance of preserving other such sites elsewhere.

Perhaps the biggest benefit of culture heritage tourism is the opportunities increase for the diversified economic and ways to prosper economically while holding on to characteristics that makes communities special.

4.4.1 Growing Importance of Adventure and Ecotourism

Cultural tourists are increasingly aware of environmental issues. It often seek out environmentally sensitive activities in order to make sure that it does not cause negative impacts to environment. Therefore, careful management and responsible stewardship of cultural heritage sites is not only a moral and often legal responsibility issue but an opportunity for tourism operators, communities, and regions. They are to support the market by sanitizing their commitment to environmental issues. Furthermore, while cultural tourists are typically older, the new cultural tourism streams are becoming increasingly attractive to the youth market. "Traditional" cultural attractions such as museums tend to have difficulties attracting youth and young adults. Yet, there is a real opportunity for cultural tourism to serve this market by offering low-cost. Dynamic options that respond to values are important to them. Such environmental sustainability, cultural sensitivity, adventure, and physical fitness will not cause destructions to the community.

4.4.2 Increasing Popularity of Short Gateway Holidays

The educated and professional people make up the bulk of the cultural tourism marketplace. They are busy with little time left over for resting and recreation. Quick, "getaway" holidays are very attractive option for this market. This trend is closely related to increasing mobility of people, money, capital, ideas, images, information, objects, and technologies. Cultural tourism operators have clear opportunity to respond to this trend. They provide overnight or weekend packages, preferably in partnership with a local hotel or travel provider. In addition, since these cultural tourists are looking for many different types of experience on their trips, the importance of packaging and partnerships are to provide their tourists with a great experience.

4.4.3 The Changing Face of Museums

Museums play and essential role in the development of cultural tourism in the tourism industry. However, it has been put in a way to catch the attentions of the tourist. The first way in which museums have changed is to empower traditionally marginalized groups. For example, the emergence of what is termed as "First Voice" museums allow specific peoples the opportunity to collect, preserve, express, display, and demonstrate their culture from their own viewpoint in their own ways, rather than in the context of a traditional arts or natural history museum. According to sources for new exhibition, we designed in consultation with the National Museum of the Philippines, is an example in Manila. Another example is a powerful new type of museum—the “idea museum”,—which deals with significant issues common to all humans, like war and peace, human rights, or globalization. These museums challenge visitors to take a stand and make a difference any time they pay a visit to these museums. They tend to see lesser object or collection based as compared to traditional museums, thus responding to the trend toward experiential as opposed to object based tourism.

According to researches the Museum of Tolerance in Los Angeles, opened in 1993, it was a popular example of a change face of museums. Not only are the stories and experiences offered changing, but also the way the tourist thinks of museum spaces. In some recent time, high profile museum projects one instantly thinks of Frank Gehry’s Guggenheim Bilbao building, i.e., to illustrate the trend towards signature architecture, that building itself becomes a popular attraction to the tourist. Such projects are also related to the city and regional revitalization projects, in which the government often plays an essential part.

4.4.4 Experiential Objective Based Tourism

Cultural tourists are increasingly seeking experiences rather than objects. Much of this trend is related to an increasing propensity for tourists to seek experiences. Those are subjective or meaningful to them personally as opposed to the destination’s intrinsic or objective significance. Attractions associated with childhood nostalgia, such as at the Museum of Korean Songs, are well positioned to advantages of this trend. Those that are apparently less well-positioned can provide innovative, interactive programming to the types of experiences sought by visitors.

This trend is also reflected in the increasing popularity of food and themed shopping experiences, interactive museum programs and exhibits, immersive environments, and “high-tech” presentation technologies such as large-format theatres. Not only are these type of experiences more fun to visitors, but each of them promotes affective learning sparks an interest to visitors. This is also encouraged them to develop that interest after leaving the attraction place, thus at the same time give them a better understanding about the museums.

4.4.5 Convergence Between Popular and High Culture in Tourism Industry

According to research, "The Rise of Museums," like the Experience Music Project in Seattle, the Rock and Roll Hall of Fame in Cleveland, and the Museum of Korean Songs, reflects a shift in our understanding of cultures. Once clearly delineated from the popular mainstream, many "high-end" cultural attractions were dealt with themes. This was once considered outside their proper scope. This trend may be partially related to the need for the museums to boost attendance levels of tourists. They broaden the market not only for the financial reasons but also to be as relevant and responsive to the community. For example, Hong Kong Heritage Museum, which has been planned and designed over an 8-year period, is an outstanding example, linking traditional Chinese culture with Hong Kong homes, toys, music, and movies.

4.4.6 Growth of the Internet and Impacts of Technology

Statistics documenting the increasing importance of the Internet in travel planning are readily available. As stated in the previous essay in this series, analysts predicted that some in the past 2003 over $30 billion in leisure travel will be booked on-line to travel to different destinations of their choices. Using of the internet is an opportunity that can easily be leveraged if effective strategies are employed. For example, tourism websites should offer prospective visitors an integrated virtual tour of the area or the attraction. It is showing the diversity of experiences that tourists can be motivated to go in. It is not only the cultural attractions but also include shopping, dining, nightlife, parks, and whatever else. The community has to offer services which are pull factors to bring in more tourists to the destination—since visitors want a variety of experiences when they travel. Furthermore, this could serve as key factors to promotion successful tourism in a destination by provide the visitor with a good experience.

The prospective tourist must find it easy to plan accommodations, car rentals, and visits to all attractions at once, i.e., "one-stop tourism shopping". The internet is very well suited to this. Websites for museums need to be planned to facilitate a cultural tourists' on-going awareness of the museum. The programs should be from before his or her visit to the follow-up programs after the visit. This will sustain visitors' interest and ensuring positive 'word-of-mouth' publicity. A cultural tourism product is created anytime as a travel motivator and the existence of attractions as a specific event or institution interactions with a tourist's personal motivator an interest in heritage, the arts or culture.

4.5 The Role of UNESCO in Promoting Cultural Tourism

The United Nations Educational, Scientific and Cultural Organization (UNESCO) is at the forefront of international community's efforts to protect and preserve. Emergency assistance is to safeguard properties in danger to long term conservation, management planning, technical assistance, professional training, public and youth education, and awareness-building. The World Heritage Centre and its partners have to develop a series of initiatives and have been actively involved in the implementation of projects, participatory workshops, seminars and training courses.

With the increasing threats to cultures, there are destructions not only by the traditional causes of decay but also by changing social and economic conditions. These aggravate the situation with even more formidable phenomena of damage or destruction. Cultures are disappearing or and becoming more unauthentic because they are used for satisfying tourists need and not preserving its authenticity. Therefore, UNESCO can play a central role in the protection of intellectual rights in traditional culture.

Furthermore, UNESCO also helps in the development of natural polices and legislative frameworks to support the development of cultural industry and initiatives in such areas. The areas are like music arts, the literary and culinary arts, fashion, festivals, theater and culture and cultural tourism. They also help in protecting the natural, tangible cultural and regional cultures initiatives. Improving of institutional capacity for advocacy and marketing of cultural products and the protection of intellectual property is also one of the keys in the promoting cultural tourism in a destination. The roles of UNESCO in preserving culture are multidimensional. These include:

(1) to encourage countries to sign World Heritage Convention to ensure protection of their natural and cultural heritage;
(2) to encourage States Parties to the Convention to nominate sites within their national territory for inclusion on the World Heritage List;
(3) to encourage States Parties to establish management plans and set up reporting systems on the state of conservation of their World Heritage sites; help States Parties safeguard World Heritage properties by providing technical assistance and professional training;
(4) to provide emergency assistance for World Heritage sites in immediate danger; support States Parties' public awareness-building activities for World Heritage Conservation; and
(5) to encourage participation of the local population in the preservation of their cultural and natural heritage and encourage international cooperation in the conservation of our world's cultural and natural heritage (UNESCO 1972).

UNESCO's work on cultural tourism seeks to help its Member States in devising strategies for long-term preservation of the cultural heritage. Another important component of tourism is the oral and intangible heritage, including music, languages, oral traditions and the performing arts.

At the field level, promoting improved tourism practices are concerned at many World Heritage sites and biosphere reserves. Tourism in coastal region features in a web-based forum on wise coastal practices, like in recent UN Atlas of the Oceans. The teacher-education materials include a sustainable tourism module in a multimedia programmer called the Teaching and Learning for a Sustainable Future. In addition, UNESCO is absolutely engaged, alongside its Member States, the private sector, civil society, and in cooperation with relevant agencies of the United Nations. They are promoting the sustainability development concepts in tourism, which we know them as preserve cultural diversity and the dialogue among cultures.

Since the International Convention Concerning the Protection of World Cultural and Natural Heritage in 1985, China has fifty-three (53) world heritage sites to date, April 2019. Of these total sites, 36 are cultural heritage sites, thirteen (13) are natural heritage sites, and the other four are cultural and natural mixed sites. China is ranking second in the world record since 2004. China has made the first large-scale renovations on six world cultural heritage sites in Beijing—the Ming Tombs, Great Wall, Forbidden City (Fig. 4.6), Temple of Heaven, Summer Palace, and the "Peking Man" site at Zhoukoudian, all of which are planned for completion before 2008.

In addition, China is a rich country of non-material cultural heritage. Several of them are described in UNESCO's list of Masterpieces of Oral and Intangible Heritage of Humanity. These are the pulling factors to motivate many tourists from different parts of the globe to come and experience the wonder of China tourism industry. According to World Tourism Organization, in version 2020, "China will become one of the largest tourist country and among the largest for overseas travel." In terms of total outbound travel spending, China is expected to be the fastest growing in the world from 2006 to 2015, jumping into the number two slot for total travel spending by 2015.

In case of Gambia, UNESCO also recognized Kunta Kinteh Island as a world heritage site in the year 2003. This site was particularly significant for its relation to the beginning of the slave trade and its abolition. It also documented early access

Fig. 4.6 Forbidden city is one of the world cultural heritage sites in Beijing, China. (Photos by Wei-Ta Fang)

to the interior of Africa Criterion. According to UNESCO, Gambia's James Island presented "a testimony to the main periods and facets of the encounter between Africa and Europe along the River Gambia, a continuum that stretched from pre-colonial and pre-slavery times to independence." The recognition of Kunta Kinteh Island by UNESCO had played a positive impact in promoting the cultural tourism in Gambia. This island had serviced as a pull factor to motivate in many tourists from out of the world to come to Gambia. In fact, Gambia is well known to slavery. It has helped in the promotion of the international roots festival. During this festivals there were different types of ethnic groups came out in their large numbers from their villages. They were well-dressed and displayed their different cultures with traditional cultures like dancing, fashion, arts and crafts, etc.

In conclusion, Italy has the highest world heritage sites in the world according to UNESCO. UNESCO had listed fifty-four heritage sited to date, April 2019 (Table 4.1). All of these heritages sites played significant roles in the development of tourisms in Italy. It was also served as a pull factor to motivate tourists from different parts of the world to travel to Italy to experience its unique sites and cultures. Most of these sites in Italy have great connections to the Italian culture. UNESCO also encouraged participations of local traditional artists and creators to identify and revitalize the intangible heritage. This included the encouragement of public bodies, non-governmental associations and local communities.

4.6 Conclusion

Cultural tourism in its widest sense provides a set of material and symbolic resources that are abundant in supplying arguably instincts, and highly mobile proofs (Rojek and Urry 1997). Culture is certainly at the basis of international tourism. It is indeed facilitated the growth of people and allowed various societies to participate in the development process of cultural tourism. It allows of sharing their norms and values with outsiders. However, in treating culture as a resource, we should not neglect aspects of agency, as the value and priority of cultures relates not only to its intrinsic worth but also to the ways that it is used. This in turn brings questions about the ownership of, and access to, the culture. In addition, it also raises issues with regard to the ways in which culture is read by particular typologies of tourists. So called 'cultural products', as Therkelsen (2003) points out: "…generate associations and meanings that are impudence by the cultural backgrounds of the potential tourist". In this sense, tourists do not encounter culture as some value-neutral form or process. Rather, they decode culture(s), in social spaces and times in relation to particular formal and informal knowledge regimes accrued through exposure to formulated tourism packages and through the normative processes of socialization.

Cultural tourism products can be those that are especially created for the visitors (e.g., museum) or they may simply be inherent in the community restaurants and streetscapes). These cultural group products may also be linked; for example, cultural institutions that are located in a heritage district or in a town. They may easily become

Table 4.1 The list of countries with more than ten (10) World heritage sites

Countries	World heritage sites
Italy	54
China	53
Spain	47
France	44
Germany	44
India	37
Mexico	35
United kingdom	31
Russia	28
United states	23
Iran	23
Japan	22
Brazil	21
Australia	19
Canada	19
Greece	18
Turkey	18
Poland	15
Portugal	15
Sweden	15
Belgium	13
South Korea	13
Check republic	12
Peru	12
Switzerland	12
Argentina	11
Austria	10
Bulgaria	10
Croatia	10
Netherlands	10
South Africa	10

Source UNESCO (2019)

the focal point for community-based festivals and special events because tourists could have easy access to these places. There are several ways in which cultural tourism products can be specified. Perhaps the most straightforward way is in terms of theme. In this case, cultural tourism products would include, but may not be restricted, to the following:

- Environmental education and nature based;
- Archaeology, anthropology and indigenous cultures;
- Heritage, history and architecture;
- Arts festivals; and
- Natural history and wild life.

References

Barton A (2005) The conceptual arguments concerning accounting for public heritage assets: a note. Account, Audit Account J 18:434–440

Boas F (1911) The mind of primitive man. The Macmillan Company, London, UK

Cohen M (2001) The grand tour. Language, national identity and masculinity. Chang Engl 8:129–141

Dieke PUC (2000) The political economy of tourism development in Africa. Cognizant Communication Corporation, Elmsford, USA

Guilford JP (1967) The nature of human intelligence. McGraw-Hill, New York, NY, USA

Hall CM, Page S (2000) Tourism in South and Southeast Asia: issues and cases. Routledge, New York, NY, USA

Hottola P (2009) Tourism strategies and local responses in Southern Africa. CABI Publishing, Wallingford, Oxfordshire, UK

Kroeber AL, Kluckhohn C (1952) Culture: a critical review of concepts and definitions. Peabody Museum, Cambridge, MA, USA

Kulich SJ (2011) Applying cross-cultural values research to "the Chinese": a critical integration of etic and emic approaches (Doctoral Dissertation). Humboldt-Universität zu Berlin, Germany

Leong WT (1989) Culture and the state: manufacturing traditions for tourism. Cult Stud Mass Commun 6:355–375

Longhurst B, Smith G, Bagnall G, Crawford G, Ogborn M (2016) Introducing cultural studies, 3rd edn. Routledge, New York, NY, USA

Mckercher B, du Cros H (2002) Cultural tourism: partnership between tourism and cultural heritage management. Routledge, New York, NY, U.S.A

Picard M, Wood RE (1997) Tourism, ethnicity, and the state in Asian and Pacific societies. University of Hawaii Press, Hawaii, USA

Richards G (1996) Cultural tourism in Europe. CABI Publishing, Wallingford, Oxfordshire, UK

Richards G (2001) Cultural attractions and European tourism. CABI Publishing, Wallingford, Oxfordshire, UK

Richards G (2006) Cultural tourism: global and local perspectives. Routledge, New York, NY, USA

Rojek C, Urry J (1997) Touring cultures: transformations of travel and theory. Routledge, New York, NY, U.S.A

Roth MS, Moorman C (1988) The cultural content of cognition and the cognitive content of culture: implications for consumer research. In: Houston Micheal J (ed) NA - advances in consumer research (vol 15). Association for Consumer Research, Provo, UT, USA, pp 403–410

Teo P, Chang TC, Ho KC (2001) Interconnected worlds: tourism in Southeast Asia. Elsevier Science Ltd, Oxford, UK

Therkelsen A (2003) Imagining places: image formation of tourists and its consequences for destination promotion. Scand J Hosp Tour 3:134–150

Towner J (1985) The grand tour: a key phase in the history of tourism. Ann Tour Res 12:297–333

Tylor EB (1871) Primitive culture: researches into the development of mythology, philosophy, religion, art, and custom, vol 1. Murray, London, UK

UNESCO (1972) Basic texts of the 1972 World heritage convention. UNWTO, Paris, France

UNESCO (2019) UNESCO World heritage centre - World heritage list. World heritage centre, unesco.org. Archived from the original on 30 March 2019

UNWTO (1998) Tourism market trends 1998 East Asia and the Pacific. World Tourism Organization, Madrid, Spain

Williams R (1961) The Long Revolution. Chatto & Windus, London, UK

Williams R (1978) Marxism and Literature. Oxford University, Oxford, UK

Williams R (1983) Culture and society: 1780–1950, 2nd edn. Columbia University Press, New York, NY, USA

Wood RE (1984) Ethnic tourism, the state, and cultural change in Southeast Asia. Ann Tour Res 11:353–374

World Tourism Organization and Organization of American States (2018) Tourism and the sustainable development goals – good practices in the Americas. UNWTO, Madrid, Spain

Xinhua (2010) China to boost cultural tourism. China Daily 21 March 2010

Chapter 5
Rural Tourism

5.1 Introduction

Rural tourism has been considered as a means of achieving economic and social development and regeneration that can benefit to local people. It does not have any demerit. More specifically, it has been widely promoted as an effective source of income and employment, particularly in rural area, where tradition agrarian industry has to decline. Recently, however, a number of established tourism areas have turned to rural tourism, in order to diversify tourism products and market. It also has to spread the benefit of tourism away from coastal areas to hinterlands to reach rural areas.

Secondly, rural tourism is seen as one of the most vital tools in the development of the tourism industry today. It can also be seeing many different forms and pursued for various reasons. There are developmental reasons to promote tourism as a growth pull such as for generations to follow agro-industry collapse or diversification of remote marginal agricultural areas in adventure or cultural tourism.

In addition, rural tourism has remained peripheral to focus on tourism research. The remaining poorly defined continues to be a general term, which encapsulates a wider range of interest groups. It is not only from tourism studies but also from economic, planning, anthropology, geography, sociology, and business studies. There has been lack of integration between their interest groups, each cultivating its own view and approach to rural tourism. Furthermore, it can also bring many benefits to allow the creation of alternative source of income in non-agricultural sector for rural dwellers. The added income from rural tourism is the contribution to the revival of lost folk arts and handicrafts. It is an ideal and natural method of rural and urban economic exchange.

The term "rural tourism" has a wide range definition. It covers all leisure and recreational activities in rural areas according to the research (Tsai 2007; Liu 2010). According to Tsai (2007) and Liu (2010), the content of rural tourism covers tourism in orchards, farms, cultures, lives, scenic areas, religious activities, food, and air

W.-T. Fang, *Tourism in Emerging Economies*,
https://doi.org/10.1007/978-981-15-2463-9_5

in rural areas. It comprises rural production, living and ecology, including production and agricultural activity oriented tourism (i.e., agro-tourism or leisure agriculture), living culture-based activities (i.e., cultural and historical tourism, or museum tourism), and ecological nature-oriented recreation activities (i.e., nature tourism, agro-tourism, green-tourism or ecotourism). It can be concluded that rural tourism is a kind of lifestyle and critical part of rural leisure industry. In order to develop rural tourism, maintenance of the natural ecosystems and landscape as well as the preservation of rural features and culture, recreational facilities in the villages should be improved and upgraded so as to increase their attractiveness and meet the needs of holiday makers into the destination.

One of the vital tools in promotion of rural tourism is helping to create employment opportunities and creating new income streams for the locals. For that reason, the notion of rural tourism covers not only tourism on the farm leisure agriculture, but also any tourist activity in rural areas that is known to be a way of enjoying rural life within the community and also, enjoying beautiful rural scenery and maintaining rural prosperity and environmental quality. Therefore, rural tourism is the utilization and integration of the unique quality and rich resources of the rural environment. This includes natural resources within the area, environment, and landscape of rural villages, such as farmland and forestry, lakes, and mountains, diversity of agriculture, farm production, harvesting fruits, gardening, venues for forestry, livestock rearing and fishery operations, local rural culture like rural buildings, historical architecture, cultural relics, temples, indigenous cultural practices and customs, elegant rural landscape, village architecture, gathering sites, plazas and squares, streets, rivers, scenes of farmers working in the field, natural landscapes and environments, and so on (Liu 2010; Botezatu 2014). It offers city dwellers or travelers a place to relax and enjoy the greenery of the countryside, and the chance to enjoy the simple joys of rural living (Fig. 5.1).

This would transform rural village into a sought-after holiday destination. In turn, it is fuelling the economic growth of the rural area. Rural tourism aims also to improve the agrarian structure, fully utilize and preserve nature and cultural resources, provide the population with a chance to experience rural living, increase employment opportunities and raise the income of farmers. All these will help to promote the overall growth of the rural areas (Liu 2010). The variety of farms, different cultivation methods, livestock rearing, and crop production that shifts with the seasons, as well as specialty farmers, all contribute to diversity of agriculture activities, weaving a rich rural landscape that is constantly changing. In modern society, agriculture and rural village contribute significantly to the overall economy by preserving the vibrancy and liveliness of the rural landscapes. Hence, rural tourism and leisure agriculture should take this into consideration and proactively support the functions of these areas, helping to ensure cultural integrity, nature conservation, and landscape protection. Its development should proceed on the foundation and framework of the overall growth of rural areas in other to promote rural tourism.

Fig. 5.1 Rural tourism is not merely organizing sightseeing trips to farms or orchards, or even just developing agriculture tourist sites or jungle theme parks and leisure farms. It is an approach to integrating the blueprint of rural development in a bid to create a graceful and comfortable living environment. This does not bring negative effect to rural areas and its locals while ensuring the preservation of nature, landscapes, and protecting the originality and uniqueness of the rural areas (Photo by Wei-Ta Fang)

5.1.1 Definition of Rural Tourism

Before we talk about rural tourism, let us look into the definition of rural tourism. This will give clear understanding about what rural tourism is all about. It will define the concept and operation views in destinations. These are some of the few definitions of rural tourism stated by different type of writers.

- Rural tourism is a discrete activity with distinct characteristics are varies in intensity and area. It discusses the differences between agro-tourism and rural tourism, and examines why there should be special relationship between tourism in the countryside and the concept of sustainable tourism (Line 1994).
- Rural tourism in Latvia is a form of tourism taking place in rural areas or settlements, providing employment and income to local population, and offering individualized holiday products to consumers.
- Rural tourism is the act of leaving the urban environment and traveling into rural regions. It is for the purpose of experiencing and enjoying natural beauty, the agricultural diversity, the quaintness of small towns, experiencing the cultural richness; and escape pressure of urban lifestyles in order to have pleasurable experiences and improve the quality of life (Irshad 2010).

To understand better about rural tourism, it is defined as any form of tourism that showcases rural life, art, culture and heritage at rural locations. It is, therefore, benefiting the local community economically and socially, as well as enabling interaction between the tourists and the locals for a more enriching tourism experience. On the other hand, rural tourism is essentially an activity, which takes place in the countryside. It is multifaceted and may entail farm agricultural tourism, cultural tourism, nature tourism, adventure tourism, and ecotourism. As against conventional tourism, rural tourism has certain typical characteristics: It is experience-oriented; locations are sparsely populated. It is predominantly in natural environments; mixed with seasonality and local events; and based on the preservation of culture, heritage, and traditions.

5.1.2 Types of Rural Tourism/Tourists

Tourism is synthesized from mass and alternative tourism. Mass tourism is characterized by large numbers of people seeking culture holidays in popular resort destinations (see Chap. 3). Alternative tourism is sometimes referred to as special interest tourist or responsible tourism. It is usually taken to mean alternative forms of tourism, plus emphasize on the contact and understanding of inhabitants 'way of living and the local natural environment'. Rural tourism attraction sites are including:

- Road-Side Fruit Stands
- Historic Sites
- Scenic Attractions
- Conservation Areas
- Cultural Centers
- Forests and Woodlots
- Walking Trails
- Garden Centers
- Country Homes, etc.

Therefore, tourists that take parts in rural tourism for those attraction sites could be:

- Farm stay visitors
- Hunters
- Snowmobilers
- Cottagers
- Cross-country skiers
- Birdwatchers
- Botanists
- Hikers
- Rally car drivers
- Bus tour groups

- School children
- Cyclists
- Photographers, etc.

The diversity of attractions included within rural tourism are heritage tourism (sometimes, it is referred to as cultural heritage tourism) (see Chap. 4), nature-based tourism/ecotourism, agro-tourism, as well as partnership-based approaches, such as scenic byways and heritage areas. Heritage tourism refers to leisure travel that has as its primary purpose the experiencing of places and activities that represent the past and this resource could also be used as a means of promoting rural tourism in a destination, such as rural tourism in European, African, and Asian Countries.

5.1.2.1 Rural Tourism in European Countries

European countries have played an essential role in the development of rural tourism in their destinations. As a result, it served as a motivating factor to motivate in people from all over to come and have a good experience at rural tourism. For example, countries like New Zealand, United Kingdom, Scotland, Italy, and Ireland, have played an importance role in promoting rural tourism in their destination. It is one of the famous kind of hospitality, and beautiful clean green environment, where people can have the chance to enjoy an authentic experience with the local.

- **New Zealand**

Rural Tourism Holdings in New Zealand has one of the largest farm stay and specialized accommodation provider. They were created in April 2001 through the merger of New Zealand's two largest farm stay companies, i.e., Rural Tours and Accommodate. Prior to this, both companies were recognized as the pioneers of farm stay in New Zealand from the 1970s (Hall 2004, 2005). According to research based on previous documents, rural tourism is a privately owned company with an extensive database of hosts throughout the North and South Islands of New Zealand.

They have reputation for providing a prompt and efficient reservations service, and focused on providing memorable lifestyle experiences for international visitors to New Zealand. As a result it has really made a great positive impact on their tourism in promotions of rural tourism.

Tourism is a vital engine of economic development in rural areas. It can bring new income and jobs into rural communities at a time when we are seeking to move away from over-reliance on primary agricultural production (Wilson et al. 2001). To realize its full potential, the rural tourism industry needs to be supported by a strong infrastructure, in terms of marketing, standard setting, and fiscal climate. This has been one of the driving tool for the promotion of rural tourism in New Zealand tourism industry.

- **United Kingdom**

In the United Kingdom (UK), rural tourism has been one of the vital engines of the economic development. Many of the rural tourism and agro-tourism development programs in the United Kingdom are driven by policies applicable in other European jurisdictions. The UK's most significant agro-tourism marketing organization is the Farm Holiday Bureau (FHB). It coordinates marketing network of over 1,000 farms providing (1) both bed-and breakfast, and (2) self-catering accommodation in the rural areas within the countryside. These establishments are found throughout the UK. The focus of the marketing program is on promoting quality accommodation with value for money prices. FHB member properties are inspected regularly to ensure that a high standard of cleanliness, courtesy and service is maintained.

- **Ireland**

Ballyhoura countryside is an attractive and typically Irish (O'Connor 1996). However, it has no outstanding tourism resource. It has relied on promoting a genuine rural experience and warm welcome. This region has only emerged as a tourism destination since 1986. Building from a low base, it now includes a range of accommodation, small scale attractions like open farms, interpretative centers, and countryside recreation facilities. The tourism strategy has two overall objectives to provide a cooperative marketing function through groups of enterprises working together and to improve the range and quality of the tourism product, on three levels in order to come up with the following factors.

- Encourage local people to start up new enterprises in rural areas.
- Support improvement and expansion of existing enterprises in the local areas.
- Attract new inward investors, piggybacking on successful development in other to set up a good rural tourism within the rural areas where locals can benefit from.

This approach has provided standard facilities appropriate to the market. Recently, the emphasis has changed from expanding the range of the products to improving the quality and performance of existing businesses (Roberts 2001). Attention is also being paid to the quality of infrastructure and services delivered by Ballyhoura Filet as a whole. Wider community involvement is seen as very important. A community consultative committee of 50 members includes representatives from each village and local societies like heritage groups. The main thrust of marketing is to tour operators and agents, providing them with a long term and quality personal service. Ballyhoura's owns website becomes a major source of individual enquiries which has also services as a marketing tool (Irshad 2010).

Careful attention is paid to the quality of the site. The keywords are used by search engines related to known consumer interests. A central reservation service is to offer independent visitors. All visitors receive a basic information pack includes a welcome letter, questionnaire, and information leaflets. All enterprises are encouraged to talk to their guests. They are asked about their experiences during their visit, so problems can be rectified before they leave. As a result, worthwhile adjustments have been made, such as introducing language training as a product. Ballyhoura Development has administered many grants. The allocation of leader resources reflects the market demand for accommodation and also focuses on quality. For example, grants are available for upgrading self-catering accommodations.

Policy objectives focus on economic benefit. For example, the Ireland Department of Agriculture and Food (IDAF) has developed a Grant Aid for Agro-tourism and Organic Farming program. The objective of the scheme is to ensure regular supply of organic produce to market by the developing organic cooperatives group companies. For each initiative, employment targets are set and monitored. Direct employment is measured by obtaining feed-back from enterprises. In addition, an economist commissioned by Ballyhoura developed a simple model for relating visitor numbers and accommodation occupancy levels to revenue generated and employment supported. Separate ratios are calculated for domestic and foreign visitors and for different types of accommodation. This has helped to check achievement against targets. The impact of tourism on local community is monitored through the Community Consultative Committee. This has also served as a driving tool to worse the development of rural tourism projects in the local communities.

- **Italy**

Rural tourism in Italy has played an essential role in development of tourism industry in their destination. As an example; Basilicata is a small region in the far south of Italy. It has two short stretches of coastline but is primarily an inland region. It has rich vegetation and a notable cultural heritage ranging from Greek temples and Roman thermal baths to Norman castles and medieval abbeys. Basilicata has a modest amount of tourism. Tourism has been slowly growing; however, it is highly concentrated in the summer season and on the coast, with the internal rural area being little visited (Irshad 2010). Rather than a comprehensive tourism strategy, the approach has been to concentrate on a specific quality initiative. The broad objectives were to:

- develop collaboration between the public and private sectors;
- understand the needs and expectations of existing and potential markets in order to promote rural tourism in the region;
- help enterprises meet these needs through training and advice so they could have good product knowledge about the area and its products and services it can offer to the tourists.

The resulting initiative was called "ÔTurismo di QualitˆÕ", a phrase applicable to the output, as well as the process. It was realized that in order to create a quality product, detailed research was required on the resources, the enterprises and their views, and potential markets, as well as current visitors. The first phase, in 1996, involved a comprehensive audit of tourism facilities, services, cultural and environmental heritage, and relevant infrastructure. The second stage involved two extensive consumer surveys. These surveys revealed a divergence of opinion be-tween people who had been to the region and those who had not. The positive reaction of existing visitors to the local culture and heritage confirmed the basis for a qualifying offer. Feedback from operators with a potential interest in the area pointed to the opportunity to package these strengths in specific itineraries and programs. This is being maintained with further customer satisfaction surveys. Training workshops and advice on business and quality improvements were held as part of the scheme. This training has been extended more widely in the region.

- **Scotland**

Rural tourism in Scotland also services as a pull factor. It has also motivated a good tourist from different parts of the world to go and enjoy the beauty of rural tourism in Scotland. One of the Scotland outstanding scenic areas with a population of 6,000 is dispersed throughout small rural communities. Forestry, agriculture, and tourism are the main employers. Scenic beauty and accessibility from major urban contribution makes it popular visitor destination. It attracts 2.2 million visits each year. A handful of sizeable attractions cater for mass market but majority of tourism businesses are small and run by families. A number of public sector agencies have long-standing commitment to developing tourism in the area. Visitor pressure and a lack of investment have led to decline the quality of visitor experience and some environmental degradation. To tackle these issues the Trossachs Trail Tourism Management Program was established in 1992 (Caffyn 2000). The overall aim is to develop tourism in a way that brings maximum benefits to local communities and local economy, while sustaining and enhancing the quality of the Trossachs environment. Key components of the program strategy are to:

- increase visitor understanding and respect for the area;
- safeguard and enhance environment assets in order to promote the flora and fauna within the rural areas;
- strengthen local tourism industry and improve the quality of visitor experience;
- monitor visitor levels and activities in order to find ways of managing and controlling their impacts; and
- increase visitor contributions to the rural area.

The tourism management program has been very successful in bringing together influential agencies with differing remits, including potentially conflicting aims like environmental conservation and tourism promotion. A community forum has been established with representatives from private, voluntary, and community sectors to discuss progress and future work of the management program. Good communication has been established with the local community councils.

5.1.2.2 Rural Tourism in Africa

Rural tourism in Africa has also played a significant role in the development of tourism industry in their countries. This result helps to motivate many people from different parts of the world to travel to African countries. The travelling will depend on the choice of destination. They should have decided to have a good tour about the unique features of the rural tourism selected in Africa. Kenya, Gambia, South Africa, Ghana, Tanzania, to name few are the examples of the countries. Tourists come to rural tourism in African contents for the purpose of rural tourism.

- **Kenya**

Kenya has several tribes and rich with culture and community homestay. Tourists can experience. Victoria Safaris has started its Rural Homestays package in Kenya

Cultural Community villages. These are the best places for those who want to volunteer their skills to various communities in different expertise. That will alleviate poverty or improve skills for rural folks while living in the indigenous communities' rural inhabitants.

The rural tourism in Kenya is really played a vital role in strengthen the living standard of local communities. This is because a good number of them have engaged themselves either directly or indirectly in rural projects they are set up in the local's communities.

- **Gambia**

The Gambia is another destination essential in the promotion of rural tourism. Tourism in the Gambia has a high reliance on package tours (Mitchell and Faal 2007). The tourists pay for flights, transfers, accommodation, and catering in their home countries before departing, denying locals the opportunities of fair benefits from tourism. Most hotels where the tourists lodge are situated close to the beaches and urban areas. Tourism in this nation, therefore, is largely urban although Gambians are making efforts to spread development of tourism to rural areas.

Cultural heritage and natural areas of different communities are being considered as key selling value propositions to the tourists. Beside the 'sun-beach' tourism in the nation has built up a reputation for Emerging Market Group 2005. Some excursions to rural areas are being organized by different ground tour operators to offer local experience to tourists, however, tourism in rural areas remain largely unorganized.

Rural tourism constituting visitations by tourists to rural areas are to experience the culture and natural environment of local population. It is considered as one of the strategies within the management portfolio of sustainable rural development. Sustainable rural development, in general, encourages locals to embark on initiatives that will lead to their own socio-economic development as well as protecting their environment.

Local Gambians are in partnership with some of non-governmental organizations (NGOs). They have started organizing themselves to offer rural tourism products and services to generate more income for the local community. However, careful consideration of the link between internal environment development with the control and/or influence an external environment factors. The local rural tourism stakeholders cannot control and/or influence the management of this development which is more balanced and strategically orientated, especially at its initial stage.

Box 5.1: Analysis of the Internal and External Environment of Rural Tourism in Gambia

The Gambia Tourism Authority, recently reformed as Tourism Authority Board, is responsible for the implementation of rural tourism policy, for example, by marketing the nation internationally and supporting tourism related human resource development projects in rural communities. The environmental impact of rural tourism projects is assessed by National Environment Agency (NEA). Community development projects, e.g., rural tourism projects, are primarily financed by donor agencies (Fig. 5.2).

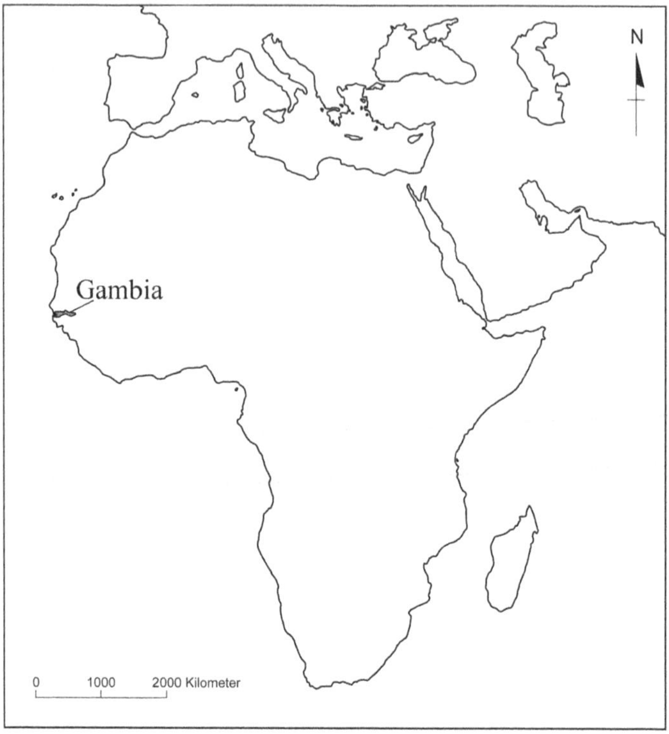

Fig. 5.2 Location of the Gambia (Illustrated by Yi-Te Chiang)

They closely cooperated with numerous non-governmental organizations (NGOs). NGOs try to support rural tourism projects by providing services like research projects, communication, training, and organizational support. As a result, it helps in upgrading the strength of rural tourism in Gambia destination as follows.

- Management implications for rural tourism offer.
- Aims and objectives of rural tourism development.
- Stakeholder analysis of the present and future rural tourism development, that is, Strengths, Weaknesses, Opportunities, and Threats (SWOT) analysis of rural tourism in the Gambia. It is one of the key factors that are used by the Gambian Tourisms Board to promote rural tourism.
- Local participation and local needs as regards rural tourism development within the communities.

- **South Africa**

The National Department of Tourism (NDT) has developed a first draft of the Rural Tourism Strategy. This is the strategy meant to ensure developmental approach upon packaging rural tourism products in South Africa. This approach is also meant to priorities spatial nodes which have a potential growth in order to stimulate growth of the tourism industry. The White Paper on the Development and Promotion of Tourism in South Africa (DEAT 1996) contends that prime tourism attractions are not located in the city centers but they are in rural areas. The Rural Development Framework states that between 10 and 15 millions South Africans live in areas they are characterized by extreme poverty and under development. In some poverty areas within the world heritage sites land found other major attractions. The strategy, therefore, further reviews product development, tourism marketing, human development, tourism infrastructure and tourism research and information for equal access of opportunities in rural tourism.

According to the United Nations, 75% of the world's poor live in rural areas, particularly in developing countries, include national parks, wilderness areas, mountains, lakes, and cultural sites, most of which are generally rural, thus tourism is already an important feature of the rural economy in these specific sites (Holland et al. 2003). In South Africa top tourism destinations, the aim of promoting tourism is to increase net benefits to rural people and increase their participation in managing the tourism product. They believe that if more tourism can be developed in rural areas, particularly, in ways that involve high local participation in decision making and enterprises, then poverty impacts are likely to be better off within the rural communities.

- **Ghana**

Rural Ghana plays an essential role in the development of tourism industry (Teye 1999). Ghana serves as a tool for poverty alleviation. The rural setting has become an attraction to older generations, who once lived there and always yearn to return to that unspoiled experience. The peace, quiet, serenity, freshness, laidback lifestyles and closeness to nature are all examples of the characteristics of the rural environment (Lukaz 2006). Since urban settings in various countries may not differ from each other due to similarities in architecture and modernization, it is only the rural that presents the cultural differences.

For example, Nature Conservation Research Centre (NCRC) is a Ghanaian non-profit organization implementing conservation initiatives in order to promote a greater awareness of and protection for the natural, historic and cultural diversity of Ghana. Ultimately the West African sub-region has helped in the promotion of rural tourism in Ghana. The following are some of key roles the NCRC put in place to promote rural tourism and protection of natural and historic sites with the local communities.

- NCRC will expand its network of community-based destinations linking tourism development with positive environmental, historical, cultural conservation, and the growth of economic opportunities for rural poor communities.
- Expand NCRC's ecotourism activities to 50 locations, involving 400 rural poor communities.

- NCRC will lead the environment and conservation community through provision of mentoring and capacity-building assistance to sister conservation organizations in Ghana and other West African countries.
- The NCRC collaborates with NGOs and corporate partners in promoting solar energy solutions in rural locations linked to positive conservation initiatives.

- **Tanzania**

Tanzania is one of African countries that help in the promotion of rural tourism in their country within the rural areas in other to make sure that the locals can benefit either directly or indirectly. Tanzania is known for its worm and friendless generosity, hospitality, and wealth of folklore. This criterion has made Tanzania a tourist appeal destination for visitors from different parts of the globe.

Rural tourism is one of the important factors to uplift tourism industry. It is set up some tourism policies in order to promote the livelihoods of the locals living in rural communities (Brown 1998). This is done to encourage the development of sustainable and quality tourism, culturally and socially acceptable. For example, in 1995, the SNV Netherlands Development Organization, began exploring various opportunities for developing a program that enables local people to offer cultural tours. This allows them to generate some extra income, while at the same time offering tourists a unique and interesting cultural experiences in the country.

During these days, the first pilot was started in Ng'iresi Village on the slopes of Mount Meru, Longido, the Usambara Mountains, and Mto wa Mbu. In all places, local residents received training and pieces of advice on how to organize tours to tourists to come to the country. With cooperation of the SNV and Tanzania Tourist Board (TTB), cultural tours were initiated in conjunction with tour operators, and tourists started to arrive shortly thereafter. Today, the country is averaging around 3,000 visitors for various cultural tourism programs, thus far reactions by both parties are extremely positive. Throughout these international exchanges, the local communities use a part of generated income for specific development objectives. The positive results of pilot phase made SNV and TTB decided to start a joint five-year program on developing and promoting cultural tourism in Tanzania to promote rural tourism in the country.

According to research, the program officially started in 1997. It is currently focused in Northern Tanzania. From the original pilot phases, programs have been added through a women's group in Mulala (north of SA River), and a tour of the Engaruka Ruins.

5.1.2.3 Rural Tourism in Asia

The Asian continent also plays an essential role in the development of rural tourism in their countries through the promotion events. The development of infrastructure in rural areas has potential for tourism. It is being supported under the existing scheme of destination development. The objective of the Asian Pacific is to showcase rural life, art, culture and heritage at rural locations, and in villages (Kapur 2016). The core competences are in arts and crafts, handloom, and textiles, set as an asset base in the natural environment. The intention is to benefit local community economically and socially, as well as enable interaction between tourists and local population. This can have a mutually enriching experience because one of the vital tools about rural tourism is interacting with the local people and visitors. Under this scheme, the thrust is to promote village tourism as their primary tourism product to spread tourism and its socio-economic benefits to rural to its geographic regions during visitation. The potentials of rural tourism in Asian region is for a major force in rural economic development. This will improve poverty alleviation to locals living in the rural areas by setting up rural tourism projects.

For example, in China, the Ministry of Agriculture and China National Tourism Administration (CNTA) signed a cooperation framework agreement to jointly promote the development of leisure farming and rural tourism. The two administrations will jointly organize a series of activities featuring "national happy rural tours". These programs have started since 2010 purposely to promote the rural tourism in China (Chio 2011) (Fig. 5.3).

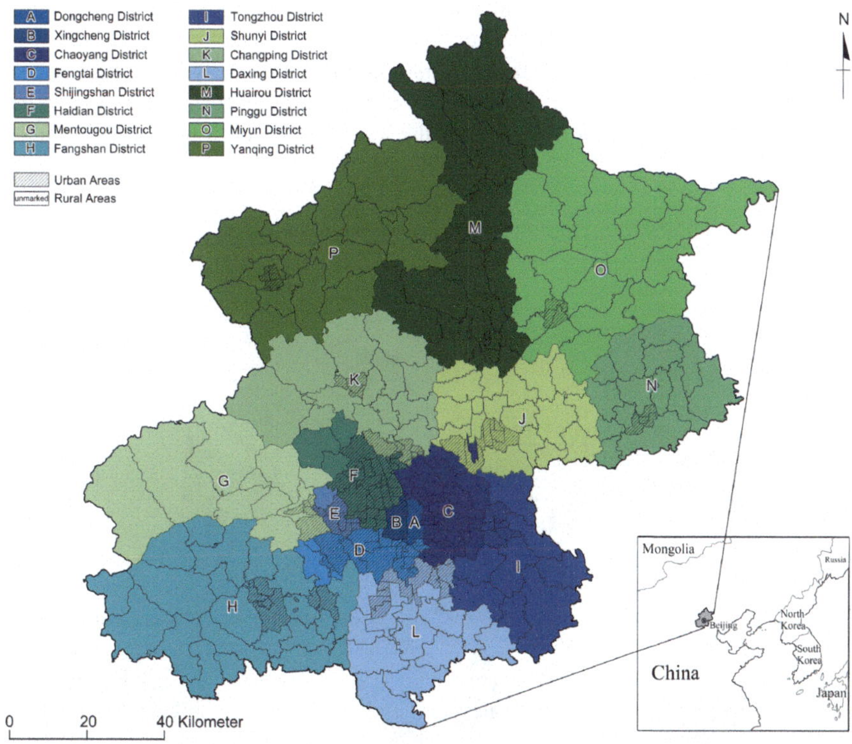

Fig. 5.3 Rural tourism in rural Beijing. In Beijing, China, you can see mountainous suburbs into a village tour beyond urban areas, metropolitan Beijing (Illustrated by Yi-Te Chiang, all rights reserved)

Two institutions have agreed to launch the leisure farming and rural tourism demonstration program across China to guide and drive the development of the industry. Through several years of effort, there would be a batch of a demonstration at prefectures, towns, villages, and enterprises, making a map for key leisure farming and rural tourism products in the country as guidance for consumers.

In Asia, rural tourism is increasingly promoted as an effective vehicle for the regeneration of rural areas. The sector has its own market and institutional characteristics, but many of these are positive, including growing demand, resource availability, and developing public-private partnerships. With rising urbanization and incomes in most Asian countries, rural tourism offers a promising response to agricultural difficulties caused by market developments and budget costs. However, the region needs to develop its own types of rural tourism, taking into account the demands of its domestic and international visitors, and its rural infrastructure and environment. Rural tourism in China has really played an important role to the growth of Chinese economy in rural areas basically from the rural enterprise.

Taiwan has also played an essential role in the promotion of rural tourism. Rural tourism has become a new type of agricultural management, which helps farmers break through current management difficulties for small farms. It has two major purposes. The first is to provide leisure and recreation to public. The second is to increase farmers' incomes. The rural tourism in Taiwan is really booming because it has motivated both inbound and domestic tourist going in for rural tourism. Due to this, it has bringing a lot of benefit to the locals living in those particular areas (Fig. 5.4).

Fig. 5.4 Rural tourism became a new type of agricultural design in Taiwan. This recycling artwork was exhibited in Tainan, located in the southern part in Taiwan (Photos by Wei-Ta Fang)

Box 8.2: Case Study Rural Tourism in India

India, also known as the Republic of India, is a country with 28 states and thousands of villages in South Asia (Fig. 5.4). It is the best country for people visiting here as tourists. People are willing to experience different aspects under a single roof. India is known as the country where the unity and diversity is seen to be strongly ruling. When it comes to tourism most people will be highly surprised to know that India is emerging as one of the best destinations in the world (Kapur 2016). Many tourists come to visit as per their requirement or preferences can find their dreams coming true (Fig. 5.5).

Fig. 5.5 Location of India (Illustrated by Yi-Te Chinag)

A lot of things seem strange to some people. They have never seen in the past but become the center of attractions as well. Most of the tourists particularly from abroad are fed-up with their city's life. Now, they want to explore the unique experience. Rural tourism in India has turned into a booming industry for its locals because it has totally shaped into one of the major sources of economic earner for the locals. It is often told to hundreds of foreign tourists that if they truly want to see, experience, or taste the country from length to breadth, they need to visit the rural villages far deep into the countryside.

Due to the increasing level of rural tourism, India has raised its position at international tourism markets. This increases the world tourism market shares. There are hundreds of things that tourists' interests are lying unexplored. They lack of knowledge on the parts of tourists. Most people will rightfully choose to spend days under the pleasant environment surrounded by hills, forests, and agricultural green fields. They can enjoy the beauty of the flora and fauna within the region. As compared with the culture shown by people in metropolitan cities like Delhi, Mumbai, Kolkata, Bangalore, and Chennai, to name few, the culture shown through various activities performed in the village look more genuine and historical. On the other hand, tourists can also have good experience in the rural tourism areas.

The cultural traditions are experienced through fairs, festivals, and religious ceremonies followed by strolling through the beautiful lanes of green paddy fields, gardens, orchards. Tourists can study and observe daily activities of villagers more directly, thus satisfy their spending holidays more frequently to the visited destinations. Rural tourism in India is having a huge potential of opportunities that can be considered as the sleeping giant. This is once awakening will bring huge economic earnings to local people and the government.

From east to west and north to south, the entire country is cemented with villages and hamlets. The country, being multi-cultural by nature, will work wonder as thousands of tribes are found in India. They can wishfully showcase their sophisticated handicrafts made up of woods and bamboos. They are truthfully nature friendly and attractive to tourists to visit and have a good experience here.

Village life is always better than city's life. People can forget the morning dawn when all the villagers are awaken up by the shrill sound made by cocks. It is indicating the time of the first ray of the dawn started. Farmers along with their oxen go to their farming fields to plough, and village women carrying water far from rivers on their head. The containers are filled full of water topped over one another. These are the amazing sights that are confined to only in the villages in India.

Most of tourists are not aware of the fact that rural tourism in India is often mistaken to be ruled by poverty and dirtiness. However, beyond that, tourists are introduced to the pure environment of nature, genuine culture and tradition of the villagers.

5.2 The Importance of Rural Tourism in a Destination

5.2.1 Objective

The main objective of rural development is to narrow the developmental gap between the rural and urban areas by improving the working conditions and living environment of the locals in their own environment, ensuring that rural population is enjoying similar privileges and quality of life, as well as their city counterparts either directly or in directly.

Along with the change of rural functions, the goals of rural development have to be adjusted properly to infuse vitality into rural areas and at the same time, during social economic growth, protect its traditions and unique styles. In addition, rural development helps to reinforce conservation efforts towards nature and the ecosystems, within their local communities, highlighting the precious status of the natural resources and unique scenery of the countryside. The efforts to preserve integrity of ancient cultural relics and buildings will further enhance the attractiveness of these areas, thus increase the number of visitors vacationing there. Therefore, the important goals and tasks of rural development include:

- Protecting natural landscapes and ecological environments as well as meeting the needs of rural tourism which could services as a pull factor in the promotion of rural tourism in the area (Liu 2010).
- Creating rural economic strength and improving the working conditions and living environment for rural inhabitants so as to promote a prosperous rural economy and meet the needs of agricultural development.
- Preserving and maintaining traditional rural features and unique styles, and maintaining vitality in the process of modernization and improvement of living conditions. It has also services as driving force in the promotion of rural tourism in a destination.

5.2.2 Functions

Rural tourism is a combination of agriculture activities. In rural tourism areas, they have many positive impacts in the development of local community. It has ventured in rural tourism. These are some of the vital functions. It could bring into the development of the village, cultural heritage, social functions, education, therapeutic, environment, economic, and recreation services, etc.

- Cultural Function: Cultural heritage preservation is to ensure the preservation and continuity of the unique lifestyles, culture and folk arts of rural villages. At the same time, rural tourism helps to create unique rural culture and cultural assets in order to promote rural tourism in the local communities.
- Social Function: To promote interactions among urban dwellers, rural villagers and the rural areas, further expand social networks of farmers, and narrow the urban-rural gap as well as elevate the quality of life in rural areas.
- Educational Function: To provide opportunities for urban dwellers to understand agriculture, educate them to crop cultivation process and livestock rearing, experience rural living and appreciate its culture and ecosystems within the local communities.

- Therapeutic Function: To provide public for leisure activities. This allows close encounters with natural landscapes and ecosystems in the rural areas, helping to relieve daily stress of work, achieving peace of mind to any person who wants to experience rural tourism.
- Environmental Function: To improve and upgrade the quality of the rural environment. It is also to protect natural landscape and ecosystems, plus a consideration to promote rule tourism in the community.
- Economic Function: To increase employment opportunities in the rural areas, improve rural income, and create a flourishing rural economy.
- Recreational Function: To provide public places for leisure activities and rural leisure tours in other to make sure that tourist have a good experience during their stay.

5.2.3 *Benefits*

- Encourages conservation and care of natural resources such as the rivers, forests, mountains as well as local fauna and flora, birds, fish and almost extinct animals. When these are valued as living assets to attract income, they care for this as result service as a way of promoting rural tourism in destination.
- Given that services are often provided by residents of the village, rural tourism benefits the village. There is greater cooperation among villagers as they operate the local enterprise provided for the tourists. This is as result help the locals living in that particular area to benefit directly or in directly to the projects that have been put in place in their community.
- Generates employment and income for the residents of remote villages. This is most especially beneficial to women who often have no new opportunities to earn income away from home, which is also one of the negative factors that as also set up a good living standard of the local communities.
- Creates market and appreciation for local products, services, and keeps them alive.
- Helps slow urban drift as it gives a better economic future for families in rural areas living in the communities.
- Provides opportunities for mutual sharing of cultures. Hopefully, these encounters will result in greater understanding among peoples and closer ties to the global family in other to promote rural tourism in the area.
- Provides incentives to preserve some of old customs and traditions, crafts, traditional festivals, architecture, food and other practices unique to certain cultures.

5.3 Planning and Development in Rural Tourism

How to invigorate rural villages, preserve their traditional flavors, and unique styles, has become an important factor for rural development. Efforts should be made to reinforce construction of public utilities, renovate village residences, improve the working conditions, living environment of villagers, preserve rural ecology and scenery, meet the need of developing tourism activities in rural villages, set up facilities for tourism, and leisure activities to create a beautiful environment for rural tourism and attractions in the rural village.

At the same time cultural relics and ancient remains in the rural village should be protected. Rural traditions should be kept and buildings of historic value renovated to preserve natural resources. The rural scenery and flavors in order to reveal unique rural features, and their significance in prolonging innate culture, should be maintained. In addition, such recreation of the rural attraction in residence and leisure, can bring society closer to rural village, and reach the goal of territorial planning to balance the development between urban and rural life. It is the real meaning of rural development serves both city life and rural life (Liu 2010). This, as a result, will service a pull factor to motivate in tourist to come in to the destination.

Secondly, one of the major types of rural tourism activity is nature-based tourism ecotourism. It is sometimes called recreation-based tourism (RBT). Recreation-based tourism refers to the process of visiting natural areas for the purpose of enjoying scenery, including plant and animal wildlife—enjoying the beauty of the flora and fauna. Nature-based tourism may be either passive, in which observers tend to be strictly observers of nature, or active increasingly popular in recent years where participants take part in outdoor recreation or adventure travel activities.

Thirdly, another major form of tourism is agro-tourism. It refers to the act of visiting working farm or any agricultural, horticultural, or agribusiness operation for the purpose of enjoyment, education, or active involvement in the activities of the farm or operation. It includes taking part in a broad range of farm-based activities, including farmer's markets, petting farms, roadside stands, and pick-your-own operations; engaging in overnight farm or ranch stays, other farm visits; visiting agriculture-related festivals, museums, and other such attractions within the rural areas. In this section, we may discuss the following focal points to improve planning and development in rural tourism. Key factors which makes rural tourism successful in a destination are follows.

- Strengthen investments in rural public service sectors and conservation efforts for cultural relics and buildings within the communities.
- Proper planning of land used within the communities to maintain a green sustainable environment.
- Creating rural attractiveness in terms of living and recreation value.
- Rural tourism enhances and revitalizes community pride.
- Protection of rural cultural assets is the basis of the attraction of villages.
- Village renewal and reconstruction of village residences.

5.3.1 Strengthen Investments in Rural Public Service Sector and Conservation

First, efforts to bolster public service sectors in the villages should be strengthened. This is an essential step towards building a better rural environment, providing convenient and quality services to local populations, as well as allowing tourists to enjoy a getaway to the countryside.

It is critical that more investments are poured into the construction of rural public services and recreational facilities for tourists. Some examples are recreational and sports activity centers, venues for ball games, swimming pools, outdoor camping hotels, children's playground, fishing, wading pools and so on. This would increase the fun and excitement quotient of rural tourism and offer urban travelers an opportunity to embrace countryside living in the rural areas.

Cultural relics and buildings in villages are important cultural assets, and symbolize the historical milestones of rural development. They should be preserved and protected. By endeavoring to protect cultural assets of the villages and high-lighting their indigenous roots and heritage, an emotional tie, are strengthened and becomes the social glue that holds the community together. Besides focusing on preserving the landscape, scenery and nature, the planning of infrastructure and public roads such as pedestrian walks and bicycle lanes should not destroy the original look and special characteristics of the place. This includes retaining the native flora and trees species that are huge attractions for tourists.

By the same measures, as required by the development of recreational areas in rural villages, houses are reconstructed; recreational facilities are added to create a nice and attractive leisure environment. Meanwhile, rural relics and historical sites are preserved, traditional or historical buildings are kept, natural resources are protected, and landscape and rural features are conserved so as to retain the unique rural style and continue their own cultural meaning. Only in this way can villages become an attraction for living and tourism.

5.3.2 Proper Planning of Land Used

Another important point is that farm stay or accommodation alone is not enough to attract urban tourists. It should be accompanied by activities that tourists will enjoy, and be part of an attractive nature and environment. Air, water, soil, grass-lands and beautiful landscapes, are the precious resources offered by agriculture and rural villages. Proper and thorough land-use planning is required to avoid pollution and the destruction of the environment and natural landscape. This relies on regulating the use of agricultural land by different departments and preventing unhealthy competition for land, as well as stopping unregulated and chaotic expansion of land for housing and industrial use. This would stamp out the problems of pollution, and prevent destruction of production conditions and living environment in the rural village. The goal of transforming rural villages into attractive tourist destinations for urbanites can only be possible with the proper maintenance and protection of the rural features, like natural and agricultural resources, beautiful landscape, and ecosystems.

This goal is achieved through proper land use planning, green efforts, and landscape planning within the rural village, as well as the maintenance and preservation of green zones and green belts in peripheral areas. On one hand, green belts such as forest, woods and open plains, are kept as oasis for wildlife and plants, while conserving water and soil resources.

On the other hand, efforts should be targeted at treating the rivers and cultivating trees and plants, and achieving a natural water system and green environment. In the end, these efforts will result in a flourishing and vibrant rural landscape, comprising rural scenery and villages, ultimately unlocking the great tourism value in rural areas (Liu 2010).

5.3.3 Creating Rural Attractiveness in Terms of Living and Recreational Values

With social and economic growth and agricultural structure changing, the role of the rural area has many functions. Agriculture and rural village are not only the food supply and raw material production process, but also they offer a beautiful landscape and natural environment. The variety, nature, beauty and individuality of rural views make rural areas attractive. Rural areas are also the vacation places, which can provide urban populations with recreation and relaxation. It is so useful to have a nature reserve and environmental protection. As a result, it is an important task for rural development by means of village renewal to infuse vitality into rural areas while at the same time providing protection for its traditions and unique style. This is a result service as a motivating factor another thing to encourage people travelling to rural villages. They can spend their vacations in the farms to support agricultural amusement. Farmers will have another source of income in the communities due to the promotion of rural tourism in the local areas.

Through comprehensive rural development planning and village renewal, we could establish a basic public facility to improve rural working and living condition (Liu 2010). Task focuses range from building and renewing houses to establishing basic public and recreation facilities in villages, from the construction of village streets, alleys, sports ground, children's playground, parks and green fields, transportation infrastructure, quality improvement with special consideration for the nature of rural areas, as well as public activities on environmental protection by protecting the flora and the fauna within that particular environment.

Furthermore, in order to meet the demands of developing rural tourism, destinations should renew and reuse traditional houses, improve leisure, and recreation facilities (Botezatu 2014). Such activities create beautiful relaxing conditions in rural areas. In the meantime, these actions will show the unique rural style and the meaningfulness of continuing rural culture. Actions such as these will recreate rural attractiveness by providing a tourist environment. This will make the whole of society more dependent on rural areas, strengthen farmers' recognition, and sense of belonging to their villages. This result could service as a pull factor to motivate tourists in the large numbers coming to the destination to enjoy the rural tourism life.

5.3.4 Rural Tourism Enhances and Revitalizing Community Pride in Rural Communities

Tourism encourages conformity to an ideal image of community. This can result in growth of personal ties and community solidarity. Thus, the basis for community solidarity shifts from shared cultural background to shared image.

Amenities play a fundamental role in shaping a community's identity and pride (Knowd 2001). Therefore, potential of tourism for improvements to facilities and amenities has positive implications for community pride, particularly rural museums as an important repository of rural culture.

5.3.5 Protections of Rural Cultural Assets Is the Basis of the Attraction of Villages

Rural development should retain traditional style and unique characteristics of the rural village. The most valuable assets for rural tourism are qualities like diversity and uniqueness of the agriculture activities, natural breathtaking landscape and environment, cultural relics and complexity, the artistic value of rural buildings, simplicity and authenticity of the culture, awesome ecosystem consisting of wildlife, flora and fauna, and indigenous animals (Liu 2010). These are what attract the urban population to rural areas; hence the process of rural development and village renewal should be based on the conservation principles of preserving the traditional, spearheading trends. Therefore, rural development should be reinforced according to unique characteristics, and style of individual regions. Through the village renewal and construction of residential areas will help farmers to create simple, clean and indigenous farm accommodation that can be offered to tourists. This can be a chance to enjoy the surrounding countryside by having a good view of the area.

The development of rural tourism does not only provide better holiday choices offering for serenity, comfort and joy of rural living, but also bring tourists for income that will revive rural economy and raise farmers' incomes.

5.3.6 *Village Renewals and Reconstruction of Village Residences*

In an effort to promote modernization of rural villages and offer to tourists, the comfort of essential public services and facilities, village renewal, should consider both the needs of the community social system and the tourists. On one hand, it should provide a blueprint for rural development while pushing for the renewal and re-construction of the buildings and residences. The aim is to retain the native rural ambience, while improving living conditions and providing tourists a clean and comfortable holiday destination (Liu 2010).

One of the approaches to village renewal, while retaining the traditional heritage, is to imbue old but historical buildings with new roles and functions (Said et al. 2013). That is, the renovation of village residences can change the purposes of rural buildings for re-utilization. This will endow old and historic houses with new vitality and fresh functions, as well as maintain, preserve cultural relics, and scenery in the rural village.

With the development of rural tourism, farmers can be persuaded to convert extra rooms or repair and remodel agricultural buildings into accommodation for tourists, meeting the needs of urban tourists who would like to experience a "farm stay" holiday. This generates new revenue to the farmers. With services ranging from sports venues, camping sites to rental of farm houses, village accommodation, cruise and fishing boats, horses and ox carts, the tourists will not be disappointed. It is essential to remember that rural tourism activities today requires a considerable initial investment to satisfy tourists' demands and meet modern standards of comfort and safety.

5.4 How Does Agriculture Related to Rural Tourism?

Rural tourism services as a dynamic role for public interest. Its contributions are fully satisfied by the agricultural activity within the rural communities. Rural tourism can also offer farmers a great opportunity the agricultural income and activities diversification of products and services.

In many rural regions, tourism is accepted as a natural part of the socio-economic fabric juxtaposed with agriculture. It is clear that rural tourism is based on rural amenities; however, it is not clear how it relates to agriculture. Are these inter-relationships of mutual benefit? That is, while rural tourism provides farmers with auxiliary funding to continue their agricultural activity, is the latter an important or even necessary component of rural tourism. According to Nilsson (2002), a farmed tourism was a subset of rural tourism based on the rural environment, farms, and farmers. This means that within the framework of rural tourism, farm tourism enterprises are more closely related to agriculture than other rural tourism operations.

5.4.1 *Evaluating the Impact of Agriculture on Rural Tourism*

Prices and sales in the tourism market are determined simultaneously by supply and demand (Camilleri 2018). The supply side reflects the cost structure of the firm and its level of efficiency. The more efficient the production process is, the less it will cost the entrepreneur to produce rural tourism services than other less efficient firm. The demand side reflects the value visitors attach to a stay in rural accommodations. Since rural accommodations vary in their attributes, we expect visitors to select different values for different attributes—one of them being the existence of a working farm (Fleischer and Tchetchik 2005).

Profits can increase if visitors are willing to pay higher price for an attribute or if the business is operated more efficiently, i.e., at a lower cost for the same level of production. Rural accommodation owners had to evaluate both sides of the market, supply and demand. Thus, we had to evaluate both sides of the market, supply and demand. Rural tourism accommodations vary widely in their attributes. Some of these attributes are similar to those of a hotel, e.g., the level of luxury of the unit or a special view, but some are unique to rural tourism. Among the latter is the existence of working farm withal of its implication. In the following sections, I summarized some previous studies according to negative and positive impact of rural tourism socially, economically, and environmentally.

5.4.2 *Negative Impact of Rural Tourism*

- Not enough public transportation with inappropriate roads to access the rural centers.
- The development of rural tourism could lead in to the increment of tax in the community; thus making the life hard for the locals.
- The authentic of the culture could be lost due to the high demand in the tourism industry.
- Pressure aggravation on resources and local facilities would cause to be polluted and destructed.
- Soil erosion, gathering of waste and knocked status of soil due to the development of rural tourism in the community.
- Tourist harms include expenses of bringing instruments and facilities, e.g., transportation systems, lack attention to customs, and local habits resulting in basically weakening the living method of local people.
- Local price inflation, migration of work power, destruction of local work structure, participation of women in part time and low-income works, seasonal patterns of demands, change of local culture patterns, reduction of life patterns, different types of pollutions and residues made by tourists, destroying available animal shelters by constructing new buildings, change of local culture for conforming to tourists' culture, destruction of the zone environment.

5.4.3 Positive Impacts of Rural Tourism

- Tourism economic effects can help to set the village economic security so that it leads to preservation of natural qualities of the vicinity of cities by the local people that this is one of the village tourism purposes.
- Many different parts of society through economic points of view will be benefited from tourism in a rural tourism project.
- Tourism can cause to strengthen and develop those parts of industries which are not related to tourism directly. It causes to attract other industries.
- Preserving natural resources and ecology of the region by villagers and local people are the positive effects of rural tourism.
- The development of rural tourism can make transportation system easier for the local as an easy movement of their products with the community convenience ties and other parts of the country.
- Rural tourism will cause to introduce the national heritage, culture of art, and history of country to other nations due to the promotion of rural amenity values (Johnston et al. 2003).
- Preserving natural resources and ecology of the region by villagers and local people resulting positive effects of rural tourism. It will service as a motivating factor to promote rural tourism within the local communities.

5.5 Conclusion

Looking at the discussion earlier, over view will help us to realize that, rural tourism is an opportunity for rural development. However, it may not be as great as some perceive. It may not also be suitable for every location. For example, local tourism impact varies greatly among rural regions. It depends on host of factors including work force qualifications, characteristics, and seasonality issues. Therefore, the development of rural tourism offers potential solutions to many of the problems facing rural areas. As a result, there has been an increasing dependence on and support for rural tourism in the hope of achieving the potential benefits.

The extents to which these benefits are realized remain the subject of much debated. Certainly, there is evidence to support the claim that is, as a vehicle of economic growth and diversification, tourism can make an important contribution to rural incomes—both at the level of tourism operators and local economy. Rural tourism events have been found to increase business, income, and employment. They are seen to assist the social and economic development. Tourism can be an important source of jobs for rural communities. Tourism not only offers business opportunities to local residents but also enhance local quality of life (Seba 2011, p. 108). Tourism can also support local culture in rural areas by encouraging restoration of local and regional historic sites.

However, as a general rule, rural communities are challenged to take full advantage of the tourism industry due to lack of sufficient infrastructure to support year round visitors. Therefore, in a bid to exploit the cash injection that tourism provides, with minimal additional resources, many rural communities have opted to host tourism events on a yearly basis. The fact that rural tourism enterprises tend to be small scale and supply a highly seasonal market present a number of other challenges for rural tourism development (Irshad 2010). Not all rural areas are equally attractive to rural tourists and simply providing. Accommodation facilities do not guarantee demand. The total product package must be sufficient to attract and keep tourists, offering suitable opportunities for spending. Developing and organizing rural tourism may require a significant investment either beyond the means of business owner or greater than justified by potential returns. Local communities and businesses may find it difficult to adapt to a service role.

The quality of products and quality of services must match tourists' demand and expectations. Therefore, tourism is not the panacea for all rural problems but it has number of positive attraction (Long and Lane 2000). It is one of the many opportunities that rural communities might consider improving productivity and incomes within the local communities.

References

Botezatu A (2014) Tourist motivation for rural destinations. Sci Pap Ser Manag Econ Eng Agric Rural Dev 14:49–52

Brown DO (1998) In search of an appropriate form of tourism for Africa: lessons from the past and suggestions for the future. Tour Manag 19:237–245

Caffyn A (2000) Developing sustainable tourism in the Trossachs, Scotland. In: Richards G, Hall D (eds) Tourism and sustainable community development. Routledge, Longon, UK, p 85

Camilleri MA (2018) Tourism supply and demand. In: Travel marketing, tourism economics and the airline product: an introduction to theory and practice (Chapter 8). Springer Nature, Cham, Switzerland, pp 139–154)

Chio J (2011) The appearance of the rural in China's tourism. Prov China 3:60–79

DEAT (1996) The development and promotion of tourism in South Africa, White Paper, Government of South Africa, Department of Environmental Affairs and Tourism, South Africa

Fleischer A, Tchetchik A (2005) Does rural tourism benefit from agriculture? Tour Manag 26:493–501

Hall CM (2004) Small firms and wine and food tourism in New Zealand: issues of collaboration, clusters and lifestyles. In: Thomas R (ed) Small firms in tourism: international perspectives. Routledge, New York, NY, U.S.A

Hall CM (2005) Rural wine and food tourism cluster and network development. In: Hall DR, Kirkpatrick I, Mitchell M (eds) Rural tourism and sustainable business, channel view publications. Bristol, UK

Holland J, Burian M, Dixey L (2003) Tourism in poor rural areas: diversifying the product and expanding the benefits in rural Uganda and the Czech Republic. In: Pro-poor tourism partnership (http://www.propoortourism.org.uk) PPT Working Paper No. 12

Irshad H (2010) Rural tourism–an overview. Rural Development Division, Government of Alberta, Agriculture and Rural Development, Canada

Johnston RJ, Swallow SK, Tyrrell TJ, Bauer DM (2003) Rural amenity values and length of residency. Am J Agric Econ 85:1000–1015

Kapur S (2016) Rural tourism in India: relevance, prospects and promotional strategies. Int J Tour Travel 9:40–46

Knowd I (2001) Rural tourism: panacea and paradox—exploring the phenomenon of rural tourism and tourism's interaction with host rural communities. Geography Teachers' Curriculum Workshop, University of Western Sydney, July 2001

Line B (1994) What is rural tourism? J Sustain Tour 2:7–21

Liu C-Z (2010) Rural development and rural tourism in Taiwan. Asian J Arts Sci 1:211–227

Long P, Lane B (2000) Rural tourism development. In: Gartner WC, Lime DW (eds) Trends in outdoor recreation, leisure and tourism. CABI, Oxfordshire, UK

Lukaz JYT (2006) Ghana: rural vs urban tourism in Ghana—which? part I. https://allafrica.com/stories/200610160914.html

Mitchell J, Faal J (2007) Holiday package tourism and the poor in the Gambia. Dev South Afr 24:445–464

Nilsson PA (2002) Staying on farms: an ideological background. Ann Tour Res 29:7–24

O'Connor P (1996) Invisible players?: women, tourism and development in Ballyhoura. Department of Government and Society, University of Limerick, Limerick, Ireland

Roberts L (2001) Rural tourism and recreation: principles to practice. CABI Publishing, Wallingford, Oxfordshire, UK

Said SY, Aksah H, Ismail ED (2013) Heritage conservation and regeneration of historic areas in Malaysia. Procedia Soc Behav Sci 105:418–428

Seba J (2011) Tourism and hospitality: issues and developments. Apple Academic Press, Palm Bay, Florida, U.S.A.

Teye VB (1999) Commentary: tourism plans and planning challenges in Ghana. Tour Geogr 1:283–292

Tsai HC (2007) Agricultural globalization and rural tourism development in Taiwan. Asia J Manag Hum Sci 2:1–13

Wilson S, Fesenmaier DR, Fesenmaier J, Van Es JC (2001) Factors for success in rural tourism development. J Travel Res 40:132–138

Chapter 6
Responsible Tourism

6.1 Introduction

Responsible tourism is a new concept in tourism industry. This concept was developed by Jost Krippendorf (1938—2003), one of the Swiss scholars, during 1980s. His aim was to develop and promote new forms of tourism, which would bring greater possible benefit to all participant travelers, host population and tourist business, without causing intolerable ecological and social damage (Krippendorf 1982).

The resonance of sustainable and responsible tourism within the industry has been intensified. This has led some businesses to adopt the language merely in an attempt to boost sales, playing on consumers' preferences for ethical travel without making real changes to the way they operate. Such "green washing" is often neither due to ignorance, sloppiness, or over-eager communications campaigns bolted to operation that lack on social and environmental rigor. A system was introduced to be evaluating and monitoring enable businesses to tackle these shortfalls. They can clarify both the positive and negative impacts of their operations, focusing on how the aspects can be maximized and the negatives one be minimized. This is the main focus for every country that wants to develop it tourism industry. It is the basis that the rationale for responsible tourism reporting has formed.

The growth of responsible tourism and the increase in consumer demand for it has led to the industry developing certification, labeling, and awards schemes to grade businesses in other for the locals to become more responsible (Muhanna 2006). This has resulted in literal scores of different schemes each with it own one-size fits all approach. While some of the schemes out there are useful in helping to undermine the industry's green washers, they do little to highlight the achievements of those who are really making progress and setting the pace in sustainable development. None of the certification schemes transparently reports what has and has not been achieved, nor do those priorities the criteria to reflect different business types and different operating areas.

According to sources, responsible tourism reporting is a new initiative offering a flexible and 100% customizable independent verification scheme (Goodwin 2005).

W.-T. Fang, *Tourism in Emerging Economies*,
https://doi.org/10.1007/978-981-15-2463-9_6

Table 6.1 Social–economic impacts of tourism to local communities. *Source* Durbarry 2008; Goodwin 2005; Pender and Sharpley 2004

Positive aspects	Negative aspects
Improve quality of police protection	Increase economic dependency on one sector or even one enterprise
Stimulate/strengthen local economy	Increase pressure on infrastructure and services, especially in cases of mass or high density tourism
Create opportunities for direct and indirect employment	Endangered social tensions linked to wage and income disparities, between hosts and visitors as well as within the' community itself
Diversify livelihoods	Create high regional leakage (the percentage of tourist expenditure that leaves the local economy)
Stimulate skills development	Encourage dominance by multinational companies or local 'power brokers' who appropriate all or most of the benefits from tourism operations
Generate investment in social and economic infrastructure (e.g., schools, clinics, and roads)	Drive up land prices and housing living costs, which can lead to displacement of local residents
Improve public services and amenities (e.g., transport, shopping, and entertainment)	Stimulate inflows of job-seekers (regional immigration which can increase unemployment and engender social tensions)
Stimulate local business growth, directly and indirectly	Stimulate outside business firms to come in without providing local jobs

It motivates small businesses to measure themselves against their own goals and publicly report their annual achievements. The scheme gives each business the opportunity to explain and justify their choices, setting the standards and priorities for what is important for their business and the communities where they operate. In addition, the impacts of tourism, both good and bad, are many and diverse (Table 6.1).

Responsible tourism has been recognized as one-way of maximizing the possibilities of positive impacts and minimizing the negative impacts of tourism. The tourist's contribution to responsible tourism is somewhat overlooked in previous documents (Stanford 2006). They are often emphasized as parts of the problems rather than as parts of the reliable solutions. They are also frequently accused of taking a break from responsibility during holidays. This argument focuses on the actions of the tourists, seeking to understand what kinds of responsible behaviors while tourists are taken on their holidays. In Stanford's argument, she also compared some responsible tourists' behaviors during holiday with similar behaviors at home. Moreover, the readers may also explore her arguments focused on effective communication to encourage responsible behaviors within a particular destination that the tourists went on their vacations. The impacts and constraints could be quite complex in her studies.

A fluid methodology, therefore, has been requested to require to be sufficiently structured, thus allows comparability, while flexible enough to allow for the unexpected to create a new structure from her collected data.

A multi-phase, multi-method literature research design could be used, based on comparable case studies (Stanford 2006). In her studies, for example, Stanford tried to detect two locations, Kaikoura and Rotorua Kaikoura, in New Zealand. The first phase of research, she employed in-depth interviews from tourists with industry representatives with document analysis. The initial stage of her approach allowed five actions of responsible behaviors to be identified. Stanford (2006) identified instances of responsible behavior, influences, and constraints on tourists' behaviors. She has also explored effective means of communication to encourage responsible behavior using Kohlberg's Stages of Moral Development (Kohlberg 1973).

We may find some of the international cases, quite complex, represented responsible behaviors in a range of under such circumstances, i.e., environmental, cultural, social, and economic perspectives (Fang et al. 2017). These behaviors were then applied in the visitor survey, based on a social psychological framework using Ajzen's Theory of Planned Behaviors (Ajzen 1991). You can see some cases applied with the concept of the Theory of Planned Behaviors as shown in Appendix 1.

Generic definition of responsible tourism and tourists were developed lately (Reid 2003; Spenceley 2012; Goodwin 2016). Five actions representing responsible behavior were identified which operationalized the definitions from the specific context. Influences and constraints on these responsible behaviors were also identified. These were the internal, (1) emanating from the tourist including culture; (2) values and attitudes; (3) ethics and motivations; (4) mindfulness and mindlessness, and (5) external factors emanating from the destination, including marketing, visitor management, information, and communication (Stanford 2006).

I may encourage that you can select some characteristics, like awareness, motivation, attitude, and tourist choices have been considered an important aspect of responsible tourist behaviors (Caldito et al. 2015). Yet, few tourists were aware of the context specific details. Information, therefore, is considered important in achieving responsible tourist behavior. Information may be most effective if it appeals to good citizenship and provides a reasoned and positive argument. A stepping model was developed to foster responsible tourist behavior according to Stanford (2006). The first and crucial step is for the destination to set its objectives, then, to market to the most appropriate tourists. Then, you can choose next step to optimize the responsible behavior of these tourists through visitor management once they have arrived, encouraging and facilitating their responsible behaviors.

6.1.1 Definition of Responsible Tourism

Responsible tourism can be defined from The Cape Town Conference on Responsible Tourism which creates better places for people to live in, and better places to visit. Some of the main objectives of responsible tourism are listed to:

- benefit local communities both economically and socially;
- respect local cultures traditions, religions and built heritage;
- conserve natural resources from source to destination;
- minimize pollution through noise, waste disposal, and congestion, and
- protect the environment; its flora, fauna, and landscapes.

The definition of responsible tourism will give us a better understanding of what responsible tourism is. At this point, one needs to know about the role of sustainable tourisms in our society. This type of tourism could be defined as a form of tourism that is centered on the following key concepts according to Cape Town Declaration:

- Responsible tourism can be defined as tourism which creates a better place for people to live in and better place to visit.
- Responsible tourism minimizes negative economic, environmental, and social impacts, generates greater economic benefits for local people and enhances the well-being of host communities. Responsible tourism often used interchangeably with sustainable or socially conscious tourism. Responsible tourism does not have a firm definition and can therefore be used for many types of travels. That may or may not be comprehensive in their efforts at responsibility which could lead into the following factors.
- Minimizes negative economic, environmental, and social impacts to generate greater economic benefits for local people, thus enhances the well-being of host communities to promote responsible tourism.
- Involves local people in decisions that affect their lives and life chances to worsened responsible tourism.
- It is culturally sensitive, encourages respect between tourists and hosts, and builds local pride and confidence within the community.
- Provides access for physically challenged people.
- Makes positive contributions to the conservation of natural and cultural heritage embracing diversity within the host community.
- Provides more enjoyable experiences for tourists through more meaningful connections with host communities, and a greater understanding of local cultural, social and environmental issues.

The distinguished characteristic of approach is the focus on responsibility of the role players in the tourism sector, plus the destinations at large in order to take action to achieve sustainable tourism development. Increasing numbers of consumers are looking at reputation and responsibility of the companies they purchase from; they want to have "guilt free" holidays (City of Cape Town, Tourism Department 2018). This affects their direct purchases from companies in tourism destinations, therefore influences the selections of sources from market companies. For example, the UK, other European, Australia, increasingly American companies are asking about the corporate social responsibility of their suppliers. They are introducing better check lists for which rate of the sustainability to be applied in their choices. This is because a responsible tourism is about "making better places for people to live in, and for people to visit". The other clause is important for responsible tourism.

It is a broad movement as the heart has commitment to respect places the tourists visit. They put the interests to the communities naturally. In the UK, for example, the Associate of Independent Tour Operators (AITO) also help in the promotion of responsible tourism. The local people as well as their communities get benefit socially, economically, and environmentally. These are the results which provide a lot of benefits within the local communities and the country at large to be driven (Mowforth and Munt 2008).

In addition, green tourism and responsible travel are also of the key concerns for AITO and its members. Each potential member for sustainable tourism credentials are examined before they join the association. To ensure sustainability, the local cultures and environment are treated with the utmost care and respect. AITO members recognize that the destinations where they provide holidays are the lifeblood of the industry. They need to protect them with a responsible travel and sustainable tourism policy to make sure that local people benefit either directly or indirectly (Goodwin and Francis 2003).

AITO is the first tourism industry association to incorporate into the business charter with a commitment to Responsible Travel and Green Tourism. Sustainable travel guidelines for its members based upon five key objectives rules are as below. They are to be re-defined as follows:

- benefit local communities (i.e., both economically and socially);
- protect the environment (i.e., fauna, flora, and landscapes);
- respect local cultures (i.e., traditions, religions, and built heritage);
- minimize pollutions (i.e., noise, waste disposal, and congestion); and
- conserve natural resources, i.e., biotic and abiotic factors, from office to destination.

6.1.1.1 Economic Responsibility

Assess economic impacts before developing tourism and exercise preference for those forms of development. This benefits the local communities and minimizes negative impacts on local livelihoods (for example, through loss of access to resources). Recognizing the tourism may not always be most appropriate form to local economics. The development is looking into the following factors from the Cape Town Conference:

- Adopt equitable business practices, pay and charge fair prices, and build partnerships in which risk is minimized and shared, and recruit and employ staff recognizing international labor standards.
- Maximize local economic benefits by increasing linkages and reducing leakages, in order to make sure communities are involved in, and benefit from, tourism. Wherever possible use tourism to assist in poverty reduction by adopting pro-poor strategies with the local communities.

- Market tourism in ways reflects the natural, cultural and social integrity of the destination that encourage appropriate forms of tourism.
- Provide appropriate and sufficient support to small, medium and micro enterprises to tourism related matters.
- Develop quality products that reflect, complement, and enhance the destination progress to ensure local benefits.

6.1.1.2 Social Responsibility

- Provide appropriate and sufficient support to small, medium and micro enterprises to ensure tourism-related enterprises thrive and sustainable in a way the local community can take part.
- Assess economic impacts before developing tourism and exercise preference for forms of development that benefit local communities. It is also minimize negative impacts on local livelihoods (e.g., through loss of access to resources) and recognizing tourism not to always be the most appropriate form of local economic development.
- Adopt equitable business practices, pay and charge fair prices, and build partnerships in major ways which risk is minimized, shared, recruit, and employ staff to meet the international labor standards.
- Maximize local economic benefits by increasing linkages and reducing leakages, to ensure communities are involved and benefit from, tourism. Wherever possible use tourism to assist poverty reduction by adopting pro-poor strategies.
- Market tourism should reflect natural, cultural, and social integrity of the destination. This must encourage appropriate forms of responsible tourism.
- Develop quality products that reflect, complement, and enhance the destination.

6.1.1.3 Environmental Responsibility

- Assess environmental impacts throughout life cycle of tourist establishments and operations. This must include planning, design phase, and ensure that negative impacts are reduced to the minimum and maximizing positive standard within the industry.
- Use resources sustainably, reduce waste, and over-consumption. Manage natural diversity sustainably, and where appropriate restore it. Put into consideration the volume and type of tourism as compared to the environmental capacity.
- Assess environmental impacts throughout the life cycle of tourist establishments and operations. These include the planning, design phase, and ensure that negative impacts are reduced to the minimum and maximizing positive level within the industrial standard.
- Managing and preserving natural diversity sustainably, and where appropriate restore it. Take into consideration of the volume and type of tourism as compared

to environment can support. Respect the integrity of vulnerable ecosystems and protected areas

- Use resources sustainably, reduce waste and over-consumption by promoting responsible tourism.
- Raise the capacity of all stakeholders and ensure that the best practice is followed. For this purpose consult with environmental and conservation experts which could help in preserving the environment.

6.1.2 Key Elements of Responsible Tourism

Below are the criteria for key elements of responsible tourism. Guiding principles for responsible tourism plays an important part on tourism to be learned as a key element in a following way.

- Market tourism that is responsible, respecting local, natural and cultural environments.
- Avoid waste and over-consumption.
- Be sensitive to the host culture.
- Assess environmental, social, and economic impacts as prerequisites to developing tourism.
- Monitor impacts of tourism and ensure open disclosure of information.
- Involve the local community in planning and decision-making.
- Maintain and encourage natural, economic, social, and cultural diversity.
- Ensure communities are involved in and benefit from tourism.

6.1.3 Trends of Responsible Tourism

Responsible tourism is the form of tourism in which the tourists, travel agencies, and the host destination can be maximized. The positive and negative impacts to environment, traditions and cultural heritage of the region, must be taken into account.

Responsible tourism is rapidly growing of global trend. Some nations like South Africa, United Kingdom, United States, Gambia, Sri Lanka, New Zealand, India, and Zambia are already practicing responsible tourism. They and other nations are gradually following the ideologies of these countries by introducing responsible tourism in their tourism industries. As a result of this, it has played an essential role in the development of their tourism industries through the help of responsible tourism actions and plans.

The trend of responsible tourism is about taking responsibility—individually and collectively—for triple bottom line sustainability, economic, social, and environmental in Sustainable Development Goals. Responsible tourism is about taking individual and corporate responsibility for implementing the principles of sustainable development. It is about taking responsibility for action rather than demanding someone else to do it. Therefore, you cannot outsource the responsibility. In travel and tourism, for too long, the emphasis has been on environmental responsibility. Examples of responsible activities can be discussed as in the following tourism destinations. They are the UK, New Zealand, and India from Cape Town approach addressed to responsible tourism.

- **United Kingdom**

In United Kingdom (UK), tourism sector is continually frustrated as the government does not take serious action. The UK only provides the level of resourcing which is necessary to ensure the country to maintain its international competitiveness.

According to UNWTO data, the industry continues to seek public funding without acknowledging responsibility for the public since a decrease in visitors coming to UK (Manzo 2019). To date, situation with negative clauses associated with tourism have been continuing since the last decade (Han et al. 2006; Durbarry 2008). It is only by facing up to this issue and developing holistic tourism strategies to maximize public yield, the government can see the real benefit from tourism and investment.

In many parts of UK, a reductive strategy has been adopted. Initiatives on sustainable tourism are largely limited to labeling and certification schemes. The public funding has been used to encourage local business to be certified. The bigger initiatives, that is, the development of new galleries, museums and public spaces, festivals, pedestrian walk, park and ride, cycle tracks, and local food, have often benefited tourism. Unfortunately, these initiatives have been designed to make better places for people to live in. Tourism may have been part of the justification but it was generally not the primary driver.

- **New Zealand**

A good example of national tourist board taking on the principles of responsibility tourism is the Tourism New Zealand. Their '100% Pure New Zealand' campaign, which incorporates the principles to date, is now 20 years old (Morgan et al. 2002, 2003). In order to develop responsible in the tourism industry, there are ways they can do.

According to sources, the UNWTO and the European Travel Commission (ETC) asked what countries they considered to be good at destination branding. New Zealand came out top, ahead of India, Spain, Australia, United Arab Emirates (Dubai), and Ireland. Respondents were citing the consistency and credibility of the '100% Pure New Zealand' brand, and strong imagery as contributing to the instant recognition of the brand and the strong positioning statement.

Based on the source of *Tourism New Zealand Annual Report 2007–2008*, George Hickton, the Chief Executive of Tourism New Zealand, commented that the result recognized the efforts of the tourism industry over the past 20 years in supporting the campaign and delivering on the promises made through '100% Pure New Zealand'. New Zealand realized that a successful brand is more than just a logo and advertising. As Hickton pointed out the efforts of the tourism industry to deliver quality, creating innovative products to reflect the values of the brand in tourism business have been critical to its success (Tourism New Zealand 2008).

Tourism New Zealand also won the Destination and Global Awards, Virgin Holidays Responsible Tourism Awards, in the 2008 because they had successfully harnessed their national brand and marketing campaign to support the development of a sustainable industry in New Zealand. Their market research demonstrated that overseas travelers were looking for landscape, adventure, people, and cultures. By targeting the kinds of visitors, it is likely to enjoy New Zealand, and working to ensure product quality (for example, New Zealand was the first country to successfully include the environmental dimension in its accommodation grading scheme). It has earned very high approval (90%) and recommendation (97%) scores from visitors.

This responsible tourism strategy has been applied consistently for 20 years. The Prime Minister has kept the tourism portfolio for himself, in recognition of the importance of the tourism brand to the country and the need to balance benefits and costs. As a result, the tourism industry in New Zealand has been recognized for what it contributes to the national economy and to society. It does not have to pledge in case for public funding. It is taking a responsibility approach, costs are minimized but benefits are maximized.

Box 6.1: What Is the Cape Town Declaration on Responsible Tourism?
According to sources, the UN Commission on Sustainable Development Commission, tourism was addressed in the post Rio UN process (Elkington 1994; Hardy et al. 2002). It was reasserted about the importance of development agenda. Lobbying by the British Government Department for International Development and European colleagues resulted in the inclusion of two calls for action. They are to:

- maximize the potential of tourism for eradicating poverty through developing appropriate strategies in cooperation with all major groups, and indigenous and local communities.
- work in partnership with major groups, especially at local level, to ensure active participation in tourism-related planning and development.

This signaled a change in policy orientation, thus resulted in further work by the pro-poor tourism partnership on strategies. As a result, a number of action research projects called *The Cape Town Declaration on Responsible Tourism in Destinations*, was agreed. It was declared at a side event at the World Summit on Sustainable Development in South Africa in 2002. It defined responsible tourism as having the following characteristics.

- Provides more enjoyable experiences for tourists through more meaningful connections with local people, greater understanding to local culture, and social and environmental issues;
- Generates greater economic benefits for local people and enhances the well-being of host communities, improves working conditions, and access to the industry;
- Makes positive contributions to conservation of natural and cultural heritage, and maintenance the world's diversity in order to promote responsible;

- Minimizes negative economic, environmental, and social impacts; and
- Involves local people in making decisions that affect their lives and life chances.

The 2002 *Cape Town Declaration on Responsible Tourism in Destinations* recognized that tourism could only be managed for sustainability at the destination level and that responsible and sustainable tourism would be achieved in different ways in different places (City of Cape Town, Tourism Department 2018). Only at local level can a decision be made about how tourism can best benefit local people and their environment directly or indirectly within the tourism industry.

Responsible tourism makes positive contributions to conservation of natural and cultural heritage. It is culturally sensitive engendering respect between tourists and hosts while building local pride and confidence (Sharpley 2012). The guest experience is at the heart of aspiration for responsible tourism. This provides more enjoyable experiences for tourists through more meaningful connections with local people, and a greater understanding of local cultural, social, and environmental issues. Following the Declaration, the city of Cape Town has developed a new Responsible Tourism Charter. That charter commits signatories to take responsibility for making tourism in Cape Town more sustainable. The hospitality industry, inbound tour operators, and the city have committed to development of responsible tourism policies and improvement plans. This contains measurable goals against progress that can be monitored and publicly reported.

The Charter commitment, signed on behalf of their members, applies to destination management. It is a commitment by the businesses to develop a plan for changing based on common responsible tourism principles, held accountable both for the plans and the deliverables.

Destination management systems and destination improvement plans will be developed and implemented based on responsible tourism practices. The city used planning, regulation, procurement, leases, and "whole of government approach" to manage tourism for sustainability. In addition, the roles and contribution of each council directorate have been defined. Detailed Responsible Tourism Policy for the City of Cape Town has been published. This policy outlines several priority issues which include the following:

- Local and Black Economic Empowerment (BEE) procurement.
- Water, energy, and solid waste (with a particular focus on plastic water bottles).
- Enterprise development, skills development, and social development within the local communities.

- **India**

In a nutshell, responsible tourism is a form of tourism which initiatives and responsibilities are taken by both the tourists and travel agencies (Kumar 2017). According to Kumar's studies, the purpose is to maximize positive impacts and minimize negative impacts on environment, traditions and cultural heritage of the region in India.

India has abundant scopes to be one of the emerging nations as a responsible tourism destination. The tourism industry in India, as Kumar described, has grown immensely in the last decade. It is sure that in the upcoming years, India would become a hot tourist destination in Southeast Asia. To sustain this growth process, it is necessary that all stakeholders of the Indian tourism industry, i.e., tour operators, government agencies and NGOs, work cohesively.

With the increased demand of responsible tourism awareness in India tourism industry, this country has put up several measures to protect the heritage. This in turn will benefit both the local communities and the tourists. India developed as responsible tourism destination, the International Centre for Responsible Tourism (ICRT), works in collaboration with private sector, government, and local communities (see ideas from Goodwin 2005, 2011, 2016). It is a Bangalore based non-profit organization, established in 2002 by Harold Goodwin. The main objective was to "take responsibility for achieving sustainable tourism, and create better places for people to live in and for people to visit."

Among all Indian states, Kerala has been one of the pioneers in practicing innovative tourism ideas into their host communities. It is a state rich in traditions and culture. It has started responsible tourism in the state to create awareness among stakeholders for a better environment to live and visit.

In addition, some of the successful responsible tourism project of India has been described as the "Sunderban Jungle Camp" initiated by Help Tourism Group. This project was to connect the heritage sites and protected areas of Sunderban with livelihood of the local communities through community-based, environmentally, and socially responsible tourism (Kumar 2017, p. 310).

"Awake and Shine" is another successful project detected from responsible tourism. This project was started by a Retired Indian Army General, "Jimmy" Singh in May 2006 (Kumar 2017, p. 310). Located near Kalimpong, it was to develop Samthar Plateau as a tourist destination depending on the homestay holidays from a picturesque landscape. This project, started off with just a few homestays, has helped the local community to improve their standards of living and made its mark on the Indian tourist map.

Another responsible tourism project of India is located in Anakkara Spice Tourism village in Idukki District of Kerala. Considered to be one of the best tourism spots, this project was established mainly to contribute a nation-wide tourism. It has been started in October 2004 by Women in Agriculture (WIA). WIA, targeted to closing the gender gap for development, is a group of women joined hands to preserve nature, share benefits of tourism among local populace, and boost the local economy and local development. This project has enabled ladies to improve their standard of living by providing their share of income in the well-being of their family and people within the communities. They also sell farm produces to hotels and lodges around the communities.

6.2 The Importance of Responsible Tourism in a Destination

According to the case studies in the previous documentation, we may understand that responsible tourism and sustainable tourism has similar goals to contribute to all 17 Sustainable Development Goals (SDGs). So, what is the differences between responsible tourism and sustainable tourism? Why they are so important?

The approach of responsible tourism is, therefore, similar as those of sustainable tourism environmental integrity, social justice, and maximizing local economic benefit (Niedziółka 2014; Dangi and Jamal 2016; Hanafiah et al. 2016). The major differences between the two are: in responsible tourism, individuals, organizations and businesses are asked to take responsibility for their actions and the impacts of their actions within the destination. They are working to help, the local communities benefit from it. This shift in emphasis has taken place because not much progress has been made on realizing sustainable tourism since the Earth Summit in Rio in 1992. This is partly because everyone has been expecting others to behave in a sustainable way. The emphasis on responsibility in responsible tourism means that everyone involved in tourism government, product owners and operators, transport operators, community services, NGOs, tourists, local communities, industry associations are responsible for achieving the goals of responsible tourism as follows.

6.2.1 Human Right and Tourism Concern

Tourist destinations should increasingly embrace the sustainability agenda. This includes some of the smallest and largest tour operators, hotel groups, and travel trade associations. The next challenge for the industry is to recognize that true sustainability means taking a human rights approach to tourism. In other to behave more responsibilities, this could be said as a result created a good environment for the local communities.

Tourism Concern, found by the Department for International Development (DFID) in UK, has defined that the human rights approach means "recognizing and addressing the multiple human rights impacts and issues associated with tourism". It makes business sense on several levels to achieve following goals. This includes risk management, competitive advantage, social sustainability, business leadership, and tourism ethics.

According to sources, the United Nations (UN) has now clarified and elaborated universal business responsibility to respect human rights (Beauchamp et al. 2012). This briefing introduces global baseline standard, as set out in the UN Guiding Principles on Business and Human Rights. The UNWTO urges that tourism industry to engage in other sector to become more responsible in their operating (Mihalic 2016). The UN Guiding Principles have received unprecedented levels of support from business leaders, governments, and human rights groups. However, the tourism industry has been regrettably absent from their development. The business responsibility to respect human rights applies to all businesses, everywhere including those in the tourism sector. It relates to direct activities, supply chains, and communities where companies operate. Companies can now work towards fulfilling their business responsibility to respect human rights by considering potential adverse human rights impacts and associated risks (Kemp and Vanclay 2013). This enables companies to reduce the risk of their activities harming the rights of others, and helps them to take necessary steps to avert, mitigate, or remedy potential, actual or perceived abuses.

Tourism Concern, for example, is keen to open a dialogue with industry stakeholders: small, medium, and large, around the challenges and practicalities of taking human rights approach to tourism. Essentially, addressing human rights effectively will require tourism businesses to think both in terms of their 'spheres of influence'. This has tended to limit human rights initiatives to issues of child exploitation and human trafficking, and'spheres of responsibility' (Ladd et al. 1999). This shift involves a company recognizing its potential to contribute to human rights abuses, and seeking to engage directly and transparently with those who may be affected, including communities in destinations. Therefore, the particular impacts can be understood and addressed. Lastly, this briefing issued by Tourism Concern, outlines a selection of the numerous initiatives, guidelines and toolkits that are freely available online. They are designed to assist businesses to adopt and implement a human rights approach.

6.2.2 *Engaging the Private Sectors' Responsibilities*

There is a range of reasons why it makes sense for private sector to engage. The green agenda was originally sold as cost saving. A broader range of factors drives the broader responsible tourism agenda through:

- Marketing and Public Relations—words of mouth recommendation;
- Enlightened self-interest;
- Staff motivation;
- Responsibility to stakeholders—communities;
- Local community expectation—license to operate; and
- Customer expectations.

There are increasing numbers of tour operators and accommodation providers. They make an effort to increase the contribution of tourism and their businesses to local economic development and poverty reduction. Some do it through philanthropy, through their own donations, other forms of support, and by encouraging tourists to donate. However, the increasing businesses are adapting the way they do business. The aim is to increase local economic development impact through employing local people, sourcing locally produced, goods and services, and encouraging the customers to spend money in local economy environment.

In the UK in 2000, the Association of Independent Torus Operators (AITO) committed to the principles of responsible tourism. They recognize that in carrying out responsible works as tour operators. They have taken responsibility to respect other people's places and ways of their life. Therefore, we acknowledge that wherever tour operators do business, it may be led both good and harm to environment (Pender and Sharpley 2004). We are aware, too, often in the past, the harm has outweighed the good.

Goodwin remarks that there are number of trends that drive changes. He said, "there is consumer demand for richer engagement with destinations". The communities who live there are more active and engaged holidays for better experience in his ideas while he wrote in personal web pages. These trends, therefore, may promote engagement with local communities and provide enhanced opportunities for local economic development through tourism. There is a trend towards experiential holidays and baby boomers looking for greater experiential engagement (Patterson et al. 2017). Significant market segments are looking to get more out of shorter holidays. Regarding to consumer awareness in originating markets, he/she uses purchasing power to try in fair trade in destinations locally and responsibly. People want guilt free holidays—particularly at times of maximum indulgence. There have been changes in the investment climate. Companies are seeking to raise capital in the stock market. They have needs to address their corporate social responsibility (CSR) policies and their practice in order to increase and defend brand value (Porter and Kramer 2006). Employees and people in destinations have become more assertive about the importance of responsible practices. There are many businesses voluntary actions prefer to legislation and regulation within the destination they are operating in.

Responsible travel is an emerging sector. It is at a stage where many consumers are aware that they need to do something. Unfortunately, few are currently translating this into actions. In Goodwin's comments, he acknowledged that this may be due to consumer apathy or because of confusion around the plethora of green accreditations that exist. What is important, however, is how the travel sector embraces this trend and moves forward. Could this be an opportunity for the travel sector to finally move away from the downward spiral of cost cutting we have seen over the past few years? Responsible tourism makes business sense because it facilitates the development of better experiences that encourage repeat bookings and referrals (Goodwin and Font 2012). It is also enables companies to engage in non-price competition. This enables them to maintain margins and avoid squeezing their suppliers, to provide better experience, higher value experience, and increasing the spending in local economy. This can also contribute to encourage repeat business and extend length of stay.

Box 6.2: Case Study in Gambia

Tourism is one of the exciting key economic sectors of the west African country called Gambia (Fig. 6.1). It is the smallest country on the mainland African, surrounded by Senegal on the three sides north, south, east and Atlantic ocean at the westernmost part. Records show that tourism contributes 16 percent to the GDP of this country. The tourism industry in Gambia is coming up. It is relatively infant and the country is eager to avoid error of order, older holiday resorts saw negative effects of mass tourism. There are number of organizations whose aims are to either secure equitable treatment, not only for those engaged in tourism but also concerned about the impacts on local society. Generally, the organizations are responsible to environment by promoting tourism through tourism industries. Tourism is playing a vital part in the development of the sub-region, not only in economy but also in preservation programs. It is rich in cultural heritage and environment. To develop tourism, it is indeed important to bear in mind that the whole economic aspects along sustainable guidelines are to be considered. Furthermore, they are eager that tourists give something back to the community thus promoting fair, sustainable, and ethical tourism practices. A holiday can be enjoyed while at the same time being aware of what effects from the actions and behaviors are occurring around local communities.

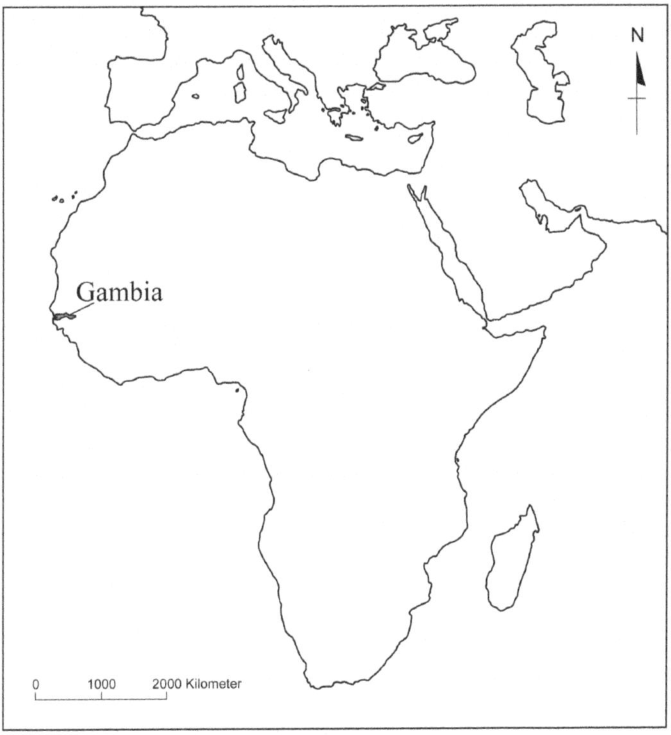

Fig. 6.1 Location of the Gambia (Illustrated by Yi-Te Chiang)

Responsible business practice is good business and it pays. It is a way of making business, not "just for the money" but also for taking responsibility to the positive social and environmental development within the destination while considering motives for CSR (Graafland et al. 2012). This is why ASSET (of The Gambia) and ONITS (of Senegal), supported by tourism stakeholders and their governments associated with local business, respectively. Gambia and Senegal signed an MOU in November 2006 to promote Responsible Tourism Development in their respected country. The two associations pledged to realize this by addressing the triple bottom line of economic, social, and environmental responsibility. They engage in product development and marketing. Almost one year after signing the MOU in Dakar, this helped make this pledge into a reality. They promoted and showcased the Senegambia Tourism Industry for the positive social and environmental development. They planned to make this as an annual event. Later, the event was going to be upgraded into an international fair held in Dakar. It was recognized as international and sub-regional world tourism trade fairs to meet for both requirements.

The objectives of the event include showcasing more developed and diverse products in order to attract people from established tourist markets. The "Mboka" Fair Committee is aware that tourists seek a variety of experience. It was expected that the Fair Committee to combine the traditional sun, sand and sea holiday market with the international event. It was increasingly competitive. The Senegambia region and its people have many to offer to international visitors. At the same time, the organizers of "Mboka" wanted to grow the business and maximized the benefits (i.e., economic, social and environmental) to the peoples of Senegambia.

- ***Key Factors Leading to Promoting the Responsible Tourism***

According to a case study, below is a list for the key factors leading to promoting responsible tourism in Gambia:

- Diversifying holiday experience for tourists through more meaningful contact with local communities, and foster a greater understanding of their cultural, society, and environment.
- Minimizing negative aspects on social, environmental, and economic effects to Gambia.
- Visitors need to consider the local cultural sensitivity, respect, and appreciate between guest visitors and the host, therefore establishing local pride, self-reliance, and assurance.
- Actively engage with local people on important decisions that will impact on their lives and future prospects.
- Create economic benefits for people in the host country and improves their well-being. This can be done through improving conditions at work and people's avenues to the industry.

- ***The Role of Asset in Promoting Responsible Tourism***

There are many institutions play in promoting responsible tourism. The Association of Small Scale Enterprises in Tourism (ASSET) was one of them established in April 2000. In the webpages from ASSET, you can see the role of ASSET is bringing and promoting a large number of small enterprises that were active in the tourism industry in the Gambia. While the ASSET promotes responsible tourism, this authority considers the local people to benefit from both directly and in directly.

The ASSET was introduce purposely mainly to help local people to come up with their own business in the tourism industry in other to make a good move in promoting responsible tourism within the host communities. These businesses include craft market vendors, tourist taxi drivers, official tourist guides, juice pressers, and fruit sellers as well as a number of small hotels, guest houses, and ground tour operators, etc. Since its inception ASSET has attracted more than eighty organizations into memberships, it has hosted an international conference on Fair Trade in tourism. The ASSET has been recognized by the Gambia Tourism Board (GTB) as an important voice for the previously voiceless, becoming a leading player in the Gambian Responsible Tourism Partnership (GRTP) program led by the GTB. This program brings together the principal stakeholders in the tourism industry (e.g., government, tour operators, ground handlers, hoteliers, and small scale businesses) in order to ensure that the adverse impacts of tourism are minimized and that local people and their communities gain maximum benefits from several informal sectors in the Gambia (Bah and Goodwin 2003).

In addition, ASSET's future goals include the training of large numbers of personnel working in tourism, the development and implementation of industry wide standards, and the marketing of the diverse and excellent services that are provided by ASSET members. ASSET won the Best award for Poverty Reduction during the First Choice Responsible Tourism Awards 2005 held at the World Travel Market in London.

- ***Aims and Objectives of ASSET in Promoting Responsible Tourism***

ASSET's aim is to assist and support small-scale enterprises in tourism to trade fairly and pursue sustainable development that contributes to conservation of the physical environment, social, and economic welfare of the community in Gambia in order to fulfill the following aspects:

- Identify appropriate quality standards for members.
- Provide assistance to members with product development, marketing, training, quality control and access to finance.
- Contribute to improving the tourism product in the Gambia and help to bring development for the country.
- The furtherance of good relations between ASSET and similar organizations in other countries.
- Protection of small-scale business interests in mass market tourist industry.
- Work collaboratively by promoting networking and joint activities amongst members.

- ***Gambia Tourism Concern***

The Gambia Tourism Concern is one of the private organization that help to promote responsible tourism in the Gambia. Its main function is to publish *Concern Magazine*, commissioned an in-flight video with other UK based organizations like Voluntary Service Overseas (VSO), Tourism Concern, and the Association of British Travel Agents (ABTA). The video is shown on First Choice flights on Air 2000 to the Gambia which features advice for tourists about sensitive tourism issues. By showing a short, entertaining in-flight video, tour operators can encourage greater awareness among tourists of the needs and wishes of people in Gambia. It is hope that the video will help the tourists be more responsible to local environment and community while staying in Gambia.

In conclusion, the Gambia green education tours are to promote responsible tourism into an interior of the Gambian communities. These tours are based on the taking of tourist around local communities and small-scale projects introduced by ASSET. These tours are organized mainly for the tourists to meet the local people and see how tourism can have an impact on a local community and how important it is for tourists to supports the projects. We can learn about community involvement, environment development, and poverty reduction initiatives through tourism by ASSET (the Association of Small Scale Enterprises in Tourism) and study lessons to be getting involved with the National Environment Agency (NEA) to see how they are working in hand with responsible tourism to protect the environment in the Gambia.

6.3 International Trend Towards Responsible Business Practice

According to sources since the Rio Earth Summit in 1992, economic growth has been promoted globally in terms of economic, social, and environmental sustainable development—the 'triple bottom line'. In tourism, however, the environmental aspects of sustainable development have tended to dominate international playing fields. This is the moment besieging its attention devoted to the development and promotion of 'ecotourism'. There has now been a change in thinking, the tourism industry should now put more emphases on poverty alleviation and socio-economic aspects of sustainable development in order to promote responsible tourism with a particular destination. Responsible tourism addresses this shift in private sector focused by giving equal weight to these three tiers of sustainability: (1) economy; (2) society; and (3) environment from Cape Town Declaration. "Responsible Tourism" incorporates the ethic of respect for culture and environment but importantly does not confine itself to these aspects of sustainable development in previous discussions.

There has been a concurrent international trend in favors of increased corporate social responsibility (CSR), as the private sector recognizes its mixed role in sustainable development. As we discussed before in this chapter, CSR deals not only with philanthropic activities, but also challenges the private sector to empower economically marginalized groups and communities through activity centers on employment, equity, and entrepreneurship to worse responsible tourism.

6.4 Conclusion

Responsible tourism makes positive contributions to conservation of natural and cultural heritage. It "is culturally sensitive, endangering respect between tourists and hosts, and builds local pride and confidence." from the 2002 Cape Town Declaration on Responsible Tourism in Destinations. Guest experience, therefore, is at the heart of aspiration for responsible tourism. It provides more enjoyable experiences for tourists through more meaningful connections with local people, and a greater understanding of local cultural, social, and environmental issues. In many destinations, tourism officer's role is simply to promote the destination online and offline in the hope of gaining more visitors. Public funds are spent to attract visitors who bring costs to the public purse and the benefits are captured by a few local businesses.

Any destinations wanting to flourish have to take a good approach to worse responsible tourism with the help of responsible tourism. It could have great positive impacts. All government, private sectors, and local communities could benefit either direct or indirect because the reforming support structure in place for tourism will free the industry from its current constraints. It is one of the few sectors of the economy which is currently creating employment opportunities. This can contribute a lot more if all people involved and take good responsibility approach.

Secondly, there is strong case to make a contribution. Tourism can make a difference to sustainable development of urban, rural areas, and communities. That case needs to involve many ways. Responsible tourism is most easily defined as being about creating "better places for people to live in and for people to visit." The order of those two aspirations is important; it answers the question about what purposes should be served by tourism.

In conclusion, responsible tourism can play a great role in the development of tourism industry. It is also good, if people take into consideration and put it into good parties. It can bring a great positive result to the economic. At the same time, all sectors in tourism industry can benefit from either direct or indirect only when the destination take it into responsibility, then put in some actions plans to make sure all sectors abided by it. Therefore, destination management systems and destination improvement plans can help in development of tourism destination. When properly implemented, it should facilitate and enforce responsible tourism practices. The city will implement their planning, regulation, procurement, leases and 'whole government approach' to manage tourism for sustainability. The roles and contribution of each individual with tourism sectors at large will increase the size of local market. It is a considerable benefit to businesses, thus gain economy for the locals from visitors.

References

Ajzen I (1991) The theory of planned behavior. Organ Behav Hum Decis Process 50:179–211

Bah A, Goodwin H (2003) Improving access for the informal sector to tourism in the Gambia. PPT working paper no. 15. Economic and Social Research Unit (ESCOR), London, UK of the UK Department for International Development (DFID)

Beauchamp TL, Bowie NE, Arnold DG (2012) Ethical theory and business, 9th edn. Pearson, Longdon, UK

Caldito LA, Dimanche F, Ilkevich S (2015) Tourist behaviour and trends. In: Dimanche F, Andrades L (eds) Tourism in Russia: a management handbook. Emerald Group Publishing Limited, Bingley, West Yorkshire, UK, pp 101–130

City of Cape Town, Tourism Department (2018) Responsible tourism in Cape Town. City of Cape Town, South Africa

Dangi TB, Jamal T (2016) An integrated approach to "sustainable community-based tourism". Sustainability 8:475. https://doi.org/10.3390/su8050475

Durbarry R (2008) Tourism taxes: implications for tourism demand in the UK. Rev Dev Econ 12:21–36

Elkington J (1994) Towards the sustainable corporation: win-win-win business strategies for sustainable development. Calif Manag Rev 36:90–100
Fang W-T, Ng E, Wang C-M, Hsu M-L (2017) Normative beliefs, attitudes, and social norms: People reduce waste as an index of social relationships when spending leisure time. Sustainability 9:1696. https://doi.org/10.3390/su9101696
Goodwin H (2005) Responsible tourism and the market. Occasional paper no. 4. International Centre for Responsible Tourism, University of Greenwich, UK
Goodwin H (2011) Taking responsibility for tourism. Goodfellow, Oxford, UK
Goodwin H (2016) Responsible tourism: using tourism for sustainable development. Goodfellow, Oxford, UK
Goodwin H, Font X (2012) Progress in responsible tourism. Goodfellow, Oxford, UK
Goodwin H, Francis J (2003) Ethical and responsible tourism: consumer trends in the UK. J Vacat Mark 9:271–284
Graafland J, Mazereeuw-Van der Duijn Schouten C (2012) Motives for corporate social responsibility. De Econ 160:377–396
Han Z, Durbarry R, Sinclair MT (2006) Modelling US tourism demand for European destinations. Tour Manag 27:1–10
Hanafiah MH, Azman I, Jamaluddin MR, Aminuddin N (2016) Responsible tourism practices and quality of life: perspective of Langkawi Island communities. Procedia—Soc Behav Sci 222:406–413
Hardy A, Beeton RJS, Pearson L (2002) Sustainable tourism: an overview of the concept and its position in relation to conceptualisations of tourism. J Sustain Tour 10:475–496
Kemp D, Vanclay F (2013) Human rights and impact assessment: clarifying the connections in practice. Impact Assess Proj Apprais 31:86–96
Kohlberg L (1973) The claim to moral adequacy of a highest stage of moral judgment. J Philos 70:630–646
Krippendorf J (1982) Towards new tourism policies: the importance of environmental and sociocultural factors. Tour Manag 3:135–148
Kumar P (2017) Marketing for hospitality and tourism services. Mc Graw Hill India, Noida, India
Ladd GW, Birch SH, Buhs ES (1999) Children's social and scholastic lives in kindergarten: related spheres of influence? Child Dev 70:1373–1400
Manzo GG (2019) Travel & tourism economic impact 2019. World Travel & Tourism Council, London, UK
Mihalic T (2016) Sustainable-responsible tourism discourse—towards 'responsustable' tourism. J Clean Prod 111(Part B):461–470
Morgan N, Pritchard A, Piggott R (2002) New Zealand, 100% pure. The creation of a powerful niche destination brand. J Brand Manag 9:335–354
Morgan N, Pritchard A, Piggott R (2003) Destination branding and the role of the stakeholders: the case of New Zealand. J Vacat Mark 9:285–299
Mowforth M, Munt I (2008) Tourism and sustainability: development, globalisation and new tourism in the third world, 3rd edn. Routledge, New York, NY, USA
Muhanna E (2006) Sustainable tourism development and environmental management for developing countries. Probl Perspect Manag 4:14–30
Niedziółka I (2014) Sustainable tourism development. Reg Form Dev Stud 3:157–166
Patterson I, Balderas-Cejudo A, Rivera-Hernaez O (2017) Changing trends in the baby boomer travel market: importance of memorable experiences. J Hosp Mark Manag 26:347–360
Pender L, Sharpley R (2004) The management of tourism. Sage, London, UK
Porter ME, Kramer MR (2006) Strategy & society the link between competitive advantage and corporate social responsibility. Harv Bus 84:78–92
Reid DG (2003) Tourism, globalization and development: responsible tourism planning. Pluto Press, London, UK

Sharpley R (2012) Responsible tourism whose responsibility. In: Holden A, Fennell DA (eds) The Routledge handbook of tourism and the environment. Routledge, New York, NY, USA., pp 382–391

Spenceley A (2012) Responsible tourism: Critical issues for conservation and development. Taylor & Francis, Abingdon Oxford, UK

Stanford D (2006) Responsible tourism, responsible tourists: what makes a responsible tourist in New Zealand? Unpublished doctoral dissertation, Victoria University of Wellington, New Zealand

Tourism New Zealand (2008) Tourism New Zealand annual report 2007–2008. Tourism New Zealand, Wellington, New Zealand

Chapter 7
Pro-poor Tourism

7.1 Introduction

Pro-poor tourism is a form of tourism geared and operated to results in, well again, disposable benefits for poor people to benefit from tourism. It operates within their societies and communities as a whole in both developed and developing countries. Looking at pro-poor tourism in its definition, it is not a specific tourism product on its own or niche sector (Hall 2007; Sarah 2015). It is an approach to tourism development and management. In a way, it should diversify tourism as a destination with solely aim to local people tapping benefits from tourism.

Pro-poor tourism develops and promotes linkages between tourism businesses for poor people in a particular destination. It is a way to enhance the mutual benefits of local people from tourism operation, so tourisms have to say to poverty mitigation in the destination. With promotion of pro-poor there is possibility to increase the wellbeing of the host community. Poor people are able to participate more effectively in tourism product development of the destination. These products can range from culture; local people with motivated well in the preservation, and promotion of the authenticity of local traditional norms and values. It serves as pulling factors for tourists to experience the uniqueness at the destination. Tourists can come and join traditional cultural festivals, shows, local drumming, traditional weddings and as well, trying the delicacy of the local foods. It is then another segment attraction motivating many tourists in search of the gastronomy of the destination, and architectural structures. This project is real for cultural and rural image of the poor. Tourists can visit on conducted excursions for half or full day trips for cross-cultural encounters.

Pro-poor tourism can start with imagination entrepreneur who wants to set up his own business to make profits. This form of tourism in simple terminology will not stand in isolation. It requires the involvement of organizations, agencies, and local community groups. All of them must work together in a common platform towards the realizations of the aspired dream. It depends on who is there for close collaboration and consideration of interrelationship and intern-dependency with many different types groups then all must give responsible commitments. This is the concern to

W.-T. Fang, *Tourism in Emerging Economies*,
https://doi.org/10.1007/978-981-15-2463-9_7

the need of the underprivileged poor people. All parties are working hand-to-hand involving employees working in travel and tourism industry, group of people, safeguarding interest of the majority residents, fixed asset holders, manufacturer of groceries, energy in facilitation of the service equipment and machinery, entrepreneurs of small scales, enterprises in tourism, related business operators, vendors, other users of tourism infrastructure (roads) and resources (water), etc.

7.1.1 Why the Poor Needs Tourism?

Tourism is one of the primary sources of foreign exchange earnings in the world today earning 49 of the Least Developed Countries recorded in year 2015 by (UNWTO) especially for the Low Developing Countries (LDCs) (UNWTO 2016). It provides a valuable source of income for many rural local economies in developed countries. In locations, where there is tourism potential, the trade also offers opportunities for small-to-medium sized enterprises (SMEs), particularly because market access is comparatively uncomplicated relative to other industries within the tourism market.

Secondly, general forecasts for travel and tourism point out to the continued growth, with 1.8 billion people set to undertake foreign travels each year by 2030, and substantial growth in domestic tourism anticipated in many newly developed countries (UNWTO 2017). In addition, the meaningful market experience rich holidays involving contact with local people continues to expand. The SMEs across a wide range of business types are potentially well placed to meet this growing demand. When managed responsibly, tourism can help alleviate poverty as well as diminish threats to local environments. It can also run direct counter to the goals of sustainable development when conducted without care to the environment and basic human rights (United Nations 2015). There are many cases of the latter but a typical situation might involve lavish water-intensive tourism facilities being maintained next to villages. The water supplies and sanitation are inadequate. Not only do unsustainable tourism practices threaten to jeopardize the health of the environment and the well-being of local communities but they can also diminish the quality and viability of tourism to the areas. Yet, such situations in numerous holidays are found in the locations within the globe.

Many evidence of analysis on journals, reports, magazines, and tourism information bureaus and offices have echoed the overall impacts of tourism in emerging destinations which cannot be over emphasized looking at things from a global perspective. You will realize that tourism has enhanced the life on millions of people globally. Their social living status is shifted by the development of tourism that takes places in destination with the construction of leisure and recreation facilities. These places are used by the poor as amusement parks. Economic well-beings are also improved under obligations to support local tourism. The livelihood of the poor is fulfilled a major success of which is the creation of countless employment for skilled, semi-skilled, and poor communities, who generate remunerations from the set-ups of tourism.

7.1.1.1 Tourism Bearing Means of Support

In underlining the effects of tourism in a three-way formula socially, environmentally, the most important of all economic impressions incur by tourism. Spawning home-grown income generation for the locals receives remunerations from official services. Rendering to industry, this kind of tourism, steering the human-resource sectors could be rewarded by proving quality, efficient, and sufficient services to both domestic and international tourists.

Tourism proceedings of income generate by the poor becomes tremendous. The poor makes high earnings through the consumptions of tourist, local obtain revenue from the selling of merchandises in the form of souvenirs. These services rendered in the form of unplanned effort in the catering and hospitality industry, recreation, and attractions. All of those today constitute some elements of community-run creativity. Governments also assist the tourism industry for its sustainable development through the acquisition of loan and financial assistance for the poor who want to venture into entrepreneurship. Tourism growth transforms underprivileged people's admittance to assets and to relate livelihood options while impacts occur on assets and other activities of the poor. In addition, the industry has being excited in creating resources, both in money and credit worthy for establishment of basic amenities for the poor. They are likely for the construction of schools, markets, good road networks, public wellbeing, centers, and other infrastructural and superstructure developments all of which reinforce sustainable development across all facets of life. Maintaining the environment around tourist areas, like keeping the beaches clean and investing in proper waste management are also important aspects of maintaining the tourism product.

7.1.1.2 Why the Poor Does Not Get Tourism Income?

Despite the fact of the tourism on developing destinations are emerging, unfortunately, these bring some adverse results to social, environmental, and economic impacts. These include the (1) inflation of good services at leading developing countries to a situation of rapid price increment on basic amenities, (2) commercialization of local culture to fit into the itinerary of tourist, (3) virtually response shows the registered control of most tourism establishments by foreign investor who have total owner-ship of hotels, restaurants, inns, motels, tour operation companies, and (4) travel agencies in the developing countries which disintegrates commercial prospects for the local poor in developing countries.

Looking at parks in India, Indonesia, Zimbabwe, Kenya, and Tanzania, should salvage and flourish pro-poor tourism development in and around the parks by giving totally management and ownership to the local people's projects. They are like craft vendors, tourist guides, and fruit and juice pressers need significant promotions and marketing by the hoteliers and tour operators increasing their income.

7.1.2 Tourism Development and Poverty Mitigation

From the previous statement, we can understand that stakeholders and practitioners are busy measuring ways in reducing the low level of living standard of the poor people through an initiative called pro-poor tourism. Tourism, however, is defined and explained by academicians, researchers, and educationalist as one of the world's largest growing industries in the world. Poverty is an issue, if not all, in most developed, developing, and under-developed countries. It is generally a concern in developing and under-developed countries and may become major issues in the top eradication agenda. It appears on list of both developing countries and under-developed countries. While poverty is much more prevalent in developing countries, it is an important concern in tourism industry.

It serves to a major earner in the GDP of many countries. It is already growing significantly in most poor countries and developing countries. These countries are today served as best tourism attractions appealing to wider audiences of international tourists, domestic tourists, regional, and sub-regional tourists.

Growth prospects through tourism industry remain at the rise in many of Less Developed Countries (LDC). It represents an important factor for general state development of the economy. In spite of this, the tourism and travel being proven for its contribution to global economic growth and development, the sector is still lack in political, economic recognition, and support by people (Chok et al. 2007). Due to less impact on the empowerment of the underprivileged, the government is acted as a mediator and facilitator between stakeholders to improve resource management and ensuring the benefits for the poor. In this section, we discuss the responsibilities for the following key players: people, government (public sector), and private sectors to mitigate poverty.

7.1.2.1 Tourism and Indigenous People

Tourism cannot work in isolation. The industry should empower local residents in one way to enhancing tourism to its pro-poor folks in the surroundings. Local people work in labor intensive part of the industry. They work in the service sectors as skilled, unskilled, and semi-skilled employees of the industry. Majority of them are low paid, therefore are not entitled and subjected to employment benefits. They are due to their low level of education, including both genders of male and female occupying the formal and informal sectors.

The efforts and steps are taken by locals to promote and support goal of tourism. This is based on accepted and traditional resources of the poor. They serve as the push and pull factor to success. The appropriate areas for poor are suitable to have tourism program for pro-poor development. They take advantage on these features and resources available at our disposal, while reducing the adverse effects of tourism on the underprivileged people in less developed countries. Various steps have been taken as causes of concern, highlighted as possible steps for change, and development towards a concise and responsible way from the tourism master plan by most governments and its allied agencies. They work in strong commercial and international interest in sustaining the industry for its long term prospect. Today, most governments, agencies, and organization involved in tourism are working on strategies and plans for long sustainability of the tourism. Therefore, many of the strategies and formulation are aimed to help the poor people. These trends led to the diversification of tourism industry leading to initiating of different forms of tourism. As examples, the community based tourism is supported local community projects; involving local people in the tourism business; giving them managerial post to run, and operate their own tourism projects. Nowadays, these kind of works can be found in many developing countries, like Gambia, Casamance Region of Senegal, Kenya, and other regions in central and east Africa.

7.1.2.2 Public Sector and Private Sector

Government (public sector) can help fulfill targets in increasing tourism expenditure by investing in tourism infrastructure, setting regulatory standards for the industry and by offering training and support to local producers. Looking to improve their products, financial assistances can be rendered to local communities for them to be committed to the conversation of natural resources and preservation of the authenticity of the local cultures. Such aids can be given through the village development committee in a community trust set-up in most developing countries. It can also be handover through the district chiefs and commissioners in other parts of the world. The emphasis on local community to get a clear share of profit given to them from the proceedings of pro-poor fund of which, should be used in support of other sustainable projects.

The state organization of the country, either direct or indirect, must involve with memorandum of understanding in tourism. It should contribute success to people. These are fall under two categories: (1) public sector and; (2) private sector. The public sector is the government, while the private sector is the non-governmental organizations (NGOs) (i.e., comprises of community organizations, foundation, agencies, and working groups to towards tourism development, promotion, advocacy, and empowerment of the poor). All of these have critical and different roles to play in planning, implementing, and administering tourism for a fair business. An extensive support is required from all angles. The players of the industry, both the private and public sectors organization, should contribute idea of reforms that will promote and empowered the vision of pro-poor tourism, the local stands in the center of significant move. They can be organized in local set-ups preserving and conserving the authenticity of their cultures, will appeal to the tourists. These cultures and other endangered attributes of the community can be packaged by organization to be sold as tourism products. These benefits will further motivate local people to construct, start natural and cultural initiative, which serves as an asset for them to tap potential benefits. This will lessen the adverse situation for locals live in the communities, villages, and provincial areas in most developing and under-developed countries around the globe.

7.1.2.3 Cooperate Social Responsibility (CSR) and Poverty Reduction

The private sector can be directly involved in pro-poor partnerships. A minimum private operator should contribute in product and sells improvement to guarantee marketable practicality. This should support local people to initiate income from tourism by small-scale enterprise operations.

The form of tourism today, being a gateway that chants a pathway to a significant transition. We put in at the core center of tourism planning and decision. Looking at ways and steps are methods to expand chances and privileges for the poor people. Most of them live in modest life standards. We universally observe in most residents in developing countries. We found out that they had little or no food, shelter, clothes, healthcare, education, and other physical means for living. They did not have quality of life but potentially being contributing to the success of tourism business.

Statistics have shown that most people in developing countries live with an average spending of 5 dollars and some live by spending less than one dollar per day. These have made depression in overall tourism destination development. Most initiatives can be supported by pro-poor activities. These initiatives are based on small-scale enterprises which are sustainable simple by all proceedings. They gain operation from hand to mouth. However, some entrepreneurs of businesses have used higher percentage of the profit. Globally, there are moves still undertaken to support the poor in some parts of the world. Some of them involve in local communities as their initiatives of promoting community based tourism projects.

They have the opportunities for the poor to open and manage their own accommodation and attraction sites in all places where tourism is significant. These are perfectly working in many developing countries. For examples, such projects are in Tumani Tenda Ecotourism Camp (the Gambia, West Africa), and the community in Senegalese's Region of Casamance. These have tremendously been effected in millions of the poor. The poor works in such initiatives with profits gain from this business to support their social livelihoods. Projects, such as local women vegetable and village school, are supported by the income gain from the operation of these local initiatives by the locals. In the following, we discuss more cases of the tourism-led poverty reduction strategies.

7.2 The Importance of Pro-poor Tourism in a Destination

Pro-poor tourism is regarded as one of the global issues but can help to generate millions of income to the poor if properly managed. The accommodation providers and catering outlets should consider buying locally grown food stuffs and local beverages where tourists can have the unique opportunities of testing the delicacies of the destination. This will pave a sustainable income ideology for the poor and as well enhance conservation of natural resources to meet the demand of tourism. Sustainable implementation of pro-poor tourism requires local participation. It involves training of local guides and conservation of wetlands to preserve and conserve endangered species. The designated land areas should be marked out by the central government which can be managed by locals to levied entrance fees which in turn are used to other community-based re-habitation works and support sustainable projects in economic, social, and environmental endeavours of tourism. In this section, we discuss the objectives, market functions, and benefits from pro-poor tourism.

7.2.1 Objective: Poor Alleviation

Tourism can be an asset to help alleviate poverty in destinations of developing countries. It can be an instrument for chanting a sustainable livelihood for the poor to make earnings and saving for disposable income. Developing countries have high turnover of tourists. This can be a tool to measure its potential ways of encouraging aids of the poor in tourism business. Linkages can be created on which different sectors of the poor are involved. The connection of embedding several enterprises into tourism should be highly encouraged like agriculture industries. It is predominantly believed that residents of developing countries live on farming for living. Tourism can help eliminate poverty to represent ultimate forces for continues growing seasons. At the initiate stage, it involves a lot courage, dedication, and motivation which might be time and resource consuming.

Stakeholders are involved in the planning, administrating, and implementing of tourism businesses, projects, reforms, and funds. These should be given clear consideration and opportunity to the poor to tap potential benefits of tourism. Sound laws and legislations should be formulated on reforming the policies. It should be deemed as necessary parts for poor tourism development plans and related businesses agendas. These can be an ability to widen the opportunities. The mobilizing resources will make tourism a gainful exploitation. It is done by creating establishments where indigenous will be deployed to work in various tourism facets their locality.

Tourism in less developed countries can be specified into different categories and forms. This will give equal rate of development for facilities and structures constructed in urban areas as well as in rural areas. Rural residents can start their development projects in their locality. They will handle and manage their own enterprises in tourism business, like eco-lodges, community parks, recreation centers, etc. These facilities should be pledged in rural area, slowing down rural-urban migration, where the poor get job or other opportunities to stay in their own settlements.

Tourism stakeholders can collectively enhance the employability of the local in tourism industry. They can fundamentally outline the opportunity of jobs to the poor into national legislation. Tourism can bring real economic benefits if properly managed and handled. It can easily create jobs and investments to open doors for hiring local poor people. These people can get valuable jobs through skill trainings. In keeping them work in the tourism industry they must be updated with tourism skills and important trends from time to time. Loan and financial aids and schemes can be enhanced to promote social, economic, and environment. The social activities can be packaged as potential niche tourism products to be sold to tourists. The poor will welcome the tourists to experience cultural exchange in their villages. The aids supported philanthropies can be generated through tourism activities, sponsorships, and other benefits.

7.2.2 Market Functions

Mass tourism travel on package holidays and reside in high standards of hotels on board's bases. The individual and adventurers will take a mile away from the usual tourism resort areas by hiking in the country side to experience different life by coming in contact with the locals. These diversify tourists are categorized into different segments on different forms of experiences, making potential markets of pro-poor. These include several categories, like package tourists, individual tourists, and travelers for pro-poor tourism. The SAVE market, currently, is one of potential market which entails student market (S), academician market (A), volunteer market (V), and educational researchers (E).

Pro-poor tourism portrays real cultural assets of the destination. It serves as the pulling factors for tourists travelling to destinations away from home to experience different life styles—unlike their normal routine life. Local destinations in developing countries like in Africa, Asia, and South America are mostly chosen to exploit these potentials. These different segments of vacationers' will have different experience of staying with the locals, in knowing how they live rather than staying in a luxurious hotel. In this mode of tourism, hey will spend less but have once experience in a lifetime of being a local holidaymaker. This will help the local people to earn extra income and at the same time the tourists will find more interesting stuffs sharing together the holiday. For the locals, they share their food, clothes, traditional stuffs, and more that they can provide for their countries. The locals will earn money by providing tourist quality food and souvenirs. Tourist will eventually make friends with local people. However, there will be a slight chance that tourists will make business offers to the locals in where they can import their products to their counties. Locals should be more friendly and approachable to tourists.

7.2.3 Benefits: Success for the Poor and Beyond

There is a variety of opportunities and benefits prevail to the poor by the tourism industry. They can exchange a lot of life experience to have access to quality and standard life which constitute their endeavor for fulfilling their daily unlimited wants.

These industries are important assets for developing economics as well as for individual who face problem of catering their needs and other social pleasure. Developments of tourism have provided a sound ecological development of landscape and wetland initiatives and core developing of projects where local people are staffs as employees.

Tourism business in developing countries has constructed strong capacity to build programs and facilitates as well as widened the scope of understanding of the poor in offering their talents and skills as sources of livings. We could be in carving a strong analysis of acquaintance from bearings of tourism proceedings on social and environmental facets of the destinations and approaches. Tourism will be regarded as a policy tool and industry instrument for positive changes, emerging in some destinations promotion. The conservation and preservation of tourism and its products will embrace nature, and enhance social set-ups of localities to improve the livelihoods of developing economics.

Social and economic resources of the poor are enhanced by these viable economic drivers. They predominantly provide skills, one of the major advantages for the poor in tourism. Tourists now travel as far and wide to learn about local traditions and local customs. These include the food, dances, languages, and music. All of them are sharpened by cross cultural exchange. They have widened the thoughtfulness of tourist expectations and perception on developing destination.

Today, the Peace Corps, USAID, and United Kingdom Volunteer Services (UK VSO) training programs in developing countries of Africa in countries like Gambia, Ghana, Kenya, Zambia, etc. For example, Gambia has benefited for small-scale tourism training and developing program in juice production. This production is also sustainable for bio-uses of the gas. In addition, the alternative uses of solar cookers save lots of trees from being cut for fuel woods. The environment is conserved for its authentic natural beauty which inhabits abundant of bird species making the place famous for ornithological tours. This provides financial benefits for some of the poor to take up ornithological guiding as their career conducting daytime excursion with groups of tourists on nature trails. Bird-watching activities have today become a famous tourist activity bring money to the poor.

In recent perspective on poverty alleviation, entrepreneurship has been a vibrant major by the poor. They work to generate their own personal income from tourism business. In recent year, emerging destination industrialists have generated their own investments of self-funded business. They have started small projects and widened their operation by plough earnings in the expansion of the entity. "*Even though sustainable tourism projects increasingly seek to address economic, social and cultural concerns, the rhetoric of sustainable tourism focuses on protection and conservation. While local populations must reap some social or economic benefits from tourism, this is principally to assure their cooperation and support for the primary objective of conservation; the benefits that accrue to local populations are a bonus*" (Department for International Development 1999, p. 2).

The social revenue generation of developing destinations where the poor are of majority could have a brighter access to dynamic and flexible forms of social regeneration of valuable meaningful gains from tourism. The potentials for participation may be greater in personal operation and community enterprises. A great of tourism registers its success on developing economies is respected of job creation. This is created from potential investment attraction by foreign investors in developing countries. The foreign exchange earnings are also looking as an important proceeding whilst the major successful cases are poverty alleviation in terms of respect for local cultures. This creates meaningful minority engagement as well as public enrichment. Tourism, thus, has paved sound cultural preservation and environmental protection as well.

7.3 Tourism-Led Poverty Reduction Strategies, Plans, and Scheme

There are many types of pro-poor tourism strategies formulated to support the pro-poor tourism ideology, ranging from increasing local employment opportunity for the aboriginal, giving them well-paid jobs so will upgrade them for a quality of life; and building mechanisms for consultation. All the organizations can involve in pro-poor tourism promotion and empowerment to support local people. This idea can be supported using any form of business organizations, like small eco-lodge, urban hotels, tour operators, infrastructure developers, community foundations and societies. The life-threatening issues are not the type of organization or the type of tourism, but increase in disposable benefits to poor people.

Poverty valuations proven that so many developing countries are collaborating with travel and tourism organizations across the sphere, comprising of World Travel Organization (WTO), Africa Travel Association (ATA), Caribbean Tourism Organization, Cyprus Tourism Organization, and United Nation World Tourism Organization (UNWTO). They are initiating and researching on key factors affecting livelihood of the poor. Tourism can affect many factors that are of detriment to the poor. It can be a prospective of alleviating and mainstreaming the positivity of tourist activities in developing countries which will further flourish the significant level of tourist grosses.

Approaches for pro-poor tourism can be divided into different groups: of local benefits inverted for poor people. This benefit can be as follows: (1) economic benefits, other social livelihood benefits, such as: physical, (2) communal or ethnic enhancements, and (3) less tangible benefits of participation and involvement. Each of these can be further disaggregated into specific types of strategies, aimed at up-lifting the welfare of the local people in less developed countries.

7.3.1 Pro-poor Tourism Initiatives

Tourism is an economic driver of a nation. It empowers understanding and mutual benefits between peoples and nature. It is an operative mechanism which eliminates and reduces the rate of poverty in communities; improve the reasonable objectives and welfare of local people, and eliminates poverty. Tourism could become more mainstream activity. Tourism can play a significant part in promoting sustainable local development and generating benefits for the poor to be successful. It must promote sustainable development by establishing a durable productive base that allows local inhabitants and tourism service providers to enjoy rising standards of living. One of the most dynamic economic activities of our time can be more effectively harnessed to address the problems of poverty more directly. Developing and committing to long-term strategies is a must to create an enabling environment within which the tourism sector can flourish. The benefits of tourism should be widely spread in society, and the poor should benefit from tourism development. Improving the quality of goods and services can increase spending on these goods and services. From goods in local crafts shops, to quality of transport, to quality of staffs in hotels, these are all goods and services. They are invariably added to the quality of the tourism product, as well as help the local populations to develop skills and improve incomes. Tourism is a rapidly growing sector that, if developed responsibly, provides an ideal opportunity for job creation and sustainable development. Tourism has emerged in this decade as a central pillar of the services economy. It can uniquely help society respond to global challenges. If its growth is managed wisely with an emphasis on ethics from poverty alleviation, the particular interests are of developing states and sustainable development.

Strategies should focus on economic benefits for expansion of employment among local people and local wages: through commitments to local jobs, and training up locals for employment.

7.3.1.1 Expansion of Business Opportunities

Expansion of business opportunities for the poor is one of the initiatives. These may be industries or entrepreneurs that make contributions like food, beverages, fuel, or building materials to tourism operations. It can also be businesses that offer products directly to tourists, like guiding services, crafts of souvenirs, bar and restaurants that serves traditional local foods and refreshments. Support can be varying from marketing, promotion, industrial and professional support to transform in operational approach, or undeviating economic in addition to guidance participations.

7.3.1.2 Helping in the Development

Second, helping in the development of accommodating community is essential. Proceeding to this project should be from the fairness payment of spending benefits of tourism to the local people, and charges and fee on tourist service. It can be used to support social livelihood project in the form of assistantships that offers by tour operators in under-standing with the government agencies.

7.3.1.3 Empowering Human Resources of the Industry

Government and Tourism Business Agencies can also empower human re-sources of the industry. They can enhance payment of local people who works for their They earn income to meet their need. With stable income, poor families have opportunity to acquire the relevant skills they need. Thus, they pass the good news from one to another.

7.3.2 Pro-poor Tourism Plans

According to the African Pro-Poor Tourism Development Centre (APPTDC), their project growths of the tourism sectors have boarded direct and indirect impact across all countries. This is true in rural and developing regions. Particularly those of developing emerging countries make pro-poor tourism well suited as a development tool extensively providing assistance to people in a vast array of locations to identify and realize the tourism potential of their areas. These processes of integrated ecotourism development are within a local, regional, and international context. This is a means by which more people from local community are employed and helps to build up the entrepreneurial skills. Pro-poor tourism should be developed in a way that encourages poor to be the beneficiaries' center at the fore front actively taking part in their own development projects by mobilizing their own needs and wants.

These plans can be vision with specific target deadline with a set goal. These can be comprised of different guiding principle, procedure, membership, and can be formed by all players to support towards the vision.

7.3.2.1 Compassionate Course of Action

Compassionate course of action can forecast agendas that permit the involvement of poor people in tourism decision making. This should allow the poor population to be aware of new decision formulations—conducted either by the public or private sector.

7.3.2.2 Broadening Decision-Making Process

There should be an enlargement with regards to the partaking by the underprivileged people in the decision-making process. Therefore, allowing that indigenous people seek advice. They should have a say in tourism procedures and law making by public and private sector organizations.

7.3.2.3 Proper Care and Concern on Local Poor

There should also be a proper care and concern on the welfare of the local poor. This will pave a pathway to the corporation of organization involved in tourism business to dedicate responsibilities on their shoulders, regardless of size, region, and scope of operation. They all should disseminate data on common ground, involving the poor in planning initiatives. Local people can have a say in things they deem necessary, whether it is economy, social, and environmental. The benefit of tourism business can operate within the dimension Economy, Social life, and Environment (ESE) within they live.

It is deemed necessary for works within the industry to attend to capacity development programs, workshops, trainings and other career developments. I will guide them to their pathway in getting well secured job where they can meet their end needs. In addition, they will help in alleviating of the environmental impact on tourisms of the poor. The management side will compete with the demands for access to natural resources between tourism and local people. This has the saying of proper planning to prevent problems. Agencies should address and concern over the natural resource. Here, tourism cannot be without those resources which include scenery landscapes, parks, museums, other historical and heritage sites. Most agencies can help in the improving the social and cultural repercussion of tours operations in less developed countries.

These approaches are aimed to minimizing the negative impact of tourism on the social, cultural, and economic and environment impacts of tourism. This can be either direct or indirect for the life of local people. Opportunities and facilities in destination should grant to local poor. They should be given privilege to have equal access to the facilities available for tourists. Among of them include beaches that are used by resident for leisure and fun days.

The public sector should lay foundation plans for training and career development programs for locals. This may be directly linked to constructing advance income generation—both long and short term plans. This can also be embedded by initiating, local owned training institutions for running course on management and service delivery of the industry. However, it may also be more for long-term indirect value, such as building management capacity of local institutions (Fig. 7.1).

7.3.3 Interventions and Scheme

Tourism has been an important viable economic business for many developing countries—undertaken by both private and public sectors. They are joining hand with the vision to increase manpower progression of the industry, as well as trying with a vision to reduce poverty in less developed countries. Tourism makes an important move in changing the life of poor women and men in developing destinations. This will directly and positively improve the livelihoods of higher percentage of marginalized population, as well as to pave a sound way for local development.

Fig. 7.1 Pro-poor tourism is as well recommended to hire local guides from the host community, as well as purchasing from the local markets. All of those contribute to the conservation of natural ecosystems, making it more sustainable on condition to generate venue for local employees. It is positive action towards cultural conservation and economic development which is an essential component of authentic pro-poor tourism (Photos by Wei-Ta Fang)

Some of the interventions and schemes have taken in developing countries with a vision for pro-poor tourism.

7.3.3.1 Playing a Vital Role in Poverty Reduction

Government laws and rules regarding tourism should be clearly stipulated to involve more stakeholders. This scheme will be taken more rooms for improvement and common grounds for partnerships to co-exist between the stakeholders to share resource information in promoting mutual understanding and benefits for the poor to equally involve in all possible viable economic sectors.

7.3.3.2 A Mean to End Rather Than an End to a Mean

Tourism in all destinations should be economic for all, regardless of social statues, age, gender, origin, and nationality. It benefits should be for all through expanding business opportunities for the poor to start their own proprietorship business in their localities. Sole proprietorship business can be operated to start and run by the poor remarkably. They can start operating food and beverage business, lodging, and camping sites. These can be initiated by means of growing employment opportunities for the poor, and improving civic revenues. This is not only done by aforementioned schemes to support tourism but also benefited to personal-earning poor. They are supported themselves through disposable income from tourism.

7.3.3.3 Customer Service Oriented from Business Sector

This service is regarding as the internationally fast growing sectors. It should be as well enriched the service delivery sector of the industry by offering training and capacity building programs and schemes for the poor. They can be enlightened on the various etiquettes, reception services, care procedures, and skills. These service industries can be resulted for them to better qualify for well-paid jobs to work as hotel and tour managers, travel consultants, hotel receptionists, as well as managers in Meeting, Incentive Travel, Convention, and Exhibition (MICE) industries, as well as in meeting and convention planners in non-economic benefits (Chap. 10). These help to alleviate cross cultural impression that poor community the perception of tourism. It will be broadening the scope of understanding social and cultural effects of tourism, to better be acquainted it ideas and skills to minimize the negative impact of tourism and maximize the positive benefits.

From the perspective, business sector marketing and promotional campaigns can be done through electronic media and also in printed brochures and flyers. This can be distributed on organized traveler markets, shows, and exhibitions. Websites are also an important way of selling a destination where customers can conduct all their booking and reservation online without consulting travel agencies.

7.4 The Essential Trends and Initiatives That Could Have an Impact in Emerging Economies

Pro-poor tourism has registered success in developing countries through the initiation, implementation, operating of sustainable faired trade business in tourism. However, some of these projects are run on small scale basis but very substantial in conveying remunerations for the poor.

With the slogan saying with the collective combination of government and people working together towards the realization and achievement of the pro-poor tourism goal, it has been changes ourselves. We are spending responsible cooperation during vacation and consider who benefits from our spending echoes prosperity for poverty alleviation. Potential travel has now considered by making a choice of booking their vacation with tour operation. This tour operation supports local sustainable project that benefits the local of the destination: economically, socially and environmentally. Tour operators have now started bestowing the challenges through donations to charitable trusts by giving hand on wider range of activities of the poor. They have vision of increasing the location of financial supports on community based programs. They have been changing a path way to meaningful developments, education, health care, early childhood development training, partnering with environmental protection and development groups.

A wider range of single and dual in positive power that prevails both sides, the financial is attached and non-financial results in impacts which constitute the impartially decentralization of tourism benefits among the poor. The building of percentage star hotels are now being converted to establishment of the community based lodes and camps. In addition, it is a new trend of ecotourism retreats that runs and manages by locals. Local people can now hold overall management and redistribution the profits on other sustainable projects. The projects can ensure the fair security of the poor. It is gradually having a tremendous transition. There is creation of job, increasing the learning center, and other educational research institute where courses are offered in tourism.

Valuable optimistic means support of initiatives embrace improvement of aids to the poor with better-quality admittance of information technology. The first hand set-up of infrastructural facilities and fair trade in the markets are strengthened the involvement of communities and organization in tourism business with prestigious superiority, cheerfulness of full involvement of the poor in managerial decision-making with high value.

Most of the international and foreign own hotels are providing training opportunities for their staff. They are equipped with relevant knowledge and skill to meet human resource demand of the industry (like, initiatives have resulted in offering of tourism courses as academic programs in colleges and universities around the world). These have become opportunities to the poor undertaking professional and skilled careers in souvenir productions.

During recent time, many international art exhibitions and displays are on displays. This brings together artists from different places promoting their artist works and skill benefits. At the same time, knowledge is dispersed through enriching and supporting of each other's talent. In parallel, the mega events are becoming key strategies of the marketing. They are developing skills through local arts products, poor people tap benefits from theirs through such works and participating in making such events a success. In this section, we discuss essential trends and initiatives that could have positive social impacts for local voices, gender equality, job opportunities to broaden the spectrum toward Sustainable Development Goals (SDGs).

7.4.1 Listening to Local Voices

Tourism has been a major economics driver for developing countries as demonstrate meaning element for poverty alleviation. Tourism industry prevails opportunities for the poor. Diversification of culturally-based products packaging and selling has influenced the role of governments in tourism toward more systematic and inclusive, expansion of business linkages, in design of economic, social and environmental tourism geared towards a sustainable development for the poor. This includes their voice in tourism plans for long-term changes with potential opportunities subsidized by development agencies. Many speculations are echoing around the globe, asking in, what ways do tourisms are detrimental to the poor. Many researchers, academic activists, and sociologists tend to investigate on how tourists can be seduced out of the resorts into real life of the destination through having a direct contact with the local people through homestay. This staying will be consumed produces of home based on foods and beverages supporting the locals. This will involve in developing tourist-friendly products reducing the level of harassing by vendors all whom tend to rush tourists for them to make sales at destinations in developing countries.

Today, tourists have unique opportunities of having cross-cultural encounters with local people. Staying with them in local communities and participating in their daily routine life has enhanced mutual understanding between the hosts and the guests. Guests learn about local music, instruments, and taste the authenticity of the local foods. Some volunteer services are rendered during their stay with the host families. Some end of supporting families is carried out the duties by sponsoring kids to acquired formal education. As in most developing countries, education is provided by the central of federal government, parents still provide learning materials, uniforms, and transportation for children. Local people are communicated and encouraged providing accommodation to tourists for their stay in local townships, which has demonstrated potential to link tourism with poverty elimination in the developing countries.

Pro-poor tourism plays a vital important role in the understanding. This is provided preservation of cultural and touristic heritage from a business perspective, customer satisfaction, and conservation of heritage for tourism. From the economic and social development of views, communities have been generated new jobs and as well fostered mutual understanding between people of the same skills, professions, and careers.

7.4.2 Improving Gender Equality

Sustainable tourism initiative in developing countries can be an equally vision for poverty mitigation in developing countries. High magnitudes of the world population are poor with women constituting the number than men. Women's opportunities and options for working their way out of poverty are restricted by cultural norms that control their social and economic mobility. The movement of sustainable tourism in developing economics is working with women groups in harnessing to contribute more to poverty reduction, though this would require a shift in focus with training programs to capacitate their perception on the tourism industry. We can give tourists with uniquely rewarding experiences through the integration of high quality products and services. They can be offered to promote responsible and valued partnerships in tourism for the socio-economic benefit of the poor.

Poverty reduction is seen as mutually supporting through sustainable initiatives of pro-poor tourism with more emphasis on gender empowerment on eradication and illiteracy among women of developing countries. Understanding their relationship to environment plays an important role in developing solutions for more sustainable use of natural resources. Through conservation project of green gardening in home growing on ornamentals have aided in ecological conversation which is significant at a local or district level.

7.4.3 Supporting Job Opportunities

This fragile industry empowers communities that are deprived in physical wealth but rich in cultural heritage and history. These infrastructures have been utilized by their distinctive features at their disposal to be as income-generating virtual benefits. By encouraging development of multiple-use, these infrastructures have been aided the host community, including roads, health care facilities, and sports center. In addition to the hotels and high-end restaurants that cater to foreign visitors, local residents are hired as waiters and waitress, and some works as chefs. This generates linkages of diverse jobs, commencing from the hotel industries to restaurants and other catering outlets of food and beverage suppliers to leisure and recreation centers.

Tourism industry and its ancillary service sectors have enabled to form complex and varied supply chains of goods and services, and supporting a multipurpose employment market with a diversity of employments opportunity to citizen in developing countries. The residents undertake careers such as tour guides, interpreter of travel in tourism organizations, chefs in hotels and restaurants giving customers a taste of foods of the natives. In addition, the coach drivers are facilitating airport transfers and other excursion, couple with hotel managers and other service sector workers. More tourism jobs are temporarily and seasonal. It is better still provides some benefits for the poor during the short period of operations.

7.4.4 Broadening the Spectrum Toward SDGs

Sustainable pro-poor tourism projects have registered significant success in national aggregates so far. It is because initiatives are nationally supported through conducting of researches coupling with location of macro-economic sustainable development poverty alleviation. These commit to developing and introducing of creative financial schemes supporting creative fundraising appliance channelling much financial resources needed for the cause of fighting poverty alleviation. These public influences change transversely the tourism sectors, approaches differ from regions to counties.

In evaluating the benefits of pro-poor tourism, statistics have shown that only few fraction of overall population of people living in developing countries. They achieve consistent paid occupations as an outcome of pro-poor tourism. However, despite the small number, these individuals' earnings are normally satisfactory to boost their everyday needs. Though these people still persist to be deprived but they are better able to encounter day-to-day requirements. Pro-poor tourism also makes available of co-operative revenues to the local communities. They are efficiently and effectively accomplished by reinvested in things such as graduate schools, healthcare facilities, and leisure and recreation centers. Further affirmative means of supports embrace development of aids on quality improved access to information and communication technology.

Pecuniary prospects incline to be rigorous and dispersed through employments benefits subsidiary others, responding of earnings on indigenous goods and services. This current undertaking pro-poor tourism works towards achieving development and poverty alleviation, aimed to scale out this operation in other poverty stricken locations in Asia and Africa, venture of shared revenue, and significance of the commercial powers transversely to national development.

Cultural values in many local tourist destinations have become part of the tourism products. They have been used as a commodity and a marketing tool to increase the average length of stay of the tourists and maximized foreign currency earnings. Many traditional ceremonies, such as circumcision and Turkish nights, have been transformed and used to entertain the tourists, no matter what the result is on the country's image, on the business future, and on the local people with regards to their psychology and cultural reference marks (Tosun 2001, p. 297). Such benefits are minimal in entire terms from a national standpoint in comparative expressions associated with welfares of tourism to the poor. Pro-poor tourism is valuable in decreasing susceptibility for numerous reimbursements for the poor with low revenue during the off-season. This variation enables the poor to be acquainted with skills and methods needed in modern financially handling which enlighten most of them to get personal savings account with financial institutions.

7.5 Pro-poor Tourism Marketing

Pro-poor tourism is considered a potential niche product when promoted and marketed to customers. These pro-poor tourism marketing promotes the business by advertising it on media industry on radio, television, internet based on which research and surveys are conducted into the wants, needs, and spending of tourists. This can better help the poor, and effective marketing can have a significant impact on tourist demands in a particular destination. Hence, we can specialize in pro-poor tourism marketing. The purposes of pro-poor tourism are aimed at helping community ventures as well as at committing to increase arrivals, sales, and profits. This could create tourism marketing projects, reach objectives, start or rejuvenate profitable ventures for the poor.

Marketing of pro-poor tourism destination embrace the unique features of the destination brand. It serves a unique attribute for tourists to experience something completely different from their normal routine life. Through these experiences of personal authentic encounters with the poor. These marketing plans will be carried out by interesting and staying with the locals understanding and learning about different lifestyles and cultures. They are packaged and sold to expose unique and compelling experiences.

Marketing strategies should include raising awareness of products available and disseminating information, within destinations to the tourists about things to do whilst on vacation. This will be invested in developing and maintaining products, like preserving heritage or places of interest by establishing National Parks, setting up cultural excursions of diversifying products, facilitating environmental management, conservation scheme, and improving their product sustainability.

Tourist excursions are packaged to suit customers need featuring local history, culture, society, and environment. They have positive impacts on local people, through an increased in incomes, as well as investment in infrastructure through donations given by tourists. All those should promote and enhance better lives and help the locals in developing countries. Tourists are becoming increasingly responsible and well-supportive. Consequently, their search for unique cultural experiences will also increase. They will also seek out new forms of accommodation. Therefore, growth rates are expected to be strong, albeit from a relatively small base.

The successes of this tourism industry vividly rely on the respectful of the environment, mindful, cultural, and social traditions, supporting the preservation and conservation of the cultures. This helps in convey affluence and alleviates poverty in societies. Collaboration involves three main parties, all of which, responsible roles to play. These bodies include the: (1) central government, (2) stakeholders involved in the tourism business, and (3) communities which tourists visit. They have duties to influence their competitive advantages by possessing a good reputation through improving the quality of service delivered to the customers. They have duties to protect their surroundings, art, music, and literature from public sectors and private sectors. The administration bodies of the country, which is the government, have a duty to launch supportive schemes. They are by initiating and applying essential codes of practices which should be in compliance with the international tourism standards to meet the demands of the international markets.

Box 7.1: A Case Study in Gambia

Tourism has been the top driver of the economy developed in such a way to provides significant opportunities to poor people in rural villages of the Gambia. Great example in this country is the Kartong Village, designated as an ecotourism paradise. Small villages lying in the southern part of the country possess unique features of ecology for tourists to travel far, wide to the scenery beauty of the place, and observe the abundant bird species.

The village development of the community works in collaboration with stakeholders and philanthropies initiated the Kartong Association of Responsible Tourism (KART), a private and small-scale tourism association. Working with the local people. It is sensitizing them about the importance of tourism and its impacts of conventional tourism, the association enlightens the poor on the values attached to their natural environment and traditional social life. As for today, it conveys greater revenue of during the peak season through their annual festival conducted showcases of the local music (Fig. 7.2).

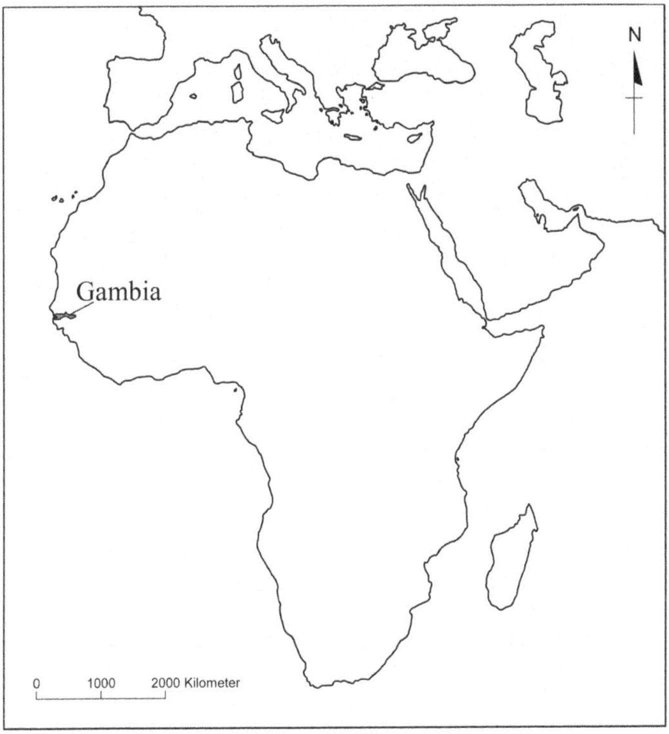

Fig. 7.2 Location of Gambia (Illustrated by Yi-Te Chiang)

Tourists travel from Europe and Scandinavia to witness the event. Majority of them like to access a greater range of complementary products during their stay. They use local guides to learn about culture, visiting traditional landmarks and cultural heritage sites. Tourists spend time and money during their stay is registering significant success on daily spending of hiking in local lodging facilities owed by villages. This is based on instinct motivation in locals to be entrepreneurial minded by operating their own business resulting in responsible tourism. These increasing businesses substantially have created perceptible benefits of income by sells of fruits, juices of "wonjo", "baobab", and guides services by local trained personnel. Small package-tour operators are managed visiting programs, such as sight-seeing of Kartong Village. During these exciting excursion programs, these guides encourage tourists to expand their gastronomic prospects outside the continental cuisines offered in hotels by promotion of local recipes.

7.6 Conclusion

Tourism has become the most vital source of foreign exchange earners for many developing countries. Globally, it is deemed, as the fastest industry after petroleum. The economic strength of the tourist business has facilitated to renovate societies, time, and again for a healthier economic growth. Despite the economic financial crisis around the world, tourism has been a developer and advantage over so many developing economies. It is an industry that is disposable at the point of production, making it directly and indirectly beneficial to the local communities. They provide goods and services consumed by industry in catering and hospitality of the tourism sectors. Tourism is a blooming international business by means of influence improving unindustrialized nations in cooperation of optimistic and harmful behaviors by its consumers. These industries remain strong even though the global challenges of political, economy, environmental, and social factors remains catastrophic by health epidemics of SARS and September 11 terrorism attack. In addition, other exploitations have raised the fear of travel using airplanes. Global financial crisis has lessened the spending power of individuals which have minimized their disposable income.

Pro-poor tourism is aimed at reducing urban and rural poverty in developing economies. It remains progressively significant, not only in emerging but also in industrialized countries. The poverty of the countries looks like invariant in relative to commercial growth. Diverse organization and theoretical frameworks have risen to deal with poverty decline in developing countries. The way to effectively achieve this objective is to clearly involve the poor at the center of tourism business. It is a comprehensive approach to deal with urban and rural poverty reduction for the entire underprivileged citizens. The developing world has now become a major growth area. In the tourism industry, simply due to its higher turnover of foreign exchange earnings and being the leading export earner for one-third of the world's poorest countries.

Emerging countries should benefit from the trade and industry prospects of both domestic and overseas travellers. Responsibility should be bestowed on the shoulders of all the stakeholders from all public, private, and voluntary organizations. All have a duty to know where their virtual benefits lie- whether it is in nature, spas, food, and beverage. When it is known the positive move can be concentrating their development approaches on that vision to increase the scope making the destination appealing to tourists. This should safeguard indigenous ownerships of development projects by the poor and support to sustain the incomes in the local communities. Ventures like community-based tourism are means of sustainability. It is also more justifiable and benefits to offer the kind of reliable experiences which tourists are seeking for.

Developing nations have demonstrated further proficient developed tourism industry with diversification in different form of tourism. All countries will be geared to help and involve the poor in the tourism business by helping them to generate money and money worth. Those projects can be running community. Based on tourism lodges and ecotourism camps, local communities will be benefited directly from the industry, establishing that tourism can be strengthening for respectable good, both in relieving insufficiency and facilitating to reinforce harmony in the countries. As saying goes, a hungry man is an angry man, the peace and security of country depends on the approaches which the system is managed and handled. With efficient and prominent proper planning of tourism, it will prevent problems to be respected the environment and places host communities at the centre of the development process. Communities should remember at all times that it is the beauty of the surroundings where they live the richness of their culture. The diversity of their wildlife attracts visitors in the first place. A percentage of the wealth that tourism generates should be spent to preserve these qualities. These can simply be accomplished if the aims and obligation lies with the governments of unindustrialized countries to safeguard that tourism develops in a sustainable way, facilitating lessen poverty and steady communities.

References

Chok S, Macbeth J, Warren C (2007) Tourism as a tool for poverty alleviation: a critical analysis of 'pro-poor tourism' and implications for sustainability. Curr Issues Tour 10:144–165

Department for International Development (DFID) (1999) Tourism and poverty elimination: untapped potential. DFID, London

Hall CM (2007) Pro-poor tourism: who benefits?: perspectives on tourism and poverty reduction. Channel View Publications, Bristol, UK

Sarah H (2015) Pro-poor tourism and its measurement in developing countries. AV Akademikerverlag, Germany (German ed)

Tosun C (2001) Challenges of sustainable tourism development in the developing world: the case of Turkey. Tour Manag 22:289–303

United Nations (2015) Transforming our world: the 2030 agenda for sustainable development. The General Assembly, United Nations, New York, NY, USA

UNWTO (2016) Tourism for development in least developed countries-mobilizing and coordinating resources for tourism with the enhanced integrated framework. World Tourism Organization, Madrid, Spain

UNWTO (2017) UNWTO tourism highlights. World Tourism Organization, Madrid, Spain

Chapter 8
Ecotourism

8.1 Introduction

Ecotourism is the best practice of tourism in recent years. Many people or tourism operators registered, succeeded, and received good reputations globally. This form of tourism includes activities and experience, mostly through sight-seeing. Tourists travel to nature related to environments or designated area of great ecologically values of fauna and flora. With particular main interest of enjoying the scenery beauty of the environment, they study and admire the biodiversity seen.

Tourism relates to fascinating and appreciation to natural environment and local cultures. It is a result of a sub-branch aspect called ecotourism. This form of tourism is emerging in the tourism industry globally, explore as new form of tourism to slow down the impacts of conventional tourism. Mass package tourism, where tourists travel in high numbers, may cause damages to environment and local communities. Ecotourism emerges to be an alternative to mass package tourism, in both developing and developed countries, around the world nurturing a strong feeling and relationship between humans and the natural environment.

Ecotourism means to end negative impacts of tourism by bonding the environment with local communities. Promoting the Ecotourism is a mean of economic development and empowerment of local communities showcasing the authenticity of the local cultures; nurturing admiration for the cultures, idea of preservation, and conservation of the natural environment for sustainable use to future generation. It is a type of tourism that calls for environmentally aware travelers before going further it is important to know who is an eco-tourist.

8.1.1 Who Is Eco-tourist?

An eco-tourist is a person who travels either individually or in small group to destinations with main motive of experiencing the biodiversity in the area. These comprise

W.-T. Fang, *Tourism in Emerging Economies*,
https://doi.org/10.1007/978-981-15-2463-9_8

the wildlife, plants, animals, and cultural legacy of the local people. They are all serve as the principal fascinations for tourist to experience a particular destination which constitute main motivations for eco-tourists;

E = means **Environment**
Co = means **Community**

An eco-tourist activity is centered on environmental aspects of the destination. It is about awareness program to learn the nature and its features; and improves the features of the local communities. They are all serve as the main attractions for the local people to experience local people on performing arts. However, diverse descriptions have being given by different academic scholars and research journals. All are trying to come up with suitable classifications on ecotourism. Therefore, ecotourism is defined as related activities to develop a wider range of investigation from several tourism experiences on the nature and biodiversity. As general, it is acceptable to mean ecotourism as a standpoint view on the following components: management of the natural resources, the nature-based experience/activity, provision of economy, and social benefits to local people in local communities from which tourism products are derived.

Ecotourism is now defined as "responsible travel to natural areas that conserves the environment, sustains the well-being of the local people, and involves interpretation and education" by International Ecotourism Society (TIES 2015). Ideally, ecotourism companies will demonstrate sensitivity to environmental concerns, try to operate within the framework of ecological guidelines, promote an ecological correct awareness, support or use local tourism services, practice long-term planning and, continue to support destination communities conservation as well as preservation programs (Sirakaya et al. 1999, p. 169).

8.1.2 Ecotourism Activities

Ecotourism empowers the experience of natural zones and the development of environmental and cultural awareness of tourist destination base. These series activities designed to give tourists unique and rewarding experience during their vacation for participating in activities. It provides with opportunities of building a strong emotional, cultural, and environmental awareness as well as giving distinctive experience for both, the host community and the tourist, for participating in activities. They provide real life experience for undertaking in the following:

- Wildlife Watching
- Bird Watching
- Mountain Climbing
- Village Tours (Cultural Encounters)
- Trekking
- River Excursion

- Storytelling
- Cycling
- Nature Trail
- Wildlife Seeing.

8.1.2.1 Wildlife Watching

Below is an outline detailed explanation of each the above-mentioned, variety of activities explore by eco-tourists. Most of these undertakings by tourists contribute immensely to environmental protection and development, and local social-economic growth of the destination in which local people are not an exception. These events experience range from wildlife watching activity.

8.1.2.2 Bird Watching

Bird watching is an increasingly popular activity in many destinations around the world. They were tourists travel far and wide to see the different species of birds as a leisure activity for tourist. This form of recreation has become a career for other people who work and major their profession in bird studies. A person who studies about birds is called an ornithologist. The person is specializing in bird watching using devices to aid the activity like binoculars and telescope. This equipment is used by both amateurs and professionals. In contrast, experts can do other identification of bird species without the use of devices but by listening to the bird sounds. Normally, the most active period of the year for bird watching in most destinations are located in temperate regions. During spring when highest species of birds can be seen, among them include male Scarlet Tanager, male Black Throated Blue Warbler, Rose-breasted Grosbeaks, Swallow (*Hirundo rustica*), Greenfinches, Goldfinches, Robins, Song Thrushes, Stonechat, Sardinian Warbler, Black Redstart, White Wagtail and Serine (Fig. 8.1).

Various bird fairs and conferences are held to host different bird watchers around the world, national and local organizations. Some of them are American Birding Association, British Trust for Ornithology, Royal Society for the Protection of Birds, and Cornell Lab of Ornithology. All of them are engaged in birding participation in national and multi-national conferences. Birdlife International being one of the main vital origination in the world is a conservation organization. Various concerns and issues are mentioned to protect and safe guard birds including to support the welfare of birds and their surroundings; avoid emphasizing birds by regulating practice of using photography, keeping back from nests and nesting colonies; and respecting private property. The British bird watching fair is the world's first and largest international bird watching event. That conference draws attention from professionals and experts around the world, geared towards supporting bird conservations taking parts in fairs held in the destination, like the Rutland Bird Fair United Kingdom (Fig. 8.2).

Fig. 8.1 Bird watching activity in Taiwan, Republic of China, *Lophura swinhoii* (Photo by Wei-Ta Fang)

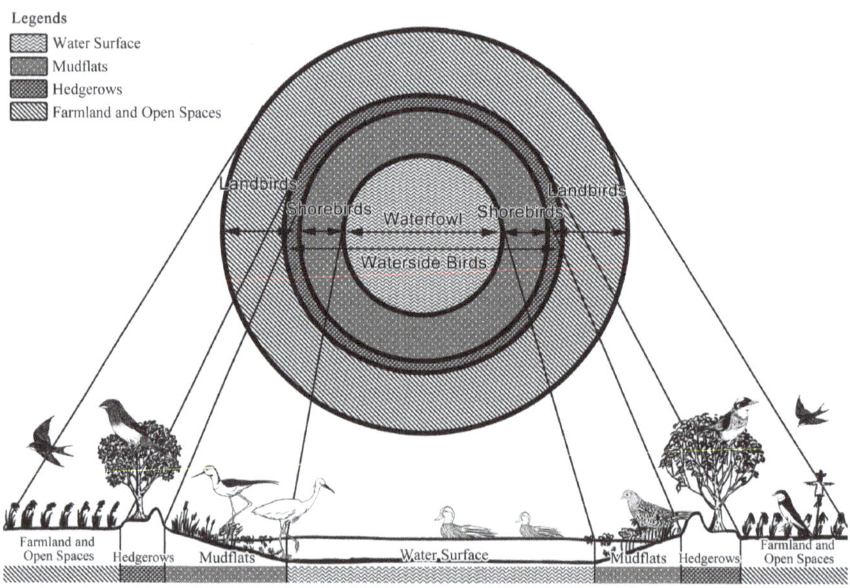

Fig. 8.2 Among of the wintering birds, numerous birds can also be seen during the winter, high number of migratory birds also travel south to their stopover locations mixed with local residents in North Hemisphere (Illustrated by Yi-Te Chiang, all right reserved)

Fig. 8.3 The climbing to the top of Mt. Kinabalu, Sabah, Malaysia (Photos by Arba'at Hassan)

8.1.2.3 Mountain Climbing

Mountain climbing is also another famous eco-tourist activity. In some regions of the world, it is known as mountaineering. It is a form of activity where tourists embark on an adventurous act of hiking a mountain with an attempt of reaching the summit. Researches have explained that mountains consist of three different parts, known as rock-craft, snow–craft, and skiing. All of these are determined by of the hiker on the journey he or she undergoes an icy area or rocky area, for health and safety reasons all of which is can be determine by the use of an experience tour guide. There are main popular types of mountain climbing which is expedition climbing and alpine climbing all of which is classy and detect by the form of route and trip the hiker took, comprising the time took to complete the journey, plus the energy consumed (Fig. 8.3).

8.1.2.4 Village Tours (Cultural Encounters)

Village tour is a unique lifetime experience in Ecotourism. It is an activity to provide eco-tourist with the opportunity to have cross-cultural encounters with the local people in local communities through participating in their daily life. This consists of a tour to the village of community development projects to understand the structure of the village. This will include the visit to schools, hospitals, parks, women's projects, and the like. The exciting part of it is the opportunity to meet the chief of the village, who explains about history talk on the initiation of the village, different families, roles and responsibilities, as well as the social events in the village. It is really an

Fig. 8.4 The Teepee Cultural Village at Hopi American Indian in Tuba, Utah, U.S.A. (Photos by Arba'at Hassan)

opportunity where tourists can spend a day or so with one of the village's families, experience traditional ceremonies, and witness a play about different cultural ceremonies, notably on festivals carried out by the locals (Fig. 8.4).

8.1.2.5 Trekking

Trekking is another ecotourism activity. It has becoming popular in developing and developed destinations where tourists take an adventurous journey on their bare foot or wearing sport trekking attire, or walking around the settlements. This type of experience is mostly common in areas where the use of modern transportation is limited. People go to places and get around on foots. Conservationist and responsible travelers support the idea as tourists used specific trails, which appeals to natural environment. This uprising activity can be done in most parts of the world regardless of geographical location. The most popular destinations for going on trekking today are the mountain ranges of South America and Asia. They are mostly done in rural settlement, where places are less busy and the trekkers can have the opportunity of coming in close contact with the beauty scenery of the nature, frequently in small groups.

8.1.2.6 River Excursions

Today, river excursions have become exciting activities for tourists to enjoy peaceful and relaxing river trips in calm water. The favorite destinations visited are like in Europe, Africa, Asia, Australia, and America. They cruise and sail in the navigable waters. This experience provides tourists with unique opportunities of seeing different array of birds' species, whales, dolphins, oysters, mangroves along the swamps,

Fig. 8.5 Boating in Colorado River from Moab to Boat Ramp, Utah, U.S.A. took on whole day. Different sceneries of Grand Canyon landscapes can be seen (Photos by Arba'at Hassan)

and different landscapes at the river banks. It is done either on modern ships or traditional dugout canoes. Both of them are available on half, and full day excursions. This excursion gives an excellent opportunity to see different form of landscape and fauna and flora environment along the river cruising. If they are lucky, they can meet local people in their traditional environments. The eco-tourists learn about the history and culture of the villagers through traditional music, dance, and foods. These are the center of experience in their excursion (Fig. 8.5).

In some part of the world people can stop at a place to watch wild animals like whales, dolphins, etc. In Africa, stopovers are done in local communities where tourists disembark from ship and go on boats at closed to be with the locals. River excursions have been the best way to explore this area by boat. From the boat, the holiday destination, tourists are able to enjoy magnificent scenery of savannah, tropical forest, and local river fishing. The fishermen are very sensitive and concerned on eco-method. There are many canoes and guided trips are offered for these types of holidays. They are completely different from the mass package experience, which is generally offered by large tour operators. These splendid Ecotourism holidays are especially favorable by bird watchers.

8.1.2.7 Storytelling

Storytelling is another ecotourism activity. It is one of the earliest type of folk art. Stories are based on the interpretation of authentic facts on events or a combination of jokes to entertaining audiences. It is mostly comprised of conveying messages through words, images, and sounds, often by the acts of using creativity. This noble activity has become an entertainment, education, cultural preservation in most countries. In some local communities in Africa, children are nurtured, and trained with traditional ethics transferred to succeeding generations. Most of them have are not written, so they are being lost after most of the best narrators die. The common one, typical of any spoken language in Africa, is the story of the *Hyena and the Hare*. Parents used it for instilling moral values to their children by narrating a story to them and interpreting the meaning of the story couple with advices.

In recent years, storytelling has become renowned interests for many travelers, who tour Africa, America, Canada, Europe, Australia, and Asia. Their main motives are to listen to local stories. In addition to the various journeys undertaken by tourists, numerous storytelling conferences and festivals flourish and draw wider ranges of audience from international community. Storytelling is diverse. In recognizing storytelling, story tellers write down stories on papers in the form of scripts to present to the audiences. The audiences will sit in a gathering to listen to the narrator. Apart from the written form, some tellers convey stories based on their personal thoughts and experience.

Other stories are collected, sometimes are changed from records of ancient remains. Folklore stories such as legends, classics, myths, and fables today, continue to be the favorites. Historically, it is believed that people in the older days used stories for communicating their ideas and issues by sitting around camp fires. This takes days. They evolve as an Ecotourism activity with many camps setting fire at night to entertain their guests and telling the stories together with dance and music.

8.1.2.8 Cycling

Cycling is another Ecotourism activity. This is an ecologically friendly activity getting around using bicycle; visiting attraction site on bikes. In most destinations, biking around green scenery landmarks has become one of gaining popularity. With tourist aspired of venturing into the adventure of their lifetime, they are most conducted in the framework of ecotourism, taking a tour through miles or kilometers on mountain bikes. Tourists snap many pictures and observe the wildlife in natural environment, witness sites or regions of unique ecological quality.

Various tour operating companies, hotels, travel agencies and locals offer bicycle rental services for tourist in destinations. Cycling holiday gives the most scope for independent exploring. They can follow any route to choose using maps and guides. The most common popular cycling destinations are in Cyprus, Austria, Spain, Taiwan, Australia, United Kingdom, France, United States, Switzerland, and Portugal. They are renowned for their cycling tours which attract lots of potential travelers through initiation of international cycling competition on long distance racing (Fig. 8.6).

8.1.2.9 Nature Trail

Nature trail is another real lifetime experience. The organic rhythm of foot travel is a wonderful way to explore and make meaningful contact with the country national parks have designated trails for used by trailers. They who want to learn about parks and wildlife use these trails to walk around the nature. It is not only confined to foot trail. Bicycle trailing is also popular form of nature trail for excitement during walking. Some tourists will be seen hiking, cross-country skiing and cycling. Destinations have different trails for different purposes. Among them are the walking trail,

Fig. 8.6 Cycling at the Yosemite National Park, California, U.S.A. (Photo by Wei-Ta Fang)

cycle trail, stairway trail, motorized trails and, equestrian trail. Signs and indications are used to instruct trailers about the comfort of the routes they are taking. Color indication is used to signify tracking difficulty: black diamond means difficult, blue square means moderate, while green circle means easy (Fig. 8.7).

8.1.2.10 Wildlife Seeing

The recent updates on ecotourism have shown indication. Seeing the natural biodiversity is satisfying and becoming a pleasure activity for tourists. Watching wildlife can be predicted as one of the most convenient form of Ecotourism. It provides pleasant activity possible and you can do it anywhere. Regardless of which region of the world one lives in, wild plants and animals can be found not far away from their residence in parks and game reserves. It has become a major attraction for tourists to destinations in bringing economic benefits to the national economy (Figs. 8.8 and 8.9).

Fig. 8.7 Tourists are walking in the forests, enjoying the beautiful scenery of the natural environment along nature trails from different attitude (Illustrated by Yi-Te Chiang, all right reserved)

8.1.3 Ecotourism Theories and Practices

Destinations are compelling series of packs with stunning wildlife, rich in diversity and well worth seeing. Most of them are managed under the supervision of parks and wildlife management. It conserves and protects the remaining biodiversity as well as the natural environment for the present and future generation. At the same time, it will create educational and leisure facilities for the present and future populations through prudent use of wildlife resources. Kenya (Maasai Mara) is the most popular packed with wildlife reserves in the world prevalent for game viewing, couple by South Africa, and Tanzania, also advent of elephant safari through the Jaldapara Wildlife Sanctuary in West Bengal. India is becoming emerging attraction in the Asian continent, with creation of hotels in amazing pleasant landscapes of Africa and creation of lakes on the tropical surfaces of Asia and bird sanctuaries in eco-friendly zones (Fig. 8.10).

However, ecotourism by virtue is of its unique distinctive approach. This results in appreciation of local cultures and simultaneously reaching the tourism benefits to the locals. The satisfaction of tourists is with the local available products on offer, above all sensitizing local people about the beauty of the environment and the fragility of nature. This means that those who implement and participate in ecotourism activities should follow the following ecotourism standards, philosophies, and practice them to have a sustainable viable business in destinations. One case study associated with ecotourism theories and practices you may read from the following journal paper in **Appendix 2**: *Environmental literacy on ecotourism: a study on student knowledge,*

Fig. 8.8 Wildlife watching activity at the Panda Research and Breeding Center in Chengdu, Sichuan Providence, China (top). Another species, the red panda, is also endemic like Giant Panda, to the temperate forests of the Himalayas, and ranges from the foothills of western Nepal to China (bottom) (Photos by Wei-Ta Fang)

attitude, and behavioral intentions in China and Taiwan (Fang et al. 2018). Some concepts and applications toward environmental literacy should be implied to the key players listed below:

- Ecotourism organizations and trusts should cover all aspect of environmental and cultural awareness or the local people. This can be packaged and displayed on promotional materials to attract tourists. On the other hand, community based sustainable projects would be encouraged, notably the traditional music industry. Today, many West African musicians leave their countries for Europe and America to search for money, all due to lack of motivation and encouragement from the authorities. Money is also the reason why many locals choose to do music and play the renowned music's like reggae, rock, rap and pop, and not traditional music. The traditional music scene needs to be taken care of by ecotourism, as well as environment education. The protection and development projects should be funded.

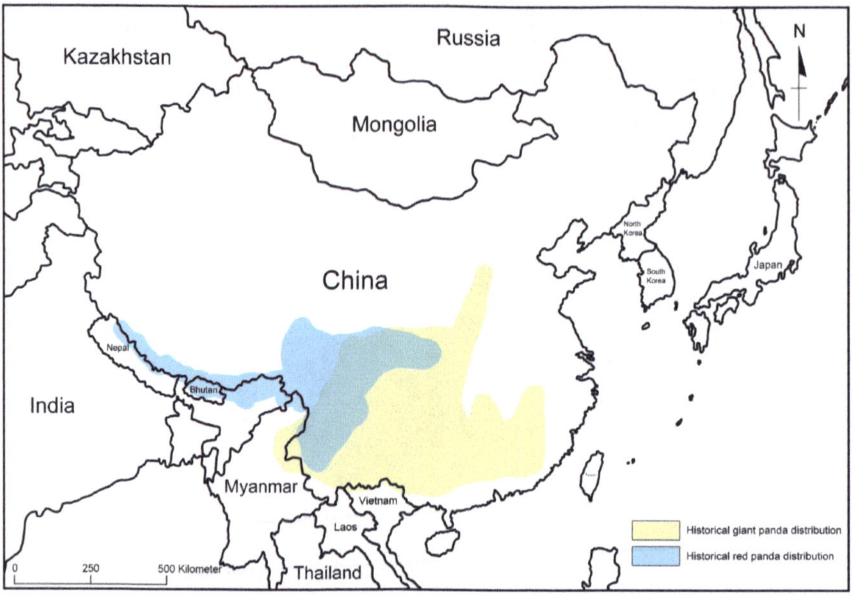

Fig. 8.9 The historical distribution of giant panda could be overlapped with the red panda which distributes from the foothills of western Nepal, Bhutan, to south-western China (Illustrated by Yi-Te Chiang)

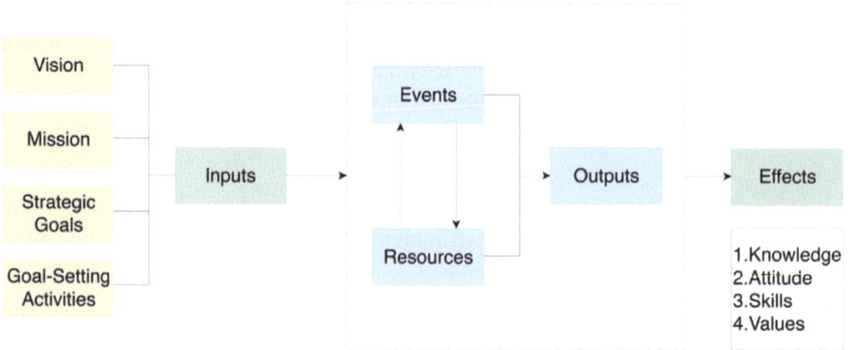

Fig. 8.10 An illustration of ecotourism theories and practices in process. Ecotourism is about ecological experience of a tourist to natural environments. It engages tourists in community activities chanting pathway to sustainable conservation of the biodiversity, learning about the local culture, focusing on the experience of natural areas and fostering of environment and cultural understanding. Its success lies on the sustainable management of the local cultures, economies, and natural resources toward better environmental literacy (Illustrated by Wei-Ta Fang)

- Projects initiated in destination should have better approaches to run direct financial benefits for conservation and preservation of both. The ecological features of the destination, authentic cultural norms, and values of the local based on which business entities should be evaluated and assessed on the corporate social responsibilities. This will be in compliance with standard and regulation to meet before certified on accountability on local employability of the locals by ensuring the profit of tourism stays with the local communities. The locals are involved in tourism planning, decision making, and administration process.
- Provision of mutual understanding and experiences should be constructing on the cross cultural encounters between the host community and tourist. They should be responsible for effective economic encouragements for supporting and improving the bio-cultural diversity and supports in safeguarding the natural and cultural heritage of destination for the future generation. The importance on inspiring individual familiarization of the local culture and awareness programs on the environment is through conducting organized tours and excursions to natural site; promotes greater understanding and appreciation for nature, local society, and culture. Therefore, the future generations may experience destinations relatively untouched by human intervention with best biodiversity conservation application.
- Ecotourism should be geared towards reducing the negative impacts of tourism. This can be done by maximizing positive impacts of the world's biggest industry. There are already destinations being affected by mass tourism developments which drained most of the wetlands and lead to the extinction of the remaining rare species. Increasing local capacity building and employment opportunities, ecotourism is an effective vehicle for empowering local communities around the world to fight against poverty. It is also achieved sustainable development by providing economic benefits and supporting the empowerment of the social livelihood to local people by creating a forum for cross-cultural experience and awareness between the local and tourists.
- Tourism is a volatile industry. Industry in ecotourism is not an exemption, grounded on the affirmed mention phrase. It is an ideal field for studying economic, political, social reformation and cultural diversity of destination to nurture sensitivity to host countries' political, environmental, and social climate. Tourism is an industry which is deep rooted on the freedom of a nation. It has a great influence on the social structure of the local communities that has an adverse change on the life of people out of the ordinary structures of their everyday lives. Some tourists disrespect local cultures and social benefits to communities. They perform bad actions to the environment. Local people must be convinced of the benefits of ecotourism before any progress can be made towards a more sustainable project. This will be instrumental in building the capacity of the local population so that they can run their own Ecotourism projects in destination.

8.2 Impacts of Ecotourism in Destinations

Ecotourism has a great influence on the host communities. Ecotourism has some portfolios of series of potential benefits. They are generated by both the nation and the local peoples in both developed and developing countries. This tourism serves as the major service industry, as well as a driver of the national economy. To simply present a brief outline of those benefits, they can simply be classified in **three** main categories available on the topic of impacts. They are mainly focusing on the benefits sides of economy, environmental, and socio-culture all. They are the major assets for ecotourism in destinations. These wider ranges of tourism activities expand from conventional tourism to ecotourism. There has been a lot of controversy on issues to the boundary at which ecotourism contribute to the positive impact of biodiversity conservation and preservation, disposing of local social-economic benefits to locals in communities, and the protection of the fragile natural environmental areas. All of those are considered as Ecotourism features making them the real ecotourism.

- Economical Impacts
- Environmental Impacts
- Sociocultural Impacts.

8.2.1 Economic Impacts

In most developing and industrialized nations, the top agenda of government plans—as well as local communities, the creation of jobs and income generating is what constitute most of their topics. However, with the advent of Ecotourism, it provides a wider range of employment opportunities for local people and residents in tourist destinations. They are all work in the tourism industry, as tourist guides, resort representatives, ornithologist, curators, based on Ecotourism helps in the reduction of unemployment and developments in destinations. They create job opportunities for locals through. They make their living from services and render to eco-tourist. Eco-tourists by virtue of their extrinsic enthusiasm are patronized existing heritage and cultural conversation projects in the local communities. They help communities with sponsorships and fund raisings for revitalizing, re-habitation and preserving historical, cultural and heritage sites for sustainability in meeting the demand of both present and future tourist.

8.2.2 Environmental Impacts

Ecotourism is a movement which is geared towards the fostering of awareness programs for environmental education and protection of endangered species. This is a

Table 8.1 Research from the "The official video of the international year of biodiversity at General Assembly Summit on Biodiversity on September 2010, New York, U.S.A." revealed that the following ecological and natural attributes were threatened with extinction by some extensive responsible people activities

Ecological natural attributes	Percentage loss (%)
Birds	12
Mammals	21
Reptiles	28
Amphibians	30
Invertebrates	35
Freshwater fish	37
Plants	70

significant form of tourism which promotes the preservation and conservation of natural environment. It may be seriously affected by the higher influx of mass tourists and their higher pressure on natural areas for game hunting. Ecotourism is the distinct paramount initiative for creating and recalling for carrying capacity to attractions and coming up with policies that helps the vision for sustainability of sites. Statistics has shown that human activities are responsible for 40% of animal extinction and pollution of farms with pesticides resulting to death zones, run-offs in lakes, thus ruining the marine life and habitat in the sea. Ecotourism raises trend to put financial supports towards the conservation and preservation of fragile natural areas in maintaining their scenery and attractiveness. This will in return provides financial benefit to local people through the visitation of tourist those sites (Table 8.1).

Experts predicted that before the end of this century more than 20% of all unknown species of fauna and flora would be disappeared. Ecotourism paved the sound practice for the wise usage of this resources and as well giving financial contribution for the sustainable use of this rare features.

8.2.3 Socio-Cultural Impacts

Ecotourism provides aids in the preservation and conversation of the local cultures. They are packages and sold to tourists on organized and conducted tours to experience the traditional local lifestyles. Tourist pays in fees to participate in these tours. At the same time, these tours help promote the authenticity of the local cultures. Local people also have some sense of pride in preserving their cultural values. Similarly, it has a potential to generate income for locals. They do it through showcasing of this cultural, norms, and values in the form of music, dance and festivals. The Eco-tourists are motivated to travel to destination with the main motivation of learning about the local cultures. Example of this is the high influx of international tourist who travels to Africa. As a part of their motivation is the comprising of some element of learning about the history, literature, and stories out of which they write books. Others take up music as a career after acquiring a vast idea in playing the local musical instruments like African singing guitar ("Kora")—Djembe among other musical instruments.

From the interaction of the eco-tourists with communities this has rose the level of understanding among local peoples in destinations. Local residents are educated about the outside world through the encounters with tourists without leaving their homes. While on the other hand, tourists do significantly acquire and know about the distinctive culture. Local communities are benefited through contribution from tourism to the improvement of social infrastructure, like schools, libraries, health care institutions, internet cafes, and so on (Fig. 8.11).

Besides, if local culture is the base for attracting tourists to the region, it helps to preserve the local traditions and handicrafts, which maybe on the link of the extinction. Socially, ecotourism equally helps in the promoting of acceptable attitudes towards each other globally. This results in global peace by decreasing undesirable views and categorizing among people. In China, red-crowned crane (*Grus japonensis*), named fairy crane, is a traditional symbol of longevity and immortality. Because of its importance in Chinese culture, the red-crowned crane was selected by the National Forestry Bureau of the People's Republic of China (PRC) as a candidate for the symbolic image of national animal of China. Fairy crane, in its noble operation has developed sense of gratitude, self-importance, esteem, mutual understanding and tolerance for Chinese culture. The opinion address by ecotourism is to promote tourism in the destination preservation so that it would both contribute to incomes generation and build respect for the local tradition and culture.

Ecotourism projects must go beyond prevailing notions of "the overlap between nature tourism and sustainable tourism" to encompass the social dimensions of productive organization and environmental conservation. Ecotourism must do more than create a series of activities to attract visitors, offering them an opportunity to interact with nature in such a way as to make it possible to preserve or enhance the special qualities of the site and its flora and fauna, while allowing local inhabitants and future visitors to continue to enjoy these qualities. They must also establish a durable productive base to allow the local inhabitants and ecotourism service providers to enjoy a sustainable standard of living while offering these services" Ecotourism: A Tool for Sustainable Development by David Barkin, May 1996 (Barkin 1996).

8.3 Ecotourism Norms, Planning, and Management

Ecotourism planner and the government should work in close collaboration with the local communities and stakeholders in destinations. They must carry-out development and management together. The government who should provide a frame work of agendas with a vision for sustainability in meeting the needs of present tourist. It comprised the need for future generation. These authorities have critical roles and responsibilities in administering and achieving Ecotourism goals. The goals should go through guideline of laydown frameworks and funding of projects in provision of

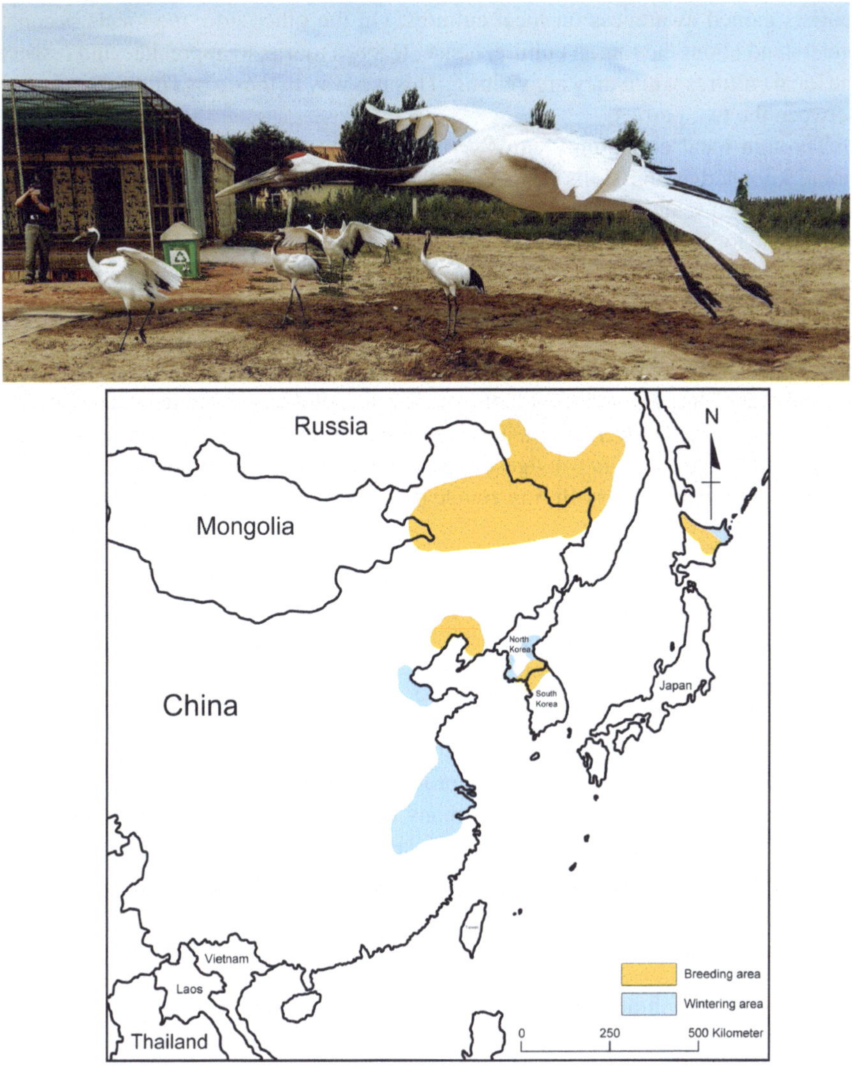

Fig. 8.11 Red-crowned cranes, *Grus japonensis*, home in Siberia (eastern Russia), northeastern China, Japan, Korea, and occasionally in northeastern Mongolia within breeding periods (Illustrated by Yi-Te Chiang and photos by Wei-Ta Fang)

skill training and capacity building programs, marketing of tourist activities, embarking and supporting of research and consultancy works in destinations in ensuring that the rationale behind Ecotourism is successful.

Ecotourism, as a new form of tourism has becoming a travel activity. It should ensure undeviating financial sustenance to local communities. The tourism activities are produced by locals but enjoyed by tourists. Through interactions with locals, tourists gained awareness on local cultures. On the other side, the locals become understand about the tourist culture, hence, teaches tourist to appreciate and respect the local cultures while they are visiting. This is a way in the cross cultural exchange between the two parties.

Through local community empowerment, local residents enhance their professional knowledge and skills in meeting the demand of the tourism industry. This supports small scales stakeholders, then benefit the locals ensuring that their money goes to the local economies by hiring local peoples. This empowerment helps locals in acquiring educational and vocational skills, the establishment of community own Ecotourism camps discourages mass tourism, mass constructions of hotels, tourism remedies and mass activities in fragile areas. The provision of community based Eco-camps and Eco-projects nurtured a greater wisdom of uniqueness and superiority in locals, hastening in development. The stakeholders aid in the developments of Ecotourism in the destination. They are through running of their entrepreneurial business in the form of Eco-lodges, camps and attraction sites, running of training and academic institutions offering course in tourism and its related fields.

8.3.1 Ecotourism in Destinations

Over the years Ecotourism was regards by destinations. It was a form of tourism for nature lovers. These emerging tourisms have then recently become a success business for many developed and developing countries. It has become not only a mere peripheral activity to support the protection of natural environment but also is being one of the major businesses of many national economies around the world, especially in the countries like Kenya, South Africa, Tanzania, Taiwan, Ecuador, Nepal, Madagascar, Costa Rica. A territory like Antarctica, Ecotourism is playing a significant role in the growth of their GDP, providing both direct and indirect job opportunities for the locals. These people engage in the administration of the tourism business, conducting guiding services for companies. They offer organized excursions, establishment of accommodation centers, and other outlets. All of those create opportunities in making life easier for the indigenous people who facilitates and make Ecotourism a successful vibrant business in destinations. Other destinations, on the other hand, continue to safe-guards the ecotourism features of the destination by enhancing the tourism development on the biodiversity (fauna and flora), local cultures. Some ecological and socio-cultural attribute and touch the heart of tourists to have potential impacts on the image of the destination economically, environmentally, socially and politically.

8.3.2 Ecotourism Code and Conduct

The way you spend your vacation in destination, interacting with the locals, and learning about the natural environment, all have some elements—both positive and negative impacts. Based on these, codes of conducts are formulated to minimize those negative impacts and maximize the positive impacts. A code of conduct is given as a guideline. It is an established guideline of out-lining the accountabilities of or appropriateness done by the Eco-tourists at the places they visit. These codes of conducts are directive of sensible approaches, engagements, involvement, and contacts on the part of the distinct Ecotourism. They are all geared toward the safeguarding the interest of the locals, environment, and equally giving travelers an enjoyable happy experiences.

Eco-tourists, as considerate and accountable tourists, will have many things to put at the back of their mind. They have responsibility to travel and leave sound footprint behind which will echoes for ages. Numerous preparations and contribution of things to do, how to do, and when to do, can be judged before, during, and after the trip. To certify the experience is in track with the principles of Ecotourism, they will have to minimize the impact on the local communities they visit or by spending through local sustainable products. Theoretically, an eco-tourist has an environmental, social, and economic influence, thus we ask him or her to help minimize any unfavorable impact on his or her holiday has on the environment, host communities and economically. The following is brief description on the codes and conducts done by Eco-tourists. All tourists must abide the on the codes and conducts during their vacation.

- Respect local cultures, traditions, religions, and heritage
- Get prepared for the journey
- Avoid displaying money in public areas
- Be flexible in your expectations
- Conserve resources
- Concern for the environment
- Apply less environmental hazards
- Protect the environment: its flora, fauna, and landscapes
- Benefit local communities: both economically and socially
- Minimize pollution: through noise, waste disposal and congestion.

8.3.2.1 Respect Local Cultures, Traditions, Religions and Heritage

Eco-tourist and conventional mass package tourists are supposed to put on appropriate dresses, accepted by the local culture and practice religion of the destination. It is mostly inappropriate for a nude appearance on the beaches and other public areas. It is considered rude and not tolerant to the religion of the local community. It is appropriate to ask for permission in the event of taking pictures—either direct or indirect—from the person you want to take the photograph or indirectly through the help of a tour guide. Observe local customs. Has an ecotourism one need to be sensitive of your personal cultural ethics and how they touch your feelings and perception about others. Being in a distant land requires a new set of life and concept, due to the difference in perceptions of personality, cultures, language, etiquettes, all of which differ from one geographical location to another, all which should not be regards as an immoral practice. One needs to feel part of the destination in order for to get an enjoyable and memorable experience.

8.3.2.2 Get Prepared for Your Journey

As an eco-tourist, one needs to familiarize him or herself with the destination on the itinerary. By searching for all necessary detail information from the internet, travel agency and even tour operators, about the destination, local people, and environment as well as its acceptable behavior for both the locals and tourists, couple with the destination's existing activities in giving the visitors an authentic rewarding experience, being it a day out on natural trail, bird watching, encounter with a local community in learning about few sayings in local and as well participating in local community festival with the interest and motivation of experiencing the local social livelihood and culture, learning about the ecological attributes, will make the tour profitable.

8.3.2.3 Avoid Displaying Money in Public Areas

Cultures and traditions are difference from one region to another. Countries are of diverse in economic; situations of which influence the life of its nationals in their daily life and reactions towards properties. Diversity in cultures also affects the way we do our routine lives in how we handle our personal belonging of money and money worth on vacations, what may be common to you in handling and dealing with your belongings, may affects the lives of thousand out there which can result them to bad acts of theft and other unlawful deeds. All valuables should be securely handled if not stored in shelve of your rooms, during visits to local communities and other public areas, donation and contributions on community initiatives can be done through registered organizations and aid agencies, giving of money may result in the encouragement of begging.

8.3.2.4 Be Flexible in Your Anticipations

Involving in travelling to destinations has both ups and downs in the fulfillment of one's desire in a destination. It might either dissatisfy or exceed the expectation of a holiday maker. Most visitor attractions are often isolated at distant zones, whether populated or abandoned, based such circumstance of changes in itineraries regarding events, will calls for an open mind of an Eco-tourist not to consider things has a disappointment, Occasionally itineraries are modified to meet the expectation of customer in line with the current available products and services with the prospect for a more profound experience coupled with a more profound enriching cultural experience in destination.

8.3.2.5 Conserve Resources

Tourism being one of the world's fastest growing industries has a lot of accountability in conserving natural environment. It can also and promotes the authenticity of local culture even though due to operations, it may habitually deteriorate the earth's assets in destinations visited by tourists. Most of these destinations are underneath immense impacts of human activity making life unsuitable for its inhabitants. Ecotourism needs to be judgmental of their vacations. They need to be responsive of the resources that are used during holidays. Therefore, they need to book their vacations with eco-friendly accommodation providers through tour operators with cooperate social responsibility. This comprises of their individual utilization of objects like energy, transportation, and specialty foods had to be imported from a foreign land. Eco-tourists need to support the sustainable and responsible use of local resources. Ecotourism is a responsible form of experience that tourists should not permit game hunting of the rare, protected, and endangered species in destinations. That such activity will lead to the loss and extinction of the wildlife, coupled with the depreciation of the local cultures.

8.3.2.6 Concern for the Environment

The environment, whether it is natural or artificial, is the most fundamental ingredient of the tourism product. However, as soon as tourism activity takes place, the environment is inevitably change or modified, either to facilitate tourism or through the tourism production process. Environmental preservation and improvement programs are now become integral part of many development strategies and such consideration is treated with much greater respect than they were during the first two–third of the last century. Relatively, researches have been undertaken within a standardized framework to analyze tourism impacts on the environment. The empirical studies have been very specific on tourisms. Studies on the pollution of air is on the positive side, the direct preservation or restoration of ancient monuments, sites and historical buildings, protection of reefs and beaches, maintenance of forests, the creation of natural parks and wildlife parks, are all of concerned.

8.3.2.7 Apply Less Environmental Hazards

The interest of promoting tourism and its related experience has invented the care for the natural environment. Without compromising for the negative actions triggered by mankind in responds for their desire fulfillment have resulted in the fragmentation of the environmental resources, leaving away the ecological responsibility to mankind. To balance their roles of development, which don't care for the natural environmental features, have to be at the vision of accessing the quality of ensuring Eco-friendly constructions of hotel accommodation and concern for waste management. In recovering the forest covers, tourists should act proper consideration on what tourists do leave behind, thus locals can feel what the effects on the nature on local.

8.3.2.8 Support Local Economies

At its best move, Ecotourism is in-charge of treks to natural locations projecting dual benefits for local communities in destinations. In one way, it preserves the integrity of the flora and fauna and generates economic benefits for local communities. On the other sides, it persuades protection of the natural environment. Ecotourism initiatives in destinations are geared towards supporting the social livelihood of the indigenous people based on the common concern of enhancing the welfare of the locals. Ecotourists are advisable to make use of the confined system of transportation used by locals during excursions.

Various Ecotourism initiatives are the establishment of full community regeneration towards more meaningful life. The long-term benefits of all Ecotourism projects consist of the improvement of local people's welfares in their communities. They are also creations of employment opportunities, community empowerment, heightened trade and industry condition, ecological reestablishment and improvement of the quality of life in destinations.

8.3.2.9 Merits and Demerits of Ecotourism

There are many potential rewards and disadvantages of Ecotourism in both developing and developed destinations. All of them depend on the way tourism is managed and handled. Ecotourism has been alternation movements to generate good quality for the natural environment. They also help in the enhancing of the welfare of the local people in the industry that stands to make a lot of money from vacationers. The vacation operators should also seek for environmentally friendly and ecologically activities. The money from Ecotourism could be used to preserve nature, wildlife, species, and earth's natural resources as well as facilitate local people with right of entry to development of quality education. However, on the contrary, if not properly handle and manage, it can also cause effects on the local economy, environment and community.

The conventional mass package tourism incurs a lot of negative impacts in so trying to shift away from that side, Ecotourism strives to minimize the adverse affects of hotels, trails, and other infrastructure by using either recycled materials or plenty fully available local building materials, renewable sources of energy, recycling and safe disposal of waste and garbage, and environmentally and culturally sensitive architectural design. Minimization of impact also requires that the numbers and mode of behavior of tourists be regulated to ensure limited damage to the ecosystem.

Ecotourism have has being a viable Eco-friendly business in regions, it construct a mutual actions toward constructing a sense green consciousness among local people all of which in way are educated on how to portray best Ecotourism practices in destinations. This is necessary capacity building programs for the local are good in Ecotourism as local updated with the necessary ethics of the industry. They are trained as linguist to be versatile in their guiding service with skills in environmental features and cultural history, environmental interpretation, ethical principles and effective communication. Ecotourism projects also help educate members of the surrounding community, schools and school children receive academic materials in some part of the world all which of promotes and empowers quality education.

In addition to the educational assistance of Ecotourism, one will also realized that it as well offers direct financial reimbursements for protection and safeguarding of the environmental resources by assisting in raise resources for environmental protection, research and education from beginning to end of are numerous diversity of methods, counting from recreational area access charges, sightsee excursions, bed and breakfast, all of which offers contributions to the welfare of the local communities in empowerment for local people in their corporate social responsibility agendas.

Ecotourism is friendlier in the preservation and conservation of local culture. It services as the one core element in aspiring tourist to destination of less culturally disturbing activities and abusive than conventional mass packages tourism. Ecotourism calls for unity towards respects of local culture. Ecotourism endeavors to be culturally courteous and have a token effect on both the natural environment and the human population of destinations that are adhering to it. Ecotourism demands a more holistic approach to travel, one in which participants strive to respect, learn about and benefit both the local environment and local communities. Part of being a responsible Eco-tourist learns beforehand about the local customs, respecting dress codes and other social norms and not intruding on the community unless either invited or as part of a well organized tour.

8.3.2.10 Demerits of Ecotourism

In the planning, management and implementation of Ecotourism in destination there are various side effects. It can be detrimental to the life of both the local peoples and the natural environment. In buttressing on the positive impacts on environment, building of awareness program and strengthening a more enriched admiration for the local culture and environment for offering a unique experience and encounter for both the locals and tourist alike. The employment cannot be over-emphasized in estimating the number of employees working in Ecotourism enterprises in destinations. Ecotourism teaches guests by on-site vacations on the subject of local political, shared public view and ecological green concerns in safeguarding the welfare of the endangered species by preserving and protecting them from the danger and extinctions.

However, Ecotourism is regarded as the successful business in destinations but if it's equally managed badly, it could result in congestion of massive buildings which are not eco-friendly in resort which will at long run pollute the natural home of the living organisms. Ecotourism paves a way for consumption of the natural resources together with learning about the natural features all which if not working in line with carrying-capacity may resulted in the abusive use of the nature. Excessive trips to parks and other attractions gradually degraded the land leading it to erosion, and run-offs. Cars no longer use the confine trails in parks, they create new trails in search of animals. They will eventually make tour operators not caring to the Ecotourism. International organizations working as philanthropies for the nature, criticized the ecological features are used as a pull factor to motivate tourists; Examples of which include construction of giant power demanding lodges in wonderful environmental locations with no consideration for sustainable development. Destinations are highly competing for the brand of best Ecotourism destination. It resulted in the development of tourist designated areas which relocates the local people isolated from coastal areas and places of natural beauty all due to serving the need of the tourist and it also unlawfully maintains rare species of flora and fauna in enclose objects to motivate and attract potential travelers by exploring the beauty of the destination.

Public sectors, Private Sector and Voluntary Sector should look back in mind that proper planning prevents problem.

8.4 Ecotourism and Recent Development

Ecotourism development in destination has to facilitate a diverse range of developments globally, leading the world into a global village through its enrichment of life and promoting of cross cultural exchange among people of diverse cultural identity. Ecotourism today, has transformed and added in the way react towards the natural environment. There has being numerous diversification in the Ecotourism all of which are geared towards meeting the demand and the needs of tourism. Among the recent development of Ecotourism are the following:

- Technological Development
- Transportation Development
- Changes in Distribution System
- Industry Restructuring
- Latest Technology
- Global Distribution System
- Concern for the Environment
- More Independent and Special Interest Travel.

8.4.1 Technological Development

Ecotourism has influenced all walks of our lives over the last few decades in a tremendous manner within the Tourism and Travel segment being one of the chief beneficiaries for local communities and natural environment. The Ecotourism product being largely intangible stands to gain a lot from technology. Ecotourism being considered as an activity of risk in the early days mainly due to the fact that the knowledge about the terrain, climate, culture etc. of the destination was limited or unknown, with the advent of Information Technology into the industry, things have changed. Tourist are now able to access information through the search engines like Google, added onto which is the use of social network life Facebook, Twitter, MySpace, and LinkedIn for marketing of Ecotourism tours and destinations around the globe were travelers also share their fascinating experience about destinations.

Ecotourism has become much more enjoyable and comfortable with the advent of information technology, and its resourceful data which can access at every doorstep. The modern day traveler has lots of options to choose from, with Internet being the favorite for the potential traveler. Places of tourist interest and scenic beauty often do not catch the attention of the customer largely due to the fact that they are not marketed well.

There are number of software which caters to the needs of the travel and tourism, including the Ecotourism sectors. Ecotourism destinations today use modern systems equipped with powerful applications that work on huge databases. Virtual Eco-tours are currently creating an entire new type of real time tour that connects historical heritages sites with geographical accurate landscapes using 3D interactive software, live videos streaming, creating of 720 panoramic photos, GPS mapping of eco-trails.

8.4.2 Transportation Development

Development of transport has taken place since the beginning of the industrial revolution and has linked the growing of economic opportunities for Ecotourism destinations. After the World War II each stage of human societal development particular transport mode has been developed and adapted including the air travel, cruise ship, railways and roads. Which have been observed that throughout history these transport has been solely responsible for economic growth linking the early stages of European expansion from the 16th to the 18th and nineteenth centuries They supported the development of international trade through colonial empires, but were constrained by limited inland access.

The first stage of the industrial revolution in the late 18th and early 19th centuries was linked to the development of canal systems in Europe and America. It was mainly to transport heavy good. This permitted the development of rudimentary and constrained inland distribution systems. And Railways stage of industrial revolution in the 19th century was intimately linked to the development and implementation of rail systems, some transcontinental, enabling a more flexible inland transportation system which allow so many people to travel for either leisure, business or on educational trips especially after the World War II, today transportation of helps in the linking of major cities to national parks, zoos, safari trips and Ecotourism camps, with the Macao's in zoos.

8.4.3 Change in Distribution System

This is a system which Ecotourism organizations, local communities, conservation projects, tour guides and resources involved in the packaging and marketing of Ecotourism product or service from destinations to consumers. The supply and distribution of these activities transforms channel and system through which products and service are controlled in a supply chain through a process of using websites for the promotion of Ecotourism destinations than using mediators like tour operators and retail agencies for marketing, though still exist the display of brochures in retail travel agencies. These modes of distribution of ecotourism products and services through the use of internet; that delivered products or services to the end customer in sophisticated supply chain systems through the use of modern technology, which have today directly, enable destinations, local communities, parks and zoos to directly link with their customers to use their products at any point in time. These have increase the speed of transactions with real time data exchange, which have eventually resulted in the increase of revenue by satisfying customer demands more efficiently.

8.4.4 Industry Restructuring

The travel and tourism industry have undergone major trends. Ecotourism is not an exemption, this transformation has made the industry to rearrange the way destination brand its products and service for the target market. The same tour companies need to reorganize and diversify their scope of operations. This will make them more competitive in meeting the demand and expectation of the customer, as well as having edge over its competitors. Such changes occur when a company or business is facing trouble with its development plans, and the brand, it will often consolidate and adjust the terms of the plans in a management and development restructuring in packaging products that suits customers. After a restructuring, the objectives on achievements are more manageable for destinations and the likelihood of future progress increases with new customers. A company restructures its operations or structure by cutting costs, such as payroll, and also through capacity building and training for its technical and skill experts. This is often seen as necessary when the current situation at a company is one that may lead to its collapse.

8.4.5 Latest Technology

What the consumers perceived they wanted or needed at the time, on the demand side is richly facilitated by the technological possibilities on the supply side. Important technological advances have had a very profound impact on the development of tourism. Among them is the application of new computer technology which facilitates the use Information technology through smart phones and intelligent devices for accessing multimedia data on accommodation, events, culture and leisure activities of destinations, applications which mainly focus on the customization of data mining techniques, intelligent, multilingual and positioning system. With destinations using the method of online interactive virtual tours that features real time sounds, images, videos with trails and a live guide guides which directs the tours by watching and listening to the video. Mapping of existing trail are done which shows which trail is to be used and which one to be avoided with appropriate signage of navigations of downloadable GPS routes.

8.4.6 Global Distribution Systems

Global Distribution Systems (GDS) have evolved over the years. It has played a major role in making online travel possible aiding in booking and reservations, with the use of Computer Reservation System (CRS) is used to store and retrieve information and conduct transactions related to travel and tours. They were originally developed by airlines, but it was later extended to travel agents and sophisticated tour operators who in addition use it as a sales channel. A functionality of CRS has been extended and what is known as the Global Distribution Systems came into existence. The major CRS operations that book and sell tourism products and services to multiple customers are usually done through the GDS. Many tourism organizations have opted to divest most of their holdings to Global Distribution Systems due to which many systems are now accessible to consumers through Internet gateways for hotels, car rental agencies, cycle rental and other services as well as excursions. A traveler or a travel agent can check out an itinerary using a GDS which is a global system interconnecting airlines, hotels, travel agents, cruise liners for daily activities on package excursions etc.

8.4.7 Concern for the Environment

The environment, whether it is natural or artificial, is the most fundamental ingredient of the ecotourism item for consumption. However, as soon as ecotourism activity takes place, the environment is inevitably change or modified either to facilitate ecotourism or through ecotourism production process. Environmental preservation and improvement programs are now an integral part of many development strategies and such consideration are treated with much greater respect than they were during the first two–third of the last century. Relatively research has been undertaken within a standardized framework to analyze ecotourism's impact on the environment. The empirical studies that have taken place have been very specific case studies such as the pollution of air, on the positive side, the direct preservation or restoration of ancient monuments, heritage sites and historical buildings, protection of reefs and beach, maintenance of forest and the creation of natural parks and wildlife parks.

The environment is such a great concern. Therefore, destinations eco-tourists are encouraged doing their part to reduce negative impacts of tourism activities on social, economy, and surely environment

8.4.8 More Independent and Special Interest Travel

Ecotourism may as well refer to the form of tourism, were tourist travels as individuals, and in groups to learn about and experience particular feature of an area. With the new trend of tourism, tourist are now becoming more flexible, independent and experience travelers, whose values and life styles are different from those of the mass tourism. As tourists become more sophisticated, they will increasingly seek tourist products that fulfill their particular needs. Accordingly, some visitors are attracted to destinations primarily for customized special interest experiences, such as education, sport or business experiences, wildlife and cultural authenticity like the Maasai Mara of Kenya, these special interest visitors are often referred to as niche markets in the tourist fraternity, today being widely observed as eco-tourist. Typically, a niche market of tourism being ecotourism is small, yet profitable, group of consumers. In a bottom-up approach, the needs of the niche market are identified and experiences developed to meet their specific needs. This approach will enable to capitalize on particular competitive advantages that it has in the tourism marketplace, attract higher yielding tourists, and promote the growth and diversification of the tourism industry as now tourist are becoming more independent and traveling on special features of a particular destination in search of discovering about the local culture of the host community and the natural environment

Box 8.1: A Case Study in the Republic of South Africa
South Africa, officially named the Republic of South Africa (RSA), is the southernmost country in Africa. This country is renowned for her varied landscape, cultural, and nature diversity (Fig. 8.12).

Fig. 8.12 Location of South Africa (Illustrated by Yi-Te Chiang)

It is The South African National Parks Organization is the leading conservation in the world and it's renowned for its incredible natural heritage, which operates, with research and management standards. South Africa offers real Ecotourism products and services, the country's 19 national parks out of which 14 of the which offers overnight tourist facilities, and a variety of good quality accommodation and safari camp sites, giving eco-tourists a real life time experience on their trip of discovery the real Africa. The country has the largest concentration of wild animals than anywhere in the world, with approximately 150 species of mammals, over 500 bird species and over 380 species of plants. Eco-tourists keep an eye for a more inclusive game drive or take part in evening safari were tourist have the opportunity to see the abundant species of endangered species of biodiversity. Today, south Africa is marketed as ecotourism destination with high profile of higher spending eco-tourists, the destination has a good number of local involvement in the development of ecotourism, benefits of which are used to reimburse back to the local community helping them in boosting the local economy.

8.5 Conclusion

Hall (1998) stated that tourism development was very difficult to sustain if it is undertaken to maximize tourist numbers and increase public mechanisms, whiles ignoring the interdependence of the local culture, ecology, society and economy.

Ecotourism is evidently identifying as a mean to end the global issues focusing on the 3 (three) elements of the millennium development goals, namely (1) economy, (2) environment, and (3) socio-cultural. It has vividly been an emerging forms of tourism that is controversial in its operation with both positive and negative impacts. This section endeavors to outline the definition, theories and practices, planning and management, code of conducts, merits and demerits, these aforementioned elements coupled with recent developments of Ecotourism initiatives in destinations. Ecotourism has been a vibrant facets of the tourism industry is sky rocketing in destinations creating proceedings for the local economy through the creation of job opportunities for the local, as promoting environmental protection and development programs in destinations. It has so far tries to preserve and conserve the endangered species of biodiversity. However, this sector has vividly embraces **ESEE** impacts: (1) Environmental, (2) Socio-Cultural, (3) Economically, and Environmentally.

Ecotourism activities provide significant benefits and understanding for developing the world fast growing service sector in emerging economies. An Ecotourism activity helps in generating economic opportunities by providing financial benefits to the local communities. It also aids sponsorship and funding scheme for conservation projects in destinations, which helps in the safeguarding of endangered species through environmental sensitization and education programs in destinations. It has led in the formulation of carrying capacity for visitor attraction centers, running sensitization agendas on parks and wildlife management, conducting environmental assessment programs in destinations alongside traffic management. Codes of conducts are designed to help enhance responsible travel in destinations with promoting a fair deal between the host community and tourists. Local people are invited in the planning and administration of tourist projects, and visitor management programs. Cultural norms and values are planned and control to the needs of the future generation. Economically, Ecotourism designs forums for creation sustainable infrastructural development projects which in turn provide employment opportunities. With the vision of safeguarding the integrity of Ecotourism; it should be embedded into academic curriculums to enhance the future generation an understanding about the local culture, environment, and historical heritage sites for them to obtain a spirit of promoting ecotourism across all facets of life.

References

Barkin D (1996) Ecotourism: A tool for sustainable development in an era of international integration? Yale F&ES Bulletin 99:263–272

Fang W-T, Lien C-Y, Huang Y-W, Han G, Shyu G-S, Chou J-Y, Ng E (2018) Environmental literacy on ecotourism: a study on student knowledge, attitude, and behavioral intentions in China and Taiwan. Sustainability 10:1886. https://doi.org/10.3390/su10061886

Hall CM (1998) Historical antecedents of sustainable development and ecotourism: new labels on old bottles. In: Hall CM, Lew AA (eds) Sustainable tourism: a geographical perspective. Longman, New York, USA, pp 13–24

Sirakaya E, Sasidharan V, Sönmez S (1999) Redefining ecotourism: the need for a supply side view. J Travel Res 38:168–172

TIES (2015) TIES: announces ecotourism principles revision. https://www.ecotourism.org/news/ties-announces-ecotour-ism-principles-revision. Accessed Mar 2019

Chapter 9
Sustainable Tourism

9.1 Introduction

Sustainable tourism is a form of tourism and tourism practice in destinations endeavoring to minimize the impact of conventional mass package tourism. With a vision to make a decrease in the effects of tourist activities on the natural environment and the culture of the local communities, at the same sustainable tourism, facilitates to create job opportunities for local residents.

There are a numerous of descriptions and introductions on sustainable tourism (UNWTO 1995). Some of them are green travel, ecotourism nature-based tourism, pro-poor tourism, and culturally responsible tourism, fair trade and ethical travel. All of those operate within the domain of sustainable tourism. Academics, researchers, and industry professionals have vividly agreed to come up with a generally established classification, which is given by the parent body called the World Tourism Organization (UNWTO), the father of tourism. They define sustainable tourism as "tourism which leads to management of all resources in such a way that economic, social, and aesthetic needs can be fulfilled while maintaining cultural integrity, essential ecological processes, biological diversity, and life support systems." In addition, they describe the development of sustainable tourism as a process that meets the needs of present tourists and host communities whilst protecting and enhancing needs in the future (UNWTO 1995).

This form of tourism aspires to certify that tourism operation and it related progression conveys an optimistic experience for tourists, local people of the host community, and natural environment of the destinations. These thoughtful and quick transformations of sustainable tourism are taken place in the destinations for the past 20 years. It has been reflecting on revolutionizing the tourism industry in the way and format it operates economically, socially and environmentally. International political and economic reorganizations have resulted in the expansion of tourism both in a spatial sense and in terms of a significant increase in the size of the tourist market.

W.-T. Fang, *Tourism in Emerging Economies*,
https://doi.org/10.1007/978-981-15-2463-9_9

9.1.1 Definition of Sustainable Tourism

Sustainable tourism is all about responsible travel to fragile, pristine, and usually protected areas. They strive to be low impact on the natural resources culturally and environmentally. Its rationale to edify the travelers and locals; offer means for management and preservation openly for assisting the financial improvement and empowerment of local communities; and then do encourage admiration for diverse traditional legacy of the destinations.

Sustainable tourism demands for cost-effectively and generally mindful tourists. Tourists, therefore, must contribute positively rather than spoiling things during their vacations. Most of these involve in unpaid assistance services, helping local people make best of use of their unlimited resources. Such kind of tourism normally engrosses trips to a destination where environmental attributes and artistic legacy are the main appeal of tourist visits. Theoretical encounters, heartening folks who explore keen on discovery and appreciating the setting of the local vicinity, sustainable tourism development, will bestow love of creatures on scenery and besides a better pleasure of our possess likely homes.

It paves pathways that lessen the unhelpful acts of mass package tours widely on offer in most destinations around the world, sustainable development enhances the awareness of both the local community and tourist about the natural features embracing environmental possessions and cultural experience of locals.

Tourism is an ideal context for studying issues of political economy, social change, and development. The natural resources management and cultural identity and expression, creating economic opportunity for local communities, energy efficiency, promoting recycling, water conservation and ecological preservations, must be off concerned and created for the economic opportunities of the local communities.

9.1.2 Definition of Sustainable Tourism Development

Sustainable tourism development has been also reconsider methodically by previous some researchers such as Barnwell and Lane (1993), Owen et al. (1993), Murphy (1994), Leiper and Harris (1995), Tosun (2001), Mowforth and Munt (2008), etc. The rationale of this critique is it may be valuable to classify several essential ideologies of sustainable tourism growth in destinations. The classification is by captivating into description theoretical opinion for progression, sustainable development and sustainable tourism development. These doctrines might be assured as follows:

- Sustainable tourism development must supply the felt requests and contentment of what constitute the tourist wish and the main motive of the travel to be fulfilled.
- It should reuse, reduce, recycle waste materials and as well lessen the use of insufficient resources in destinations.
- Sustainable tourism has to add to the surfacing of required situation in tourist destinations which will lead local people to gain self-esteem and self-efficacy to

feel free from the evils of want and ignorance from place identity theory (Goulet 1971; Wang and Xu 2015). That is to say, it should help host communities be free or emancipated from alienating material conditions of life and from social and political servitude to nature, ignorance, misery, institution, and beliefs from political choices and logical decisions (Todaro 1994; Tosun 2000).

- Sustainable tourism development should accelerate not only national economic growth, but also regional and local economic growth. This growth must be shared fairly across the social spectrum.
- Sustainable tourism development should achieve the above objectives or principles in an indefinite period of time without compromising the ability of future generations to meet their own need.

The aforementioned principles of sustainable tourism development remind the readers that the remit of sustainable tourism development is extended to consider the role of tourism in contributing to sustainable development more generally (Hunter 1997).

9.1.3 Origin of Sustainability

The word sustainable is derived from the word sustain. Sustain means to maintain and prolong tourism, safe guarding it for both present and future generation, and the like. Sustainable tourism has to do with the long-term preservation of responsibility, in safeguarding the integrity of the biodiversity, protecting of the authenticity of the local cultures and economic livelihood of the locals. They demand the responsible use, and management of tourism resources in destinations. Hence, it minimizes the impacts for sustainable tourism which emerges as an end to the negative impacts on the local economy, culture and environment.

Sustainable tourism is being a new trend of tourism development in recent years. It seeks to conserve the ecological features and attributes of the destination, helps enhancing the financial well-being of the locals and cultures of the destination. The objective of sustainable tourism in destinations is purely based on an intention of improving the platform of constructive experience for tourists, as well as helping the local people enjoying from tourism business, even if the environment is at risk. Sustainable tourism in every destination has some guideline elements. The sustainable tourism main concept and beliefs are consigned to the environment, economic, and socio-cultural aspects of tourism development in destinations. It is an appropriate sense of balance related to well-known among three dimensions to promise its long-standing sustainability in serving the need of tourism and host communities. As a result, sustainable tourism should operate in Triple Bottom Line with the following dimensions:

- Environmentally
- Socio-culturally
- Economically

9.1.3.1 Environmentally

Destinations of well-managed environment are of significant attractions for tourists. They are precious as a source of foreign exchange earnings for the destination, providing job opportunities for locals but equally ecological features. However, they are increasingly under pressure from tourism activities, leading to extinction of fauna and flora, environmental degradation, and natural resource depletion. Therefore, sustainable tourism should construct the most advantageous use of ecological possessions comprising of a key component in tourism development, maintaining crucial environmental processions, and serving to preserve natural legacy and biodiversity of destinations, for both the present generation and future uses.

9.1.3.2 Socio-culturally

Tourism developments, in both rural and urban settlements, often deter citizen of destinations from having the right of access to land, water and natural resources in need of serving the tourism industry. Land areas have been designated as tourism development areas with restriction to them but backed by punishable laws. Cultural features of the destinations, which constitute one of the most key factors that tourist wants to discover in destinations, are commercialized to fit into the itineraries of entertainments programs. In valuing this socio-cultural authenticity of the local social, set-up of host communities ought to be of great interest of all stakeholders. In helping to preserve these, they constructed source of revenue cultural heritage and traditional values, and contribute to inter-cultural legacy preservation and appreciating them with thoughts of open-mindedness.

9.1.3.3 Economically

Tourism is being an industry that relies on the human resources of destinations for the services delivery. The sustainable tourism should emerge as a potential tool for local opportunities to gain tourism benefits and support for source of revenue for the local people. They should be guaranteeing a feasible long-term economic growth, by providing socio-economic remuneration to all stakeholders that are moderately spread in communities, including constant job opportunities and income-generating openings and communal forces to host communities, and contributing to poverty alleviation within the under privileged groups of the community.

Based on the affirmed mentioned statement, International Day on Monuments and Sites (ICOMOS) further stated that sustainable tourism refers to a level of tourism activity that can be preserving in excess of the extended period for the reason that it results in a disposable advantage for the economic, social, natural and cultural environments of the destination in which it takes place (Drost 1996).

9.1.4 *Types of Sustainable Tourism*

Sustainable tourism activities are unique experience giving tourists rewarding and real life time experience. It is representing an essential cross connecting cultural diversity experience and environmental awareness program. Major symbols of this unification are noticeable in diverse personality of local traditions, starting the diverse dialect in the village, to the customs, as well as in local traditional festivals, such as:

- To enjoy nature and scenery;
- To see famous tourist attraction in local community;
- To taste food and try delicacies of the country;
- To experience a different culture;
- To broaden your mind, culture and education;
- To learn how to cook traditional food and play traditional musical instruments;
- Bird-watching;
- Conservation programs, i.e., tree planting;
- Nature walk;
- Story telling; and
- Village tour, etc.

This phrase refers to general tourism programs that present preservation, conservation and sustainability, i.e., socially, environmentally and economically. This is in compliance with sustainable tourism principle, imply by sustainable tourism a sector of tourism. Based on the main beliefs of sustainability, with accountable tours to destinations these steps forward will promote the interests of local people and safeguards the atmosphere ecological features in place, encompassing constructive contact, mutually to local community and biodiversity in protecting the endangered species.

9.1.4.1 Community-Based Program

Village tourism better known as Community-Based Tourism (CBT), with its unique features of visitations increase the attractions of the locals, and bits and pieces, offer the tourist to mingle with the community. The village tourism showcases the customs, experience a way of life of local community which are still untouched, so tourists can hand-on the experiences. Village tourism a sustainable form of tourism where tourist hike in local communities, staying with host families to experience with the local people, taste the local food, share agriculture system and social system, music, cooking, dance, and so on. Furthermore, clean scenery and environmental features are extra tips of tourist experience and expectation, which constitute part of their trip.

Organizing home offers popular of village homestay to tourists. These amenities construct them an enjoyable stay in local homes, giving them a happy, low destructive vacation with authentic experience. Homestay sustainable tourism initiative is done via numerous opinionated services, such as aiding the local livelihood of the local people, supporting the economic proceedings, and promoting environmental conservations. The lodgings in village tourism (community-based) are extraordinary. Visitors spend the nights in homestay provided in the village. It gives visitors a memorable moment because they can feel a purely village atmosphere.

9.1.4.2 Homestay Program

Experience village tourism or traditionally notorious as homestay tourism, tourists reside in isolated smarmy secured homes. These homes are easily accessible by road using 4-by-4 (4×4) vehicles; like official tourist taxi, and safari cars. Discovering homestay is an experience that is not merely distinctive but undergoes of the experience remains as a real lifetime experience. Homestay includes delightful and liberal and caring after the tourists. It is not just a tendency however appeal, amid assortment of rituals, traditional fairs and festivals, customs, craft and fascinating landscapes such are encountered.

For some reasons, it is a romantic hideaway for quiet and peaceful ideal of the second honeymoon. To sum up, an adventure travel offers you a great escape and opportunity to reveal in your freedom. Stay in homely comfort and get to know hospitality even better. Enjoy the comforts of homestay in off beaten destinations of the world.

9.1.4.3 Landscape Design

Scenery beauty of the destinations, though the beauty is longer than what encourage tourist to visit, the landscape features panorama of mountains, high rise hills, coastal sides, parks, are strongly decorated with patterns that sharp their appearance and visualized an image. All of these make them popular during the summer for camping, winter flowers, and butterfly watching, are emerging trends in destination. Certain flower species are found during particular period of the year at specific landmass with attractive insects. Tourists are interested in botanical features. They love to visit during this time of the years in exploiting the outlook of the regions with diverse features of flowers and insects, values of environmental features, are enhanced through the high participation of tourist.

9.1.4.4 Eco-friendly Camps

Sustainable tourism echoes for the tree planting and rehabilitation programs in the regions, It is geared towards sound path foremost to conservation of the ecology esteem of environmental development program of the fauna and flora. In the sensitivity of the successful sustainability, it is a mixture of picturesque good looks, enjoyable tours, wildlife scenery watching excursions, conservation of fauna and flora, appealing conduct of both host community and visitors towards ecological features, and also cultural assets of the destinations.

Eco-friendly camps is a retreat for independent travelers, couples, special interest travelers as a secrete place away from normal crowded seaside resorts. Eco-retreat camp is located in destination some in an isolated environment. However, some eco-friendly accommodations are next to seaside with complete sea view, construct with traditional objects giving comfort of home away from home, pleasing you feel at home at the similar moment resembling the local.

9.1.5 Agents of Sustainable Tourism

Looking at sustainable tourism from a clear perspective, as well as narrowing down the mainstream philosophy about tourism operation, one will realize that sustainable tourism has some driving forces. It will result in its success in the sight of realization of its meaningful objectives all of which cannot be without the involvement of key players who work toward the protection and improvement of the world's distinctive destinations in the course of sensibly administered tourism and tolerant destination. With some tourism operations that sustain or enhance the ecological features of a destination, including everything ranging from natural environment features of fauna and flora, cultural assets of food, music, local handy craft, arts, dance and lifestyle, aesthetics, heritage sites of buildings, monuments, and the economic well-being of local people, the tourism is engaging business through the provision of jobs to locals. All of which are one way or another, the main factor of tourist attraction in destinations agents. This will comprise of the following aspects which play significant roles in the operations, product development, and management of the tourism industry:

- Public Sectors
- Private Sectors
- Voluntary Sectors.

9.1.5.1 Public Sectors

Public sectors are the industries mainly concentrate on the State Bureau of Tourism Department and Ministry that operates in the congress to deal with tourism and it related issues. They include for licensing, product development, destination marketing, the quality control of tourism products outstanding to this trade, and industry formation of the tourism. The government as you might expect plays important roles that is significant in providing infrastructural and super-structural development as well as facilitating development and collaboration between the individual tourism organization and business entities. They are to build up logical tourism destinations.

The public sectors comprise of national tourism bureaus, tourism boards, conservational agencies, regional and sub-regional development organizations, and local authorities.

9.1.5.2 Private Sectors

The tourism private sector comprises of business organization within the tourism industry. It is dispersed in the midst of number of the different conventional business segments, and brought jointly and classified as tourism businesses, since they rely entirely or partly on the expenses of holiday-makers. A large majority of these are small scales enterprises to standard business enterprises. They operate with small number of employees with no specific limit. They comprise of non-state owned institutions and agencies including private tourism institute, tour operators, airlines, event planning agencies, land owners, travel agencies, hotels chains, property developers, and travel agencies.

9.1.5.3 Voluntary Sectors

Voluntary Sectors comprise of charity foundation in the destination. They work to compliment the effort of the public and private sector agencies, for example, the Tourism Concern, Ecotourism Gambia, South Gambia Foundation, fair trade, and community groups in Gambia.

9.2 The Importance of Sustainable Tourism in a Destination

9.2.1 Objective

Sustainable tourism occasionally is consigned as an action of undertaking healthy conducts experience by performance of good quality activities. Its re-sources successively support the local community business ventures. It is in-fact an approach with the purpose of doing good than to hard possessions of the destination—be it socio-cultural, environmental, and financial. All of them take place in the tourism for its success through the commercialization to meet the demand and wants of the incoming and domestic tourists. Sustainable tourism productions and operation of openings will facilitate long term benefits through the operating using these methods (UNWTO 2001). They can improve all three aspects of growth, success, and sustainability. Sustainable tourism and its developments encompass them in three main areas: (1) socio-cultural, (2) environmental, and (3) economical. These three factors are considered as the triple bottom. Sustainable tourism being a form of tourism is searched for the lessening environmental and socio-cultural harms of destinations whereas providing tourism trade and industry benefits to home societies and indigenous people.

Globally, tourism industry has been the world fastest growing industry and the main service sector in most developing and industrialized countries. The objective accounts for preservation of the natural environment, conservation of the local cultures, and couple with achieving social equity in destinations. Based on which sustainable tourism emerges as an alternative for conventional mass package tourism. This has been in years led to the deteriorations of the local cultures, exploitation of the ecological attributes and as well fluctuation of tourism proceedings from destination to foreign economy. The affirmed mention effects as convey tourism engages tourists to a wise use of environmental features and directing their spending towards supporting local projects. This type of tourism in destination is supposedly support local peoples. In such doing, tourists enjoyed their vacation by having a more authentic experience as well as the locals also gain a wider positive reception for what makes their destination distinctive focuses on the three main areas namely:

- Preservation of the environment
- Economic well-being of the destination
- Cultural assets of the local residents.

9.2.2 Functions

The success of tourism business in every destination is heavily depends on the growth, used of natural resources, historical cultural heritages sites, human resources in the local environment as tourist attractions, and destinations in which sustainable tourism is not exemption, factors of which serve as the pull features of attracting tourist to experience the destination. However, multi-functional endeavors to constructing tourism sustainable are requested.

9.2.2.1 Economically

In economic broad tourism industry is one of the most viable sectors. It is tap-ping lots of potential benefit for nations. This industry should ensure trade, feasibility, and antagonism of going to places of interest of emerging destinations with the interest for novelty and discovery. Such ongoing activities are able to be of significant advantage to local economy. They are bright to prolong affluence generating remuneration for communities which are sold as tourist products. This will facilitate grant maintain financial benefit to the local villages, more especially to the local people. Tourism has been the fastest growing businesses making the most giving of economic success, counting the quantity of sightseer expenses that engaged in the vicinity. Sustainable tourism connects and sanctions local people in setting up and formulate decision concerning the organization and prospect progress of the industry in diverse location of the destinations utilizing different available recourses.

Poverty alleviation through the creation of job opportunities for local resident is on the move from strength to strength, reducing the unemployment rates in destination, and produce excellence confined works. They are maintained via tourism, counting the altitude of compensation of service rendered, in accordance with the nature of service delivered. They are also depending on the diversity of the work on their availability to all devoid of intolerance and no segregation by origin, gender, and disability or in other customs. These have encouraged local source of revenue and handwork of the local people in destinations.

9.2.2.2 Socio-culturally

Socially sustainable tourism enhances community fairness getting an extensive and pale allocation of financial and community remuneration from tourism. All of the beneficiary communities work to enhance the development of tourism, devoting their skills, and tirelessly pump efforts for recuperating prospects, revenue and job opportunities accessible to the underprivileged community residents. With promoting cross cultural exchange and unique experience between tourist and host communities in safe and congestive environment, enjoyable and satisfying experience for tourists, and thus obtain all exclusive of bias. Tourism preserves and fortifies the excellence of living of local communities, respecting and improving the remarkable traditional legacy of the host communities. They are all pay esteem to tourism, authentic culture, traditions, and distinctiveness of host communities, including social structures and access to resources, basic facilities and living bear method, and steer clear of several misuse and miss conduct of cultural values of social humiliation.

9.2.2.3 Environmentally

Promotions of local culture for tourist experience have helped in the preservation and conservation of the local cultures. The sustainable tourism, be it any locations of the destination, adheres for quality landscape, feature of natural scenery beauty, visitors are highly motivated to discover places of good quality, resorts of sound environmental programs with destruction to the vegetation.

Sustainable tourism works in close collaboration with biodiversity features of the region. It is strongly sustaining the management of natural areas, and natural home of living organisms of flora and fauna. Tourists in large number or in small groups are motivated to discover natural homes of wildlife in safari trips, bird sanctuaries, zoos, nature reserves, and low impacts to the visitor attractions. Recently, statistics have shown the increasing number of destinations which have at least done through the creation of book keeping center, botanical garden, bird sanctuaries, and national park which are homes to diverse range of wildlife. Today, such ventures are increasingly become noble investment in Europe, America, and Africa, some part of the Caribbean, Middle East, and Asia.

Low energy electrical and electronic appliances are installed at their accommodations, thus reducing the consumption of inadequate and non-renewable possessions of energy supply. The consumption of fuel wood will lead to the degradation of forest cover in most developing countries, while most women of living in developing or under developed countries, depend on the consumption of wood as fuel to cook and prepared their daily meals for the family.

Environmental greening program on the roofs of the buildings serve as covers will reduce the greenhouse gasses of atmosphere, rainfall, and soil acidity. Places at destinations are now complying with this new initiative ways of greening. They plant flowers on roof tops, engage in floriculture to beautifying places, discard wastes to minimize and promote sustainable environment in sustainable tourism schemes and tourist awareness programs.

9.2.3 Benefits

Good quality sustainable tourism is a procedure of satisfactorily, resolving and appealing, to the long term progresses of the destination concerning the tourism products and services on offers. Best sustainable tourism practices involve maintaining and constant relationship with local communities tied with good cultural, historical heritage everywhere prior remains of long-established legacy of local residents and their performances culturally, with an analysis to achieve pleasure and support of their possessions. Whether tourists take part in uncomplicated nature hike preserve revolve interested in an expedition of chronological bequest. In contrast, it is competent to demonstrate, extra than several manuscripts might feasible accomplish in portraying the tradition and authentic culture and scenery beauty of the destinations.

Providing good sustainable tourism development and supporting the initiatives should be a number one priority for any company or business entity involved in tourism and its ancillary services in very simple terms without them we would not be able to achieve the aspired dreams of sustainability.

In simple terms, it is vital for small and large businesses to have in place proper guidelines for staff members to follow. If a tourist on vacations leaves the destinations satisfied with a feeling, he has been provided with excellent service and a desire to return or back in the future—coming back for vacation at similar place of the initiative. High-quality sustainable tourism development supplies potential benefits to a business from providing a consistently high level of customer service include.

9.2.3.1 Satisfied Customer's Need

The first benefit of good tourism development is the dual meanings of local's satisfaction in preserving the ecological features, contributing economically and culturally;

coupled with which is the satisfaction of tourist expectations. Happy visitors become repeat customers. This means competent, compliant destinations are needed to handle questions and issues of global voice on sustainability. For examples, such nations which excel at achieving the vision goes a long way with the resolution, rehabilitation, realization, and achieving of the sustainable goals. Destinations that strive to help local communities can ultimately lead to better customer satisfaction. The customer is more likely to be a source of repeater business and to recommend the destinations to friends and family.

9.2.3.2 Consumer Loyalty on Destinations

Vacation as a edifying and judiciously explore is a key indicator of quality sustainability in development. Destinations with more common concern of the triple bottom lines are partnered with thoughtful tour operators, although these are rarely conscious needs on the part of the tourists. The piece of evidence remains that they explore for places, which still have something to offer at the cultural and emotive level. It occurs when a product or service is consistent in the formality, it is supposedly to do the job. These products and services are precisely embraced the physical and ethereal good services in destinations. They are preserved, appreciated and improved, finished obvious, and brought up-to-date. Yes, sustainable tourism can make you your holiday in destinations an experience some undiscovered things in your life, as well as contributing to the well-being of the host.

Consumer loyalty means what it says. It is when someone remains a repetitive customer of a product or service. The result can be a steady revenue stream. If a repeated customer is consistently satisfied about the service provided, he will be in line to remain a customer, and he or she only tells one or two people about a positive experience. This will lead to more sales from their own repeatedly business and from the new customers generated by their recommendations.

9.2.3.3 Less Cost Needed to Attract Potential Customers

Quality destination program on sustainable talk shows and programs relieve pressure to attract new tourists with intensive marketing. Statistics show that the costs are more to gain new customers, it is, therefore, to retain existing ones. The benefit of good initiative and destination programs comes into play here. A thorough sustainable tourism program helps maintain existing tourist base rather than lose a percentage of it. Special interest tourists are attracted equally to the conventional mass package tourist besides stay on board. Destination higher display out-performs other competitor in sustainability. It is more likely, destination with strong programs remain one of the leaders in the industry with good reputation. Interestingly, if a destination offers a product or service at a higher quality than places of mass destruction to local communities and social livelihood of the locals, it has a positive public image that will generate more sales through attracting high turnover of higher spender tourist.

Today, destinations embark of the staging of various events in the form of festival, carnivals, and the like. This promotes the local cultures of the residents but others will generate economic benefits for them. Celebrations, the most admired fair of them all being cultural music by local residents, administered to replicate and express completely this exacting unify ethnicity of the inhabitants. Widely observed destination has further than its reasonable split of usual festival festivities. They signify remarkable grouping of ritualistic basics, enthused in turn by dedicated team of organizers who are trained at local level to manage and run the events successfully.

Tourists will engage partially or fully in vacations with a vision of contributing sustainably. In one way, it is back by cultural encounters coupled by enjoyment of traditional storytelling and traditional set-up of the host.

9.3 The Advantages and Disadvantages of Tourism Impact

Tourism is considered as the act of management, planning. Though, tourism adversely contributes immensely to the benefits in destinations. On the other hand, it leads ecological harm through the felling of trees for constructing the lodging and accommodation. This gradually results in erosions and deforestation, demolition of sandbank, sand beaches, noise pollution, and toxic waste by sun bathers. These services have resulted in depressing activities of tourism. Effluence is becoming an alarming for local people. Additionally, it possibly will put off holiday-makers from further travelling to those destinations simply due to the bad publicity on the image. Subsequent to this, tourism will weaken traditional customs, norms and values, by commercializing them for the purpose of making income. This will habitually associate with growing waste, drawing, destruction, and sound pollutions resulting to the depreciation of the culture, as a result tourists do not constantly esteem traditional cultural legacy of the locals, ensuing to the vanished of such cultures.

9.3.1 Disadvantages of Tourism

Outline list of some of the disadvantages of tourism are listed as below. Today, sustainable tourism has caused both positive and negative impacts. It is also responsible way that the actions should intervention such acts to reduce the unwanted impact.

- Tourism in the mass package tourism results in environmental degradation noise pollution and social damage through tourist encounters with ecological features as well as the commercialization of cultural assets depreciation the authenticity of the local cultures.
- Investments in most destinations are foreign owned; proceedings of which did not support the locals. Tourists spend more money during their vacations all of which are used to pay for tax of foreign imported food stuffs, and beverage, as well as

spending on importation of foreign human resources, all of which are considered as squander than required.

- Tourist activities on the beaches and town settlement, coupled with the construction of hotels, lodges, and motel constructions along the coastal sites, have being detrimental resulting to waste disposal and accumulation of waste materials along the seaside. This is making life unsuitable for the marine life, dirtying the environment with synthetic hand baggage, plastic bottles, artificial crockery, tin cans, batteries, and cigarette counterfoil, all of which are used and irresponsibly disposed by tourists at anytime and anywhere they want.
- Transport and means of transportation with the use of coaches, buses, cruise ships, airplanes, are today be becoming a threat as result of their emission of smoke leading to carbon ocean emission and air pollution, noise pollution leading to ocean layer depletion.
- The accommodation industry has being heavily replying on the intensive usages of water and energy utilities for the lighting of the rooms reserved by customers. This is results to the higher cost of energy charges for the locals making less number of citizens to afford it. The uses of higher-powered generators are producing noise pollution for the surrounding vicinities.
- The event industry comprising of meetings, incentives, conferences, exhibitions and carnival outcomes in gathering litters, noise pollution, a large amount of power consumption, and water utilization, are eventually breaking down the supply of water and energy to areas of the less developed countries.
- The marine lives in the sea, rivers, ponds, oceans, are affected holding to the tourist activities on recreational activities. The uses of sun cream, sun oils, during sun baths on the beaches, are disposed as garbage into the water. However, the dramatic growing interest of tourist leisure activities in the oceans and other navigable waters with motor boats universally agreed, will endangered the environment of marine life, leading to displace of ecosystem.
- The ecological features which are under threat through the environmental greenhouse effects, mass debris throwing away; tourist threat on and tear losing protected plant life and extinctions of the wildlife (UNWTO 2004).

9.3.1.1 Environmental Impacts (Negative)

- **Erosion**: Erosion is adversely being regarded as negative impact of mass package tourism, especially when higher level of visitors' influx in destinations. Their coming is greater than environments ability to cope with, normally due to the number of tourist exceeding the carrying capacity of the destination or attraction. This results to the overcrowding and consumption of the resources. The uncontrolled mass package tourism activities bring about potential threats to many natural areas. This can put enormous pressure on an area and lead to impacts like sand erosion. Construction of hotel, leisure and recreation centers all consumes sand dunes which serves guards to protect the seashore.

- **Pollution**: Transport by air is continuously increasing in response to the rising number of tourists and their greater mobility. Tourism industry in its mass form; now accounts almost about 60 percent of air travel and it can contribute to several air pollutions. The uses of rail, car, bus, and coaches are increasing in respond to mass tourism and cruise transport, as well. All of these sources of pollution are rapidly growing as tourism is developing and increasing. Tourism has brought great negative impacts in causing annoyance, stress, and even hearing loss for humans. It causes distress to build life and can cause pollution from air emission, noise, soil waste and littering, oil and chemical pollutions, and even visual pollution. Sustainable tourism initiates fund and certification for green travel, as well as awards for eco-friendly accommodation providers.

9.3.1.2 Economics Impacts (Negative)

- **Seasonal Jobs**: In most destinations, tourism industry operates on a seasonality basis. Most often, it serves for only six months and the other six months off season. In the off season, most people working in tourism industry do not and cannot have any other work to earn living. They cannot sustain comfortable life because they cannot keep the six month wages for all year round. During this off season, they always facing as compare to those people having permanent and long year round position. As consequences, people who rely on tourism found to be high numbers in the society. Most of the young boys stop school and all the other work to support their living. Unfortunately, some of them stray on the beach, go around the hotel begging from tourists, or else to bég for money. This has resulting a big number of young people dropping out from schools.

9.3.1.3 Social Impacts (Negative)

- **Sex Tourism**: Sex tourism is one of the negative growing impacts of mass package tours within the industry. Prostitution is getting more rampant and obvious. Tourists follow more blacks for sexual purpose. They are influencing young children into this system because of money. This system is exploiting our young children and youths in doing prostitution or doing sex business, even it against most religions and cultures. Sustainable tourism has being an end to such acts underlying cultural awareness program.
- **Drugs**: Local people use drug and abuse it. It is common in the tourism industry. Most of the tourists involved in drug use usually come with it and introduce it to local young people in destinations. Most of the youths hang around the hotel and use it through the influence of tourists. Drug abuse is becoming an alarming issue in destinations, government and local policy formulators are laying down strict laws and fines of anyone dealing with such substance during their vacations. In certain countries, like Malaysia, Singapore, and others, people found possessing

with certain weight of drugs can be put into death. It is very scary scenario but people are taking easy as drug selling can earn huge amount of money.
- **Cultural Exploitation**: The attitude of tourists like prostitution, drug abuse, inappropriate dress conducts, can lead to cultural exploitation. It is because, if it is against the cultures, norms, and customs, it is not protected and preserved for the future generation, all will one day be lost. The authenticities of the local cultures are depreciating due to the commercialization local cultures to meet the demand of the tourists mostly are modified to suit the taste of the tourists.
- **Inappropriate Dress**: The display of social affection has been an issue in destination, couple with inappropriate dress, and nude appearance at beaches, most of the tourists, especially females are dressed full-naked or semi-naked, or almost look like naked. They lay under the sun on the beach. They can sometimes found in town with similar kind of dress which is actually against or forbid the local or/and traditional culture of most societies. It is in-fact violate religion practices in some destinations likes Muslim communities in Saudi Arabia, Brunei, Indonesia, Malaysia, Africa, and Muslim countries elsewhere. They are concerned with daily dress codes. Unfortunately, it has affected the future young generation and can influence others in tourism industry.
- **Public Display of Affection**: The display on affection of parts and many activities and events done by the tourists in the tourism industry can affect public. The place taken of random activities in the industry can affect the public directly or indirectly. Offending local customers and religious, most activities of tourist in the industry, can affect local customs and religious. Activities and habits of European like prostitution, drug abuse, in appropriate dresses actually offend local customs and religions.

9.3.2 Advantages of Tourism

Tourism is an experience of charming, appealing, and sterilized expected travel to destinations. However, it is an experience of tourism philosophy on the pathway of ancient remains, environment and economic benefits for the community. Sustainability is incorporating local communities and local assets of physically value astonishing reserves, countless of such resources being bare to the respect of tourist in the plentiful exhibition halls and center, traditional houses, and museums, mostly managed and controlled by the national center for arts and culture in destinations. The nations realize the potential gains of tourism. This is one of the motivations why many emerging economies see the growing necessitate of globally acknowledged initiatives of tourism that appeals to the economy, culture and environment, excluding locally tailored principles of sustainability in supporting local people in destinations. Nations like Taiwan promote the aborigines, Kenya promotes the Massai Mara, and South Africa's Zulus dance. Advantages of Tourism could be environmental, economics, and social impacts positively.

9.3.2.1 Environmental Impacts (Positive)

- **Conservation**: The quality environment, both natural and manmade, is essential to sustainable tourism. However, tourist relationship with the environment is complex; many activities can have adverse environmental effects. Many of these impacts are link with the construction of general infrastructure like roads, airport, and of tourism facilities—including resorts, hotels, restaurants; shops etc. Sustainable tourism has benefit effects on the environment by contributing to environmental protection and conservation and it is a way to raise awareness of environmental values and it has served as a tool to finance protection of natural areas and increase their economic importance.
- **Preservation**: Sustainable tourism is essential—directly or indirectly—in preserving the natural environment like forests, wildlife, nature parks etc. It embarks on series of environmental awareness programs that helps to call the attention of the international committee on such features and how they can be maintained for the future generations. Sustainable tourism industry has great impact in preservation of natural environment by increasing the environment values and financing protection of natural area as tool which can even create an economic importance.

9.3.2.2 Economics Impacts (Positive)

- **Jobs for Local People**: In many destinations, the tourist industry serves as an industry where many people provide with job opportunities to enable them earn for their living. Many resorts are built accommodations like hotels and lodges. It provides direct employment for the people associated with occupations in bars and hotels. This has increased in higher magnitude and even restaurants, bar, and hotels that creates job opportunities for many peoples. Self-catering services are taken places, game and theme parks, natural resorts and local bars are opened. Businesses around hotels, the creation of all these facilities appear to be sources for job opportunities and thus reduce unemployment in the country. Sustainable tourism in its pure form accounts for a significant contribution of job market for local people at destinations, alleviating the level of poverty.
- **Taxes and Economic Benefits**: Most developing and developed economy are fully relying on tourism. Firstly, many countries depend heavily upon travel expenditures by foreigners as a source of taxation and income for the enterprises. Ideally, tourism development is frequently an approach to support specific area for the rationale of growing business potentials through tourism as an industry. Lot of taxes are received from tourism industry. Every tourist will pay tax individually as an outsider. All organizations pay taxes to tourism industry like hotel, tour operators, travel agencies all pays their dues. Travel agencies and tour operators and airlines and all the others in tourism industry do the same. The government relies on these taxes to maintain country's economy. Tourism is now the largest

growing industry and the engine that drives the economy. In addition, tourism is the backbone of the destination's economy through taxes paid.

9.3.2.3 Social Impacts (Positive)

- **Supporting Rural Livelihood**: Sustainable tourism paves opportunities like regenerating handcraft. It is one of the importance impacts of tourism industry in the destinations, coupled with environment and awareness of the modern life, as well. Encouraging women run organization with the development of tourism, many tourism organizations open for business but many services are needed in the tourism industry. This brings about women running organization through the influence of tour-ism such initiatives are an example of entrepreneurial entities or small and medium size enterprises.
- **Language Opportunities**: The world has been a global village with more and more people becoming multi linguist and bi-linguist. Many people can speak couple different European language like English, Swedish, Norwegian, and Finnish. Other people can widely speak foreign languages in tourism industry because of their interaction with tourists and the interest that their work required. It is common in tourism industry people speak several languages that tourists can understand. For example, receptionists are bi-linguist or multi-linguist. Sustainable tourism facilitates the recommendation of people to learn about the basic greetings words to communicate with locals in destinations.
- **Cultural Exchange**: Many tourists now want to have cross cultural exchange and encounter with local people. This booms the event industry aims at practicing out cultures, thus aware of it existence. Other example, the Gambian would like to practices European cultures. In this situation tourist exchange their culture for Gambian and exchange for European cultures.
- **Education for Local and Tourist**: Sustainable tourism industry plays an important role in educating local people, communities, destinations and tourists. This brings the training programs of local people in most of the local societies to get the benefits and be able to preserve products like tomato paste, Garry from cassava, soap making etc. These activities support their livelihood ventures. Sustainable tourism has been a global issue which has left no stone on turn. The tourists also can train from them. To be culturally aware and understanding the local communities, we have to drag international funding aids. It encourages sponsorship for local students to acquire training programs and establishment of training centers in destination, most of which runs fulltime or part-time professional and services programs.

9.4 Planning and Development in Sustainable Tourism

Natural resource planning and development further enhance ecological design of resorts, scenery design and management. Since tourism has some positive and negative impacts, we should pay attention for carefully planning and development in sustainable tourism. Tourism is as being the act of management, planning and implementation of travel, tours, and tourism services in destinations. The main principle is not merely for leisure activities, but stipulation of services to tourists, as well as the promotion of mutual benefit among two conflict impacts, positive and negative.

In the following, all of these show the positive effect that tourism conveys for a destinations and the citizens. However, the auxiliary thoughts on the negative effects of tourism demonstrate to facilitate that tourism is catastrophic economically, environmentally, and culturally. In this section, we discuss planning and development from global to local perspectives.

9.4.1 The Role of World Tourism Organization in Promoting Sustainable Tourism

The United Nations World Tourism Organization (UNWTO) is a tourism body under United Nations agency. It is responsible for the promotion of sustainable and universal accessible tourism within the industry. Being one of the main pillars in tourism industry and a leading international organization in the field of tourism, UNWTO promotes tourism as a driver of economic growth, inclusive development and environmental sustainability. It offers leadership and support to the sector in advancing knowledge and tourism policies to many destinations worldwide.

The UNWTO encourages implementation of Global Code of Ethics for Tourism (GCET), to maximize tourism's socio-economic contribution while minimizing its possible negative impact. It is committed to promoting tourism as an instrument in achieving the Sustainable Development Goals (SDGs), geared towards reducing poverty, and fostering sustainable development. UNWTO generates market knowledge, promotes competitive, and sustainable tourism policies. They foster instruments, tourism education and training, and works to make tourism an effective tool for development through technical assistance projects in over 100 countries around the world. Currently, the UNWTO's membership includes 158 countries, 6 associate members and over 480 affiliate members representing the private sectors, educational institutions, tourism associations and local tourism authorities.

9.4.2 Sustainable Tourism in Less Developed Countries

Less Developed Countries (LDC) feature bustling seaside resorts, natural and man-made resources making the community well-known for sightseeing and cultural experience by the people. They experience it at very first time. It is also an important part of a culture, encompassing the countryside and natural environment including flora and fauna, scientifically known as biodiversity. Between the main rural settlement and urban town features is a thick tropical palm forest of baobab trees and cashews with a variety of animal life in Africa, science parks and national parks in Asia and theme and pleasure parks in Europe.

Sustainable tourism in less developed countries are aimed to extend the optimistic impacts, in the course of a special focus on management, of both the fauna and flora, fetching economic and environmental benefits for the local inhabitants, and the education of visitors who want to learn about nature.

Successful sustainable projects have being recorded in most regions around the globe, notably small and mediums sized businesses. The entities accomplish endeavors through the establishment of community based initiative projects boasting bee-keeping center, tie and die batik factory, carving and souvenir shops, women groups in cultural village and also working so hardly to have our own horticultural garden.

Sustainable tourism organizations award tokens; single approach of pleasing sustainable tourism initiatives that is accurately meeting the terms and standards as laid by the principles of sustainability with these aspirations is by benevolent awarding them with plausible exterior credit.

9.4.3 Sustainable Tourism in Local Communities

Sustainable tourism is a kind of tourism corporations and partnerships involving local communities in tourism operations, planning and management at local level; communities with heaps cultural heritage sites and biodiversity. They are encompassing the destinations, with enriching for sustainable tourism and ecotourism. Both are empowered and trained for sustainable tourism initiative, running and managing the accommodation and historical heritage sites. Local village settlements are encouraged to exploit tourism resources being at seaside, ecological features, historical heritage sites, and the cultural assets. Local lifestyle is completely adverse from the way of life in urban settlements. This makes things so interesting, people are still lives in tradition mud houses, and makes the sleeping room separate from the restroom and kitchen.

9.4.4 Sustainable Tourism in Rural Settlements

Sustainable tourism in rural settlements aims to have least unhelpful effects on the life, economy and environment of local people. In order to reduce destruction through the strengthen of trade and industry benefits of tourism, in both rural and urban settlements, people need to sustainably sightseeing experience, typically carry out in mostly safe and sound natural regions. People have a propensity to be mutually ecologically and socially sensitive. Tourism rural experience is much interesting to familiarize the local life in rural settlements. For African women, they wake up early morning, rush to the well to get water, cook and prepare traditional rice porridge make out of sorghum or corn powder for family breakfast. After breakfast, they go to their local gardens to grow crops like tomato, cabbage, carrot, bitter tomato, onion, potato, and other kind of vegetables. The culture of this beloved experience includes social values and traditions, practices, spiritual beliefs and artistic expression that interpret the backdrop of socio-economic, and ethnic values of the rural settlements.

Tourist can join the early morning trip to the market make shopping for the meal, cook lunch for the family. The whole family will join hand in one big basin before using in a big calabash to eat lunch, Traditional believe quote says when the whole family eat together there will be "no extreme dislike" in the family.

9.4.5 Sustainable Tourism in Tourist Contribution

The major contribution of tourism and tourist in every destination are to encourage the use of public transports, using local taxis, buses, motorcycle, and bicycles. They are mostly available for rental in the destinations. It can also be booked or reserved through retail travel agencies. The purchases of these items should be done with consideration of buying the materials that support local people, as well as to contribute to social livelihood of the host communities. Is it the way that tourists think, act and speak (Fig. 9.1)?

Attitude can be of a significant influence to the lives of the hosts, however, only if you want to change it. Attitude to host communities during vacations have to be properly checked in the way and manner. We consume water and energy in destination, most of which deter local from such facilities. Do always switch off your lights, air-condition, and other electrical appliances, if not in need of them, or else choose accommodations and services that use alternative energy.

Attitude towards local people in destinations can be spotted through your appearance. It is indeed, how proud you feel to be associated with the locals. People who make effort to look presentable, by widening their connections and tolerance, with each other will generally feel positive and welcome by the local communities—in developed and developing destinations. For sustainable tourist in search of responsible, travel in one of contributing and experiencing your vacation is to think globally

Fig. 9.1 Environmental protection and development as being a cause of concern, regarding waste disposal tourist purchases, should be confined to materials needed to avoid environmental littering. In this case, I did not see any littering in Switzerland (Photos Wei-Ta Fang)

and act locally. In doing so, there will be a raise in your awareness on the culture, norms, and values of the locals.

9.4.6 Sustainable Tourism Certification

Sustainable tourism organization does awards qualifications as a form of official recognition. This means of motivation in so doing it's an encouragement for their long term successful aims for sustainability, and here we describe how it works. The awards are in the form of certification for recognizing the effort and services made by enthusiastic community groups, charitable trusts, foundations, accommodation providers, and tour operators, all of which operates with the goals of sustainable tourism. Considering sustainability in tourism and tourist experiences the activities are sold by tour operators either independently and as well on package tours, sustainable tourism certification is done looking at the triple bottom line in three main aspects:

- **Socio-culturally**: Sustainable tourism relates on the doings that does not hurt the local traditional legacy of the local communities, and rejuvenate the community formations of way of life of the community. They are done in the form of traditional musical entertainments, food, dance, music and musical instruments, helping training through teaching the children in our cultural values.
- **Environmentally**: The ecological features of destinations serve as the pull factor, attracting tourist to discover destinations, sustainable tourism activity curtails any harm to the surroundings. Its ecological features comprise of biodiversity that constitute key elements for tourist visitation. This idyllically proceedings will promote environment protection and preservations programs in destinations. Annual tree planting programs are done in destinations, prohibit of killing and hunting of endangered species, are considered crimes in destination. They will definitely be fined or penalized when found guilty. Pleasure garden and other leisure centers are created in destinations. Wide tours are conducted to experience wildlife spotting and game drives. Kenya, South Africa, Tanzania are well famous for the sustainable tourism initiatives and conservation of the nature, drawing wider audience of tourists across the globe for safari tours, ornithological tours and game drives.
- **Economically**: Sustainable tourism adheres for economic supports, enhancing the livelihood of the local people and communities, excursions, souvenirs, entertainments and other tourism product and service offered in local communities are geared towards generating economic proceedings for the local. This will prolong to the payment of the economic welfare to local community alleviating poverty. Small and medium size enterprises are run in destination by entrepreneurs, providing job opportunity to locals, thus provide financial to the investors and development for the community through the consumption of the resources.

9.5 Sustainable Tourism in Recent Era

The modern growth of tourism development globally, philanthropic the impression of comprehensible vision is in progress expansion of tourism. In both developing and developed nations, it is subsequently a pro-growing paradigm. By yearly, the tourists increase in arrival quantity. The painstaking is a gauge of achievement and careful sign of accomplishment of understudy realization of affluence convey to underprivileged local residents. Sustainable tourism has been the top agenda of the development issues. It is perceived as principal to supervision of possessions in such a manner that appeals to the destination socially and economically. The desires of tourists are met and satisfied whereas upholding the enriching of cultural authenticity, vital environmental development, natural multiplicity maintaining of the endangered species.

9.5.1 Tourism in Local Constraints

Today, sustainable tourism in its purest form has underpinned tourist experiences on the triple bottom line. It is pleasurable, particularly for the excursions situated in exterior of big metropolitans. Over here, the excursionists boast opportunities to get together and cooperate with native residents. Accommodation centers and other facilities are located in rural areas. The excursion is conducted and family visit is supervised by the community council of elders, namely Village Development Committee (VDC). The overnight accommodation, excursions of countryside, or ecological attractions rely on the high-quality determination of their local community for their continued existence. Statistics have shown those tourism initiatives do not do well to local residents. The appreciation of natural scenery possessions often drops down. The shore up of society contributions for its sustainability has resulted the bureaucracy and the community having conflict in the project. Local residents and home-grown community populations, whose common structure is attractiveness for tourists, stand as natural preparing assets sighted by tourist on organized excursions, either in groups or independently.

9.5.2 Tourism About Cultural Identities

The UNESCO World Heritage Sustainable Tourism Program focuses on linking up national and local authorities. The tourism sectors and local communities should create an integrated approach to sustainable tourism within the context of the World Heritage Convention. A critical element of this initiative is providing stakeholders with the capacity and tools to plan develop and manage World Heritage responsibly and sustainably, based on the local context and needs.

Development and delight of this kind of tourism is not merely maintaining the heritage sites and cultural features. It can even increase and improve the image of the destination whilst local people have pride in their culture as well they have a better living. Offering challenging ideas and performance of traditional activities result in a prosperity and trendy of local culture. Tourists contribution in the well-being of local citizens, sustainable tourism enlarges the scope of admiration of affection visitor attraction, will be motivating locals to return roots of cultural diversity.

Sustainable tourism experience boasting cultural uniqueness, diversity in dress, music, food, and dance, will suitably be reconsidered. The acceptable approach will make it understandable to everyone approaching them through the showcasing of such attributes in its sustainable way. Hence, it offers an impression of lively performance, authentic and natural beneficial and proficient contribution for the satisfaction of tourism felt need on environment. Commencement of sustainable tourism initiatives starts at their doings of tourism business in destinations. The philosophical study needs to be conducted and developed by a specialized consultant, either locally or by a foreign consultant. Strong pulled destinations will abide and devote upholding cultural uniqueness, natural beauty and economic concern shall be upheld by all stakeholders.

9.5.3 Tourism for Concerning Stakeholders

The involvement of consultants as well as all stakeholders in sustainable tourism development is highly required as they enlighten local entrepreneurs and investors about the dos and don'ts. Based on such stipulation development agendas and plans simultaneously are linked to diverse affiliations functioning within the tourism industry in various destinations. Periodical era of exertion are crucial tourist development and administrations, essentially to be specific, measurable, achievable, realistic and timed backed by obvious objectives, technical know-how on how to accomplish agendas back. This will be done from the support and involvement of the local communities, acquiring local supports, coupled by the living cultural and environmental assets, and tools. They are used in the development and growth of tourist to a new dimension, increasing responsiveness.

Sustainable tourism initiatives in destinations do local training citizens to the uppermost stage of all folks contributing within the progression of sustainability, precisely: local communities, investors, small and medium scale stakeholders, youths and women groups. With aspiration of constructing a brighter prospect of tourist in searching the need of present tourist likewise is preserving it for the future generations, through the preservation and conservation of culture, environment, and economic livelihood of the locals.

In its purest form, sustainable tourism is the likelihood of development, joint collaborating locals and tourist administrators in development programs. Therefore, it is important to produce an integrated offer, lessening interactions to general travellers and cultural initiatives. Marketing and promotion of destination cover wider range of benefits. When a destination is advertised to potential travelers, it is equally presented its attractiveness of tourist attractions, craftworks, local cuisines, its atmospheric ecological features, and added with popular events of which tourist might engage in during their vacation. Portraying destinations to tourist with wider ranges of activities to suit their interests, for the potential of assimilating sightseeing inside the range of sustainable growth of tourism, for restoring a locality's true identity, its historical, architectural, ecological, agricultural heritage, and thereby what is to come is people to have a look.

Box 9.1: Sun Moon Lake in Taiwan

Taiwan is located in the western Pacific Ocean about 160 km off the southeastern coast of Mainland China. As one of the islands between Japan and the Philippines, this island straddles the Tropic of Cancer between 124°34′E and 119°18′E longitude, and 21°45′N and 25°56′N latitude (Fig. 9.2). With an area of about 35,961 km^2, is a little larger than the combined states of Massachusetts and Connecticut, or is about the same size as the Netherlands.

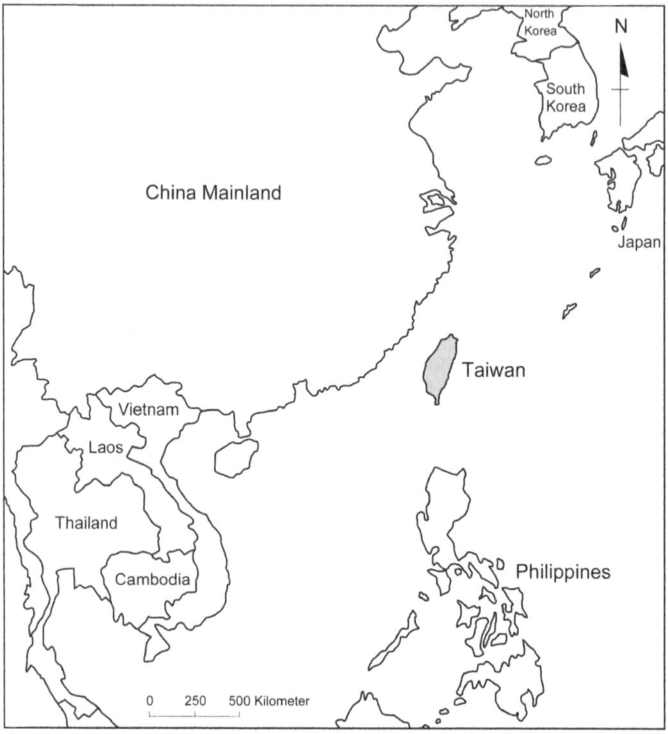

Fig. 9.2 Location of Taiwan (Illustrated by Yi-Te Chiang)

Recently, Taiwan is home to wider variety of tourism developments and projects. Being an island destination, Taiwan has something to offer on both interior and exterior. This magnificent destination is endowed with vast lots of natural and artificial attractions. They are ranging from aboriginal cultures and pestle dance at Sun Moon Lake, to the wider collection wild animals at Taipei Zoo and Hsinchu Zoo, all under the sustainable management of local people. Tourism in the current century is geared towards authentic experiences of local life of the destinations. When and How? Both domestic and international tourisms are fascinated with local lifestyle of the host community. As for example, it is the largest turnout of people during night markets of Taiwan. People like to taste local traditional cuisines, like the stinky tofu and "shabu shabu" restaurant.

Famous for Chinese tourists, Sun Moon Lake is situated in Nantou County's Yuchi Township near the center of Taiwan. It is a true example of sustainable tourism project in the Asian continent, lying in the magnificent island of Taiwan. The land possesses unique features of pristine attractiveness of the natural environment supported by abundant enriching historical legacy and cultural diversity of the aboriginal peoples. These aboriginal peoples still maintain their cultural authenticity of the region (Fig. 9.3).

Fig. 9.3 The exotic island of Sun Moon Lake has attached a sense of pride between the traditional communities, ecological features to the modern state of earth tourism, giving tourist a unique reward experience, in such a way that appeal to the environment and economy (Photos by Wei-Ta Fang)

The island has been an excellent point of conserving the forest cover of Taiwan. It has been preserving various ecological assets; ranging from plants to animal life. The birdlife is second to known in Asia. Taiwan is a paradise for migratory birds during winter season. Sun Moon Lake is homed to most of the species. The cultural feature of the pestle music is still in its pure form. Various tours are today sold to eco-tourists. Cycling tours, hiking tours, boat trips, and natural trails are offered at the lake. The aboriginal people who live around the lake benefit from the proceedings of the lake. They, the tourists, pay some amounts during the visits to their communities, promoting cross-cultural exchange, and social interactions among local people. Local people who live within the vicinity of the lake benefits from employment opportunities, the showcasing of the local cultures generate fiscal remunerations for the habitants.

9.6 Conclusion

Sustainability do focus on the experience of natural area. They are also fostering of environment and cultural understandings among locals and tourist, paving a wider appreciation for conservations, enhancing the linkages between tourism businesses and poor people; so that tourism's contribution to poverty reduction is increased and poor people are able to participate more effectively in product development.

Sustainable tourism development is a pattern of resource use that aims to meet human needs while preserving the environment. These needs can be met not only in the present but also for generations to come, if we are aware with what is going on to all everyday activities.

Benefits of sustainable tourism in every economy cannot be overestimated. An often-quoted definition of sustainable development is defined in the report as “development that meets the needs of the present without compromising the ability of future generations to meet their own needs”. Looking at from the above mentioned quotation, you will realize that sustainable tourism embraces proceeding of three key areas. They are all of significance interest to the global world, as a millennium development goal to achieve environmental suitable sites, with high economic statues, and authentic cultural assets. The sustainability of these features have been one of the primary assets of the developing nations. They are working tirelessly to promote sustainable tourism initiative which has today being one of the main aims of tourism development and global vocal point for policy makers.

An overall loom to tourism development, the enlargement and administration aspired at generating prospects for the local communities and residents. With those, they will achieve profit from tourism that leads to the management of all resources in such a way those economic, social, and aesthetic needs can be fulfilled while maintaining cultural integrity, essential ecological processes, and biological diversity and life support systems. The main thrust is on environmental sustainability. Social benefits are just one aspect of sustainability.

Sustainability adheres to instrumental constructing of capacity building pro-gram of the locals. Hence, they can entrepreneurial minded in operating their own small scale enterprises in a sustainable manner. Conducting policies that safeguard the interest heritage and remaining site as visitor attractions, constructing carrying capacity to the attraction sites, raising awareness program at local level with code of conduct, the local and visitors are supposedly working together harmoniously. They will be employing locals in resorts as tour guides, resort representatives, supervisors and managers gains of which stays within the community.

Accommodations, attractions site, tour operators and destinations are currently operating on small scale to be sustainable for long term. This will initiate intended growing in size substantially. It is also for the measurement of behavior change (in response to community based initiatives) which would lead to improve the culture and promote entrepreneurship to comprehensively reduce unemployment.

References

Bramwell B, Lane B (1993) Sustainable tourism: an evolving global approach. J Sustain Tour 1:1–5
Drost A (1996) Developing sustainable tourism for world heritage sites. Ann Tour Res 23:479–492
Goulet D (1971) The cruel choice: a new concept on the theory of development. Atheneum, Berlin, Germany
Hunter C (1997) Sustainable tourism as an adaptive paradigm. Ann Tour Res 21:850–867
Leiper N, Harris R (1995) Sustainable tourism: an Australian perspective. Butterworth-Heinemann, Oxford, UK

Mowforth M, Munt I (2008) Tourism and sustainability: development, globalisation and new tourism in the third world, 3rd edn. Routledge, New York, NY, USA

Murphy P (1994) Tourism and sustainable development. In: Theobald WF (ed) Global tourism: the next decade. Butterwoth, New York, USA, pp 274–290

Owen RE, Witt SF, Gammon S (1993) Sustainable tourism development in Wales: from theory to practice. Tour Manag 14:463–474

Todaro MP (1994) Economics for a development world. Longman, New York, NY, USA

Tosun C (2000) Limits to community participation in the tourism development process in developing countries. Tour Manag 21:613–633

Tosun C (2001) Challenges of sustainable tourism development in the developing world: the case of Turkey. Tour Manag 22:289–303

UNWTO (1995) Lanzarote charter for sustainable tourism. UNWTO Declarations 1995(15). UNWTO, Madrid, Spain

UNWTO (2001) Tourism satellite account: recommended methodological framework. World Tourism Organization, Madrid, Spain

UNWTO (2004) Sustainable tourism in protected areas—guidelines for planning and management. World Tourism Organization, Madrid, Spain

Wang S, Xu H (2015) Influence of place-based senses of distinctiveness, continuity, self-esteem and self-efficacy on residents' attitudes toward tourism. Tour Manag 47:241–250

Chapter 10
MICE Industry

10.1 Introduction

Today tourism is account for its high quality, and high volume. It ranks number one among the global popular uprising diverse businesses, like MICE. The abbreviation, MICE, is defined as the combination of Meetings, Incentive Travel, Conferences, and Exhibition. It is a new trend that describes an emerging industry of the travel and tourism industry, more especially in the event industry (Schlentrich 2008). MICE Industry is an organized category of tourism where a large number of people in groups travel to destination together. In discussing certain key topics of their interest being it national issues, trade and industry, concern on academic topics, it has contributed to the development of tourism industry around the world. The events are designed glowing in progress, with specific theme, aims and objective for some particular purposes. MICE comprises of diverse events all of which operate on different objectives, time and dates.

10.1.1 Definition of MICE Industry

The MICE Industry has various definitions depending on which way a writer will understand it. The MICE planning involves the process of marking plans for Meetings, Festivals, Ceremonies, Incentives, Conventions and Exhibitions, planning are the acts and set of policies, and strategies for particular events. It covers a wider range of programs starting with date and schedule preparation, budgeting, site selection, site reservation, permit processing, transportation, accommodation, health, and safety checks as well as contingency plans. The importance of MICE has created big positive impacts of the tourism industry. It fosters unity among industry stakeholders resulting in the mutual exchange of ideas and corporations on projects. Through, the hosting and participating on events are the industry stakeholders; organizations integrate and build partnerships together on the promotion and advancement of specific

W.-T. Fang, *Tourism in Emerging Economies*,
https://doi.org/10.1007/978-981-15-2463-9_10

goals. MICE cover the assortment of methods, strategies, imitation, implementation, staging and launching of the events. In the following sections, we discuss the definition of MICE Industry.

10.1.1.1 Meetings

According to various scholars and experts, meetings are classified as gathering of people, for specific purpose in addressing a common concern of the attendees, target group or a general public. The number of participants may range for two to many. That has been assembled for the reason of achieving a common objective in the course of oral communication, like allocation information such gatherings take places in diverse form being it face to face and as well through the use of modern day technological facilities at our disposal such as, video conference, Skype call, Yahoo, MSN and telephone.

Listed below are several types of meetings for this area:

- Forum
- Clinic
- Seminar
- Workshop
- Symposium

The overall purposes of organizing events, such as meetings, are becoming common phenomena trying to the venue of the following:

- Meetings have been venues where business entities demonstrate their new products and services they have to offer to general public.
- Increase of awareness on specific issues, products, and services are the core elements of meeting industry. It is a forum where different people with diverse thoughts gather to share their perception about the topic (agenda) of the meeting, like a scientific committee meeting (see Fig. 10.1).
- Today, the meeting industry brings together people of different caliber with the purpose of exchanging ideas and views.
- Different meetings have different motives among the objectives staging meetings, raising funds constitute one of the main elements of most meetings happening around the world, and funds are raised from the registration of the participants to sponsorship aid from organization who donates money or money for them to be promoted.
- Delivery of materials like journals, reports, and statistical data, are among the few evidence released during meetings. The delivery of academic journals is done in educational meetings, while business, health, government meetings present reports, resources, and equipment during such gatherings.
- With the modern psychological view of people where everyone is liberal in their own thinking and reaction towards issues, meetings have become important asset to encourage people about issues, and meeting at places to substantial messages.

10.1.1.2 Incentive Travel

Incentives are rewards given to staff (employees) of an organization for achieving a specific objective, mostly as the best staff of the year, graded on performance, personal appearance, hygiene, customer satisfaction, and overall the dedication and hard work at the office. Today, incentives are awards given in a form of travel package to destination for short and weekend break at the expenses of the company.

10.1.1.3 Conference

A conference has a broad definition depending on the type of topic to be addressed. The topic is in simple explanation. The conferences are meeting of people to talk about specific subjects. Today, various conferences will address on academic, trade, security, government, and other related interest. This will draw participants' attention at local pivot being domestic and international.

Fig. 10.1 Before conference, we organized scientific committee meeting for the 10th INTECOL International Wetlands Conference in China during 2016. The International Association for Ecology (INTECOL) supports all collaboration in the international wetland studies (Photos by Wei-Ta Fang)

10.1.1.4 Exhibition

Exhibitions are common phenomena but formal events. They gather people together on prearranged arrangement of shows, witnessing the display of a selection of items being commercial, arts and others. In practice, exhibitions usually occur within museums, galleries and exhibition halls, and World's Fairs.

10.1.2 Origin of MICE Industry

In discussing certain key topics of their interest being it national issues, trade and industry concern of academic topics, MICE has contributed to the development of tourism industry around the world (UNWTO 2012). It usual events are designed glowing in progress, with specific theme, aims, and objective for some particular purpose.

Historically, the development of MICE industry can be traced back during the gathering as far back as in the ancient era for the purpose of attending meetings, conventions, and expositions since the ancient times. They are mainly for social, sporting, political, or religious purposes. Notably, *Bible* comprises of various evident with mentions of the merchant trades (Publishers et al. 1995). From the ancient civilization of ancient people to the modern day third generation people have uphold ceremonies. This will portray the sense of meetings for discussion of communities and other social welfares, tradition festivals, shows and carnivals. They are used to promote high sense of mutual understanding and relationships among people of diverse ethnicity. MICE otherwise is called special events as defined by Goldblatt (2002, p. 6). He further defined MICE "as a unique moment in time celebrated with ceremony and ritual to satisfy specific needs" at the commencement of the meetings where the early religious gatherings of Mecca, Rome, Canterbury, and Santiago de Compostela occurred.

Publics have assembled to join gatherings, treaties, and discussions since the ancient times, primarily for the purposes of sporting, political, social, or religious. Following the world civilization from the modern day of Egyptian, countries started to organize conferences, discussed on issues and development agendas, whilst academician offered it as a forum for knowledge sharing, with the industrialization of the globe trade, prosperity become a noble venture organization, and business companies started to give incentives as a token of appreciation and staff motivation for the excellence job performed. Meeting, Incentives, Convention and Exhibition have the following parts of ancient day gatherings to the modern day travel and tourism event industry.

10.1.3 Structure of the MICE Industry

The structure of the travel, tourism and MICE industry consists of various organizations that form different components of the industry. They work together as an umbrella body towards having the overall event and tourism product in meeting the demand of the travel, tourism, and event industry. These organizations work very close-linked operating together in different sectors to the form the MICE industry to fulfill their customers' needs and expectation. The structure includes the following.

10.1.3.1 Private Sector

These are commercial businesses that directly or indirectly support meetings, incentives, convention, trade shows, and exhibitions. They operate as ancillary service providers in the travel and tourism industry and are mostly operated by private individuals who are highly motivated. The current trend in the industry, which gives them motivation and courage, has started up business in meeting the service requirement of the travel and tourism with aim solely to get profits. Today, there is numerous meetings and event planners across the globe organizing and staging different kind of events, ranging from cooperate, private, to mega events. Such private bodies include:

- **Travel Insurance and Finance Services**: The structure of the travel and tourist industry is concerned with persuading and assisting organizations to cross the nation boundaries in promotion and development of goods and service satisfying the customers. This also is to ensure their welfare, in recognition travel insurance are made available to both organization and customers for compensation in case of damages and loss of personal belongings. The organizations also offer loan on credit base to the needy with a vision for them to get publicity for a profitable operation of travel and tourist business. Travel insurers provide covers for goods, services and people. Money exchange bureaus are available for tourists who want to change their money from the hard currency, the local currency is used for payment of goods and services in a destination. Today insurance companies cover risk and damages on conference and other gathering facilities and personal belongings.
- **Marketing Support Services**: Travel, tourism, MICE products, and services that go into the market by delivery to the consumers have to undergone some forms of marketing. They are sending the information about products and services to the right people, at the right place, and the right time. They are also keeping informed the customers by having access to competent quick and easy access to information on the products and services through internet. The booking and reservation are made easy. They are based on effective marketing to customers. They help customers to get wider access to the overall MICE products and services with ease and convenient. The marketing sector services consult organizations in the structure provide them with marketing strategies.
- **Travel Trade Press**: With the structure, you have travel trade press through organization. They are set up and provide up-to-date articles and documentary on the

event industry, the event happens both of the past and the future, news in the form of newspapers, magazines, television shows, and Gazettes. News press informs potential customers and investors about the current trends and issues affecting the industry, as well as about the future programs in places, example trade fairs, conference, and symposiums and other travel and tourism related issues.

10.1.3.2 Producers

These are the main providers in travel, tourism, and MICE Industry. From the structure, they sell products and at the same times advertise the products and special offers to supply customers they want. Producers produce and distribute the supply, information and guides of the travel—tourism products and services like hotel, accommodation, attraction at the destination, and the mode of transportation from their originating place to the final destination and back. These services are sold to tourists either through direct, travel agents, or tour operator who are best describe as wholesaler of tourism. These include:

- **Carriers**: These are the transport providers in the structure. They are those who operate major form of transport. They take people from one destination to the other. This comprises the airlines, cruise line, rail companies, road transports, and coach hire. Transport providers will take tourists from one destination to another destination. The aviation industry has eased the movement of people of third generation, mostly local and international airlines for facilitating transportation. Event planners, today, consider the convenience of accessing the destination that constitutes the vital element for destination market organizations. Airlines operations might be on chartered flight moving at any time when they get passengers or scheduled completed. They can also operate using on fixed timetable from regional and national airports. All operating in various cities will link event delegates and participants to their receiving destination.
- **Accommodations**: Accommodation is a fundamental part of both domestic and international tourism. Travel and event products are being major parts of the event, plus the audience decision of whether or not to attend a specific event. The primary purpose is to allocate the availability of accommodation, and, to provide it sufficiently to everyone is another matter. The accommodation often includes some forms of meal arrangement, can be either the bed and breakfast, half board or full board—a form of all inclusive board. The provision of overnight accommodation for people staying away and the provision of sustenance for people eating away from home are important. The accommodation types are large variables, based on location, market, and ownership. They are small family run businesses or multinational chains. All of those provide different level of services in meeting the demand of the industry. These are all logistics prearranged well in advance before participants and delegates arrivals. This can be either through event organizers or independently through booking at a convenient accommodation.

- **Attractions**: The attractions, whether a tourist is visitor or local, will experience attraction in the destination. He or she will play a great importance in MICE industry. A tourist will be attracted to the sightseeing will be motivated in his or he traveling. This will be based on the destination which can offer particular features, being it manmade or natural. In fact, it is a reason for tourists in paying a visit to destination, thus joining that specific event of cultural and educational reasons. Attraction has made it possible for everyone involved in the event planning and organization to appreciate how important is the attraction in tourism. Nature will attract people to make it possible, open opportunities in transport, catering, accommodation, guiding, souvenir production, and so on. Attractions are classified based on primary purpose of which to allow public access for entertainment, interest or education rather than being a primary retail outlet. These can be ancient monuments, stone circle, Theme Parks, Disneyland, Amusement Park, stately homes, castles, temples, historical buildings, wetland, and so on.

10.1.3.3 Public Sectors

Authorities made many decisions in MICE development. These are the policies and planning initiatives taken at national and regional level ensuring that some of the more tangible aspects of the MICE overall products are readily available to the consumers. National government often set policies that stimulate MICE development plans and implementations. They coordinate the developments that are taking place in the country.

Today, many countries with an established MICE industry increase national income by having a separate department of the government. They will be responsible for the MICE development programs and National Tourism Organizations. They do the marketing for the destination, research and corporate planning on product developments, finance and administration, and also setting the quality control and licensing body. Many of these organizations are non-profit making. They set up provides services to supports local businesses and visitors to an area.

- **National Tourist Organisations**: These are bodies set by government in having a framework to regulate MICE growth. It monitors products and services, as well as development of promotions, compulsory registration of accommodation, and also other services providers. The major players in the Meeting, Incentive Convention and Exhibition Industry (MICE), which makes the industry to lay down the sector on solid foundation, are the Convention and Visitors Bureaus (CVBs), meeting planners and their clients, convention centers, specialized services, and exhibitions centers, tourism planners and administrators. In addition, a national tourist organisation regulates the quality of hotels, transportation and others by making they have the license of operation. They pave sound polices and agendas for MICE promotion and developments. Government in the countries heavily depends on tourism. They ensure that those coming into contact with tourists have the necessary skills to deal with them. The staffs must be competent, polite, and friendly as they are the first impression to visitors. Therefore, it is important to the country's long term image in tourism industry. Moreover, it is strongly applied to the MICE industry, through the certification and training of MICE professionals.

- **Regional Tourist Organizations**: These are sub-bodies of the national tourism organization. They involve directly with providers within their region. They also promote their regions for customers to be attracted through special events and occasion that might be of interest to tourists. They have information centers, conference and exhibition halls, centers with accommodation, attractions, and restaurants for event planners. The more famous they are, they become easy to attract more tourists.

10.2 The Importance of MICE Industries in a Destination

MICE Industry, is visualized as the youngest fastest, growing industry in travel and tourism industry, has being extremely large in expansion over the years. Meetings, Incentives awards, Conventions, Conference, Expositions and Exhibitions, are the order of the day, happening in every corner of world. It is based on outstanding economic significant, national agencies, and taking up the lead for the development and promotion on the MICE industry. Today, the most widely observed and stage events of the MICE industry around the world is through conferences, held in small cities and help in mega venues. Conference are mostly staged and organized to serve in the form of forum for national development, trade shows, exhibitions and bilateral cooperation to academicians, thus gaining reputations and growing in high interest among the cosmopolitan population of America, and Europe to the metropolitan regions. The MICE industry is differentiating by four high growth potential of diverse dimensions, like economical, educational, and social benefits.

10.2.1 Economical Benefits

The meeting, incentive, convention, and exhibition (MICE) industry have been dynamic business for so many years but widely perceived as event industry. The MICE industry, a service business sector that has been uniting together of different sectors of trade, transportation logistics, investment, and the movement of people from one destination to another. It has a vibrant business, which have been dynamic in Australasia, Asia, North America and Europe for over a century. All supports implementation of MICE initiative and serve in some aspects as development goals to be achieved in high added-values, and extremely valuable novelty. The great benefits yield by the MICE industries are the immense benefit of generating job opportunities for locals and international natives in destinations. The diversification of the industry brings attention of the international experts to come up with the short form of word namely MICE, Today, MICE has been a vibrant economic sector fetching financial support for economics, as well as raising awareness about destinations.

Various international and multinational organizations are dedicated to the welfare of the overall MICE industry. They formulate some guidelines and regulation of what constitute a conference, meeting, exhibition, and so on. The International Congress & Convention Association (ICCA) is one of the organizations dealing with affairs of MICE. Statistics have shown that a total of 12,558 international association meetings held across the world annually (ICCA 2017). ICCA is coupled by the Union of International Association (UFI) which formulates criteria of what should comprise to be at a conference, the duration of the event, and the number of delegates or participants.

The UFI reveals that once a year, the MICE industry yields significance economic benefits for governments and private nongovernmental organizations records. The UFI records show that about US$1.16 trillion (including US$400 billion are for conferences and US$760 billion for exhibitions), of financial proceedings are incurred by the MICE industry, specifically through exhibitions (UNWTO 2014).

MICE may perhaps be added up on to convey giant trade, industry and financial advantages to the countries and cities that host them by meeting the requirements needed by the organizing committee and as well winning the bid as most event are now concurrent of bidding system were the winner become the host of the prior event by hosting it, destination complete with having well-structured infrastructural and super-structural developments.

Box 10.1: MICE Industry Development in Asian Continent

Emerging Asia demonstrated the most remarkable growth in industry development on manufacturing and services, such as Southeast Asia, China and India (OECD 2018). Countries around the globe are developing and formulating sustainable development plans for economic growth and attracting foreign investors, as well as tapping funds on certain projects. The promotion of genuine administrative guidelines is initiated to help and achieve those aspired collective national development goals. In so doing, Asia is not left behind. As developed and developing countries are globally cheerful about MICE industry development, exhibition space will continue to grow in the years ahead. This is particularly The MICE industry in Asia-Pacific, where the revenue of $229 billion in 2017, and is expected to grow to reach $441.1 billion by 2025 (Sable 2019).

Sable pointed out in her Jan 2019 report Asia-Pacific MICE industry by the momentum of production relocation to Asia from the rest of the world. The Asian meetings made over US$159.4 billion in revenues of the Asia-Pacific MICE industry in 2018. There was a growth in the quantity of exhibitions and a growth in available exhibition floor space in China, Hong Kong, India, Taiwan, and Thailand.

Local investors have demonstrated a giant milestone in the successful planning and administration of MICE event happening in the Asian continent. They did it through the contribution of the money or money worth for the successful registering of high growth meeting, incentive, convention and exhibition industry with high values and quality imaginative and liberal development strategies and actions. Among the key elements of the prestigious growth and rapid expansion of the MICE industry in Asia was convenient air access for both domestic, sub regional, and international airlines making Asia accessible to the rest of the world. The good magnitude technological development also served as a plus, in addition to the existing good number of standard exhibition and convention hall supply as a pulling factor for the organizers.

Box 10.2: Case Study of the Status of MICE Industry Development in Taiwan
Tourism and worldwide travel destinations have shown significant developments of the industry. This is actually the birth of the MICE Industry, one of the 21st century booming industries. It is tremendously increasing in good interest in the United States of America and European countries. Countries around the world strive to recover the market as MICE had gradually moved to Asia. Countries in Asia have constituted giant economic power. Among of them are the Taiwan ROC, China Mainland, Singapore, and Hong Kong. Other countries are at the conscious of economic developments and benefits from MICE Industries. Many have launched all kinds of programs and projects.

MICE Industry in Taiwan experienced key of development. These development plans symbolize an inauguration or strength of mind in 2005. The Department of Commerce, under the auspice of Ministry of Economic Affairs, have put their talents together then instigates the MICE Industry Development Plan. This plan is vital. It is considered as the "ten major construction projects" that led to Taiwan's economic miracle in the 1960s. From the overall MICE development and implementation projects in Taiwan, the plan is auxiliary separated into five operational strategies. Overall plan for promoting and supporting the MICE Industry, the plan is aimed for cultivating and certifying MICE Industry Personnel, assisting MICE Industry Management, and thus establishing a MICE Industry Information Network. In order to improve the economic country, Taiwan's MICE industry with the stakeholders, launched a four year MICE development plan, assisted from the Ministry of Economic Affairs. The agencies reside over the project development and implementations (Fig. 10.2).

Taiwan is in the heart of Asia with well-structured, developed MICE industry country. Taiwan has been a venue for numerous MICE events of the Asian continent. Statically in 2017 Country & City rankings (ICCA 2017), Taiwan alone had hosted almost 141 international meetings and its related events especially in events like the Taiwan Exhibition & Convention Industry show; the show for MICE professionals EXCO TAIWAN, Taipei International Food Show FOOD TAIPEI, Taiwan International Hotel, Restaurant, and show on Catering Taiwan HORECA. The Taiwan Ministry of Foreign Trade did launch the MICE project office which duties are to stage, launch, and successfully bid for meetings, incentives, convention and exhibition to be host in Taiwan. The office in close collaboration works was to promote Taiwan as a global events destination for the rest of the industry professionals.

In doing so, Taiwan with the meeting as progress was considered as a significant manifestation for appraising national affluence, coupled with high recognitions at international stage. As an emerging globalized service industry, MICE is especially valued by national governments for its three concepts. They are high development potentials, high added-value, and highly beneficial innovations. The other three large area are: (1) large output, (2) large opportunities for employment, and (3) large industry associations. Moreover, the other three advantageous are: (1) the advantage over other industries in human resources, (2) technological know-how, and (3) the efficient utilization of assets.

Fig. 10.2 Location of Taiwan (Illustrated by Yi-Te Chiang)

Good numbers of developed and developing nations are making used of the MICE Industry as development approaches for the new generation. Taiwan's MICE Industry has been around for only a quarter century. It is a much younger industry than its more matured counterparts like Hong Kong, Japan, and Singapore, though not without its own competitive edge. There is no disputing the fact that the MICE Industry in Asia is currently experiencing a period of great expansion and growth. The regional competition has only just begun. Subsequently, the growing the interest of Taiwan as a MICE destination led to the formulation of a development plans for overwhelming images in the global event industry. The nation has successfully drafted the development plan for training and certification of MICE Industry professionals, promoting Taiwan as MICE destination, Implementation of MICE projects, and the overall aim and objective. Therefore, the framework created is promoting and developing Taiwan as MICE destination.

Upon launching the MICE Industry Development Plan. Department of Commerce has established Taiwan MICE project Office. It is the single window MICE Industry development. As for "providing inquiry services to domestic and foreign businesses" the MICE project office has organized 24 different seminars or forums in the past years with a total of 1,134 attendees are 241 inquiries of domestic projects and 194 international projects. As for subsidies and funds for international events that have attracted nearly 140,000 visitors in the past for years. According to Tang (2014), the economic value of the convention industry in Taiwan was estimated at approximately of NTD $8.7 billion (US$300 million) in 2012.

The project staff in the office also works hard to bring more events to Taiwan. Every year the office sets aside funds and provides assistance to related agencies in earning Taiwan the rights to host events. They aim to enhance Taiwan's image globally. The office provides assistances in the areas of international bidding site surveying provision of promotional materials getting assistance form peripheral businesses. They are coordinating to related core industries governmental funds, subsidies, and assistance for visa application into Taiwan.

According to statistics successful events can attract over NTD $71.7 billion output value of economic benefits from the MICE industry (Chen et al. 2012). Also "Meet Taiwan" website is established to showcase Taiwan's MICE image (Taiwan Business Topics 2016). The department of commerce has laid a strong foundation for the future development of Taiwan's MICE industry. It has also made the first step the efforts and achievements in the past four years have opened up an optimistic future for the industry.

In this case, since Taiwan computer shows establish a forum for computer manufacturing companies, wholesaler and retailers to showcase and exhibit their products to potential customers. In contrast, the lantern and dragon festivals create awareness about the cultural features of Taiwan. They are popular events in Taiwan today.

10.2.2 Educational Benefits

The overall MICE industry has shown a tremendous progress of development over the recent years in various destinations across the globe. The construction and establishment of conference are like exhibition halls around different continents, more especially in the Asia Pacific, American, and Europe. However, few successes have been registered in the African continent even researches indicate the growing interest of young intellectuals. The scholastic ambition for acquiring compressive knowledge about the MICE sector is through the international body (agency) dedicatedly responsible for the affairs of the MICE. Most often, they deal with other organization. The ICCA (International Congress & Convention Association) information on figures has shown that overall sum of 400,000 conferences and exhibitions are held worldwide annually at the total outlay of US$280 billion (ICCA 2011; UNWTO 2014; Zheng 2018).

The MICE industry has a wide range of opportunities for people with relevant skills, knowledge and qualities to work in various department of the industry—both national and private institutions. They are currently providing training like at the national hotel school, travel and tourism, schools and colleges, for people who want to make their career in tourism, travel, hospitality and MICE management.

The structuring of the travel tourism and MICE Industry requires professional training and certification program based. Such needs in academic institution are coming up, providing training and giving qualifications, and training for aspiring tourism professionals. The meeting planners and experts for people, who will be either directly or indirectly working in the industry, will equip them with technical and academic skills needed to deal with their customers effectively and efficiently. These skills are needed for effective service delivery in meeting the man power demand of the industry.

10.2.3 Social Benefits

Corporate Social Responsibility (CSR) is the responsibility and dedication that the event organizers bestow upon themselves to some interest and welfare of the local community. CSR Promotes to empower the usage of local public transport services. It is also taking care of the interest of the locals in nativity and congestive environment by managing waste and other litters. Fund and money are raised from the organization of the event. The support by local charity organizations, schools, hospitals, and others seek to serve the local people and whole community. The money raised at the event should be used to support them especially students and their welfare in a sustainable way.

10.3 MICE Planning, Design, and Marketing

MICE planning, design, and management depends on the types of the event and the particular individuals and organizations. Normally, the MICE planners consider following points to include planning and selection of the theme, date, key invitees, and other logistic. They will embrace coordination and location of support groups for electricity, multi-media, food and beverage staff, health safety and security, and other utilities.

It deals with ideas of running an event successfully to achieve the aims and objectives of the event. The planning of events starts with a creation of having objective of what you intend to achieve at the end of the event, and what is the theme, location, human resources, logistic, budget resources, and participants.

10.3.1 How to Organize a Successful Conference?

The MICE events like any other organized event happening currently in most of the developing and developed nations. Its economic prosperity is staged and backed by well-established plans and guidelines. Among them include the following outlines to organize a successful conference.

- Agenda
- Site or Venue Selection
- Theme Décor
- Floral arrangement and table top décor
- Catering which entail food and beverage management
- Lighting design and operation
- Sound system and enhancement
- Video production
- Audio visual effects
- On-site management
- Budget preparation
- Health, safety and security check
- Contingency plans.

10.3.1.1 Budgets

Ever successful conference is backed by sufficient financial and monetary budget. Today, we witness the harsh financial crashes. Many event planning organizations are enhanced and supported by financial proceedings of the conference registration fees, and sponsor-ships. Conferences cannot be held without financial support. They are based on the organization, planning, development and implementation of special events. They need and require funding and saving financially in making the event a worthwhile program. Standing on well financial plan will cover cost for expenses of the event. Conferences should be cost effective in demanding for payment from attendees, participants and sponsors.

10.3.1.2 Venues

Venues for conference are very vital and dictate the success of the conferences. Establishment of conference halls, therefore, with modern days' technological facilities are first to be known. Conference development and administration required logistics and planning of event planners to effectively consider the location, cost of facilities, utilities, as well as in bidding for suitable congestive venues which will appeal to the attendees.

Conferences of different participants should be located in ideal venues. The venues must be accessible either by road and air by majority of the conference attendees. Conference locations are supposed to be close to the target audience. Currently, it is realized that many people have to travel on long or short haul to be present at various conferences.

Today conferences are stage in difference locations depending on the winner of the bid. However, a well meeting quality standard should be based on attractions, accommodations, access, public services and facilities, security, and hospitality. They are held around the world in small and large cities. The venues range from Meeting Rooms in small hotels to large Convention Halls of different countries. These are few names among widely popular venues with pristine reputation like Taipei Convention and Conference Hall, Taipei World Trade Center (Taipei, Taiwan, ROC), Hyderabad International Convention Centre (Telangana, India), and Queen Elizabeth Conference Centre (London, United Kingdom), and the Scottish Exhibition and Conference Centre (Glasgow, Scotland).

10.3.1.3 Facilities at Site

Convention and conference centers are designed to handle mega events. The venue supports the event at reasonable capacity for all facilities required during the conference. The seating arrangement and all facilities at the conference hall play important roles to run the conference smoothly. This perhaps appear like an undersized element to somebody who has not at all been occupied in conference consultation preparation. However, the conference room is very important as it is the main place of meeting and serves as forum for all events. For conference electronic equipment involving computer, facilities like LCD and large display screens, PA-Systems, and so on are compulsory. They are huge selling point in conference organization. Various event planners reimburse additional bill just for the expediency of not having to carry their personal display and projector. The Acoustics and the Seating Arrangements also factor into the room set up.

10.3.1.4 Accommodations Obligations

Every conference proceedings have dedicated places for lodging and catering for the audiences. A place where delegates stay overnight will provide them with food, and beverage or/and other refreshment. The hotel accommodations have facilities the provision of excellent conference rooms and meeting halls. If the carrying capacity of the conference hall is insufficient to cater for all, that means there are not enough hotels in town for all of the attendees. It does not make sense to hold it there. While many attendees would be willing to drive for several miles to attend, many would not participate in the event. The accommodation plays a vital role in event industry indeed.

Box 10.3: Type of Conferences

Today, among the entire event industry conferences are the most highly organized event happening around the world. This single fraternity of the MICE industry requires a special attention for industry stakeholders. Event organizers have drawn people for diverse academically and intellectual diversity. Conference attendees are fully ready, well advance about the significance day of their participation in such gathering. Therefore, it requires some elements of professionalism and dedication and responsibilities. In doing so, all conference organizers must have specific goals, time frame, and theme for the conference.

There are many types of conferences as many as the venues. However, the International Association of Conference Centers (IACC) have initiated and provided some guideline about quality assurance on the facility and equipment of conference centers. As for the basic, what kind of things should constitute a conference? So far, there is diversity in the nature of conferences and conference hall depending on the location and locality as well as the format and approach also dictate the pattern and design of hall. The organizing committee chairs are overall concerned to the planning and organizing of the whole program. While all of the forums regards as conferences types share some resemblance, there are some distinctions explained as below:

- **Academic Conferences**: These can be held for two main reasons. University scientists hold them to announce for research results. There is also a "call for papers," where newly published information is presented to a group of students or/and researchers. Many conference-planning companies specialize in these types of gatherings for colleges and universities. These companies help in publishing the research or paper, as well. Many of these conferences are held on campus of a large university.
- **Medical Conferences**: Thanks to science, new advances in medicine are announced all the times. Medical conferences are ways for doctors and other medical professionals to keep up with these advances and to earn the needed continuing education credits to keep their license. Often, professional organizations, like the American Medical Association, hosting an annual conferences to give doctors the opportunity to earn these credits.
- **Business Conferences**: These can be several types of conferences or meetings. (1) Annual meetings. Many companies hold this type of meeting for shareholders to announce the company's accomplishments of the past year and to highlight the future. (2) If a company has many divisions around the world, often the management teams will get together to talk about their business' direction and to share best practices. This may be the only time they get together in the same town. Often these conferences are held in the same town in the company's headquarters.
- **Training Conferences**: Large organizations like the United Way, conducts these types of conferences several times in a year. This gives every-one opportunity to learn new information and share their best practices with their colleagues. Often training conferences will be presented by a marketing company and will be opened to anyone from any company. These types of conferences require a classroom type of setting and easy access to restaurants, if lunch and dinner are not included in the session.
- **Internet Conferences**: These virtual gatherings have grown in popularity as travel costs have increased while company budgets have not. Even though, they may seem

as simple as just getting online, these conferences require planning—particularly if the participants are in different time zones.

Conference management may seem easy to an outsider: find a place, find a caterer and set up the room. But it is much more than that. Conference planning involves every little detail down to making sure there are enough water glasses on each table to making sure the sound system works. These tasks are usually distributed to several people and vendors but are usually over seen by a conference manager, who carries a huge responsibility. Any small mistake can put a damper on the entire event.

10.3.2 How to Design a Successful Event?

Planning and design of MICE is one of the key processing works before preparing event launching and receptions. It entails the creation of event time, determining the goal of the event, and actions that need to be accomplished during and after the event. Considering the establishment of partnerships, with funding agencies and sponsors, selecting a suitable location, and a well congestive ideal place for audiences, need to be heavily considered. MICE planners today use the unified systems of designing events determined by the events itself, most case event is dictate by the standard procedures, various gathering warfare were also done by chiefdoms, kingdom and empires.

Standard Operation Procedures (SOPs) to plan events are set of guidelines, rules and governance that help organization: event planners and event managers. Event planning is where the design and implementation of themes, setting, consumables, services, and suggests to provide unique experience for participants to have a unique and remanding experience of attending the event. On the other hand, the SOPs, guide the planners should give them idea of what, when, how, and where to do. If focused on the principle of having the event at the right time, at the right place, it should be done it at right way and format.

Event design is the creation, conceptual, development, and design of an event to maximize the positive meaningful impact. It is for the event audience has an event manage working for Sheraton Hotel designing a reception, welcome party, and an outdoor field trip the following will be put made the design Catering, Content, entertainment, environment, production, programmer, and theme. The Catering-Theme will be a provision of food, refreshment, drink, and snack. Available during the course of the event, will be allocated to the participants and audience at the event operated. Run under the proposed aspects, and contend at a strategic ideal environment, will accommodate every attendant. Entertainment will also be crucial element of being part of the event.

- Theme—appealing to all the senses.
- Layout—Creative use of venue
- Staging—in the theatrical sense

- Entertainment—Programming and the talent
- Catering—quality service to all
- Above-mentioned key points will be the guidelines to operate run the event of the welcome party, reception and international truth convention.

MICEs are event of any kind that drams together peoples for specific purposes. The event industry being a growing industry is as well diverse expanding in two separate ways namely Cooperates and Private events.

10.3.2.1 Cooperate Events

The event industry is being a part of global dynamic business of performing commercial activities at local and international areas. This has resulted to more formal approach with emphasis on professional career development and trainings. Today, the MICE industry is presented bright strategies of the events. These events facilitate to support travel for tourism. In general, an event industry, either external or/and internal, as well as apprise and inform publics is about the firmness and endure of organization within the event industry. Among few general portfolios uphold by these segments of the industry are the most operational approaches to event organization, staging and implementation of various kind of events successful as prescribe by the objectives notably. Cooperate events shall embrace music shows, parties, game shows, business parties, family parties, private parties banqueting and product launching etc.

In a wider spectrum, a commercial event is occasionally has to do with scheduling, preparations of meetings, incentives, conventions, conferences, exhibitions and other major and minor occasions that are all meant at achieving the inclusive objectives of the business and trade enterprises. Several types of events are listed in a comprehensive list of a corporate event packages emphasis on internal essentials and gatherings. To mention, travel, tourism and cooperate event organizations embarks on several staging and hosting of events what serves as a medium of information and knowledge sharing and entertains serves as a venue for passion and fun for other audience.

10.3.2.2 Private Events

The word private signifies those events that are not officially and formally in the staging and organization. It is considered that occasion is fallen within the casual gatherings of celebrating. The considered gatherings are like birthday, marriage, and wedding anniversaries. Sometimes, the other events will include participants comprising of closed-friends and relatives. Mostly, these are events that marked as significant moment of real memory in one's life. Such events are commonly celebrated by individuals and organizations deficient to reminisce warm-heartedly of their existence, they always linger mind because of the gorgeous setting fashioned by decorations, lighting, entertainments, beverages and food in diner form. It has becoming a custom by most private individuals and private organization to mark the anniversaries annually mainly to raise awareness, fun and as well expanding their network connections.

Non-governmental organizations (NGOs) is privately owned, operated, and managed to serve the interest of the members of a particular society or group. The vision deals with the aim of NGOs as long term plans. They want to achieve to the benefit of its members, either to raise profit, product awareness sensitized on campaign for a particular, or else.

10.3.2.3 Mission

They want to achieve, how and why, it deals with the question of why is organization operational that is its aim, what it does, and what does it do for the benefit of the community.

10.3.2.4 Budget

Budget means the required resources needed to achieve it aims and objectives. Budget also talks about what resources are needed in cash on kind to achieve lost the short term and long term plans.

10.3.2.5 Personnel

Personnel deals with required skills and technical human resources needed to operate the organization the qualification, selection and recruitment process need for the trained staff.

10.3.3 MICE Marketing

MICE industry like other sector does the marketing to put the message across. It is be a promotional campaign to constitute one of the development plans from policies makers in the government. Normally, after drafting the MICE project plan, marketing, and promotion will be the third most vital aspect to be considered.

Several countries are completing to be the best MICE destinations. Basically, the marketing and promotion are the processes and the way of persuading potential event planners. The meeting, incentive, conferences, conventions, exhibition, and exposition at a particular venue are the most suitable for the community. The satisfaction in launching a successful event will embark the effective promotion at the right media channel. Part of competitive edge destination marketing and promotion personnel helps visa application process for individual and participants, as well as helping in investment, awards programs and market overview. Basically, the MICE and its related events have diversification in marketing and promotion of its event. It is usually has special products and services for various gatherings of social, cooperate and private events. The promotions by MICE planner and authorities can make certain event is transmitted throughout the aspiring tools in the community.

The wonderful colorful beautifications, foods, invitations, websites, newsletters, refreshment, amusement, and taking pictures, on condition that is a permanent illustration of the magnificent occasions of popular, and specific event are done at the appeal of both the organizers, hosts, and participants.

MICE like other commercial sector deals with element of the marketing tools to persuade its potential target market to come forward and utilize the products they have are offered at the destinations. This is further breakdown into subgroups for ease comprehension for the units the four Ps, such as products, price, place, and promotion.

10.3.3.1 Products

The product aspects of MICE marketing and promotion deal with specifications of actual goods or services. Basically, this comprises of the event venues, accommodation, and ease of transports at the gateway. The scope of this element generally includes supporting elements like warranties, guarantees, support the planner, and organizers chosen venues. Various MICE products have different appealing factors depending on what to offer. They are for example, between 12th and 13th century, this period has been the era for the commencement. The commencement was at the first exhibition site and center. The trading was centered among some major nations. These nations were expanding their economic statues at that time. They were France, Netherlands, North & Central Europe and finally, Italy, also parts of the trading countries at the time. The products were the motivating factors for their gathering. They were wine, silk, metal goods, furs, cattle, and pigment. This was sharpened the exhibition industry with modern exhibition-taking place in London in the United Kingdom.

10.3.3.2 Price

MICE event depends on the planner who dictated whether it was a profit making event or not. Despite the fact that most event runs on saved budgets but yet still participants and delegates have to pay in sum amount of charges. This can be either as conference registration fee, accommodation, flight and airport transfer to the venues, food and refreshment. In doing so, token that is levy is the price; the MICE industry expert has a unique system of setting special charges for attending and participating in the events. They are mostly including discounts on early registrations; in some instance this outlay need not be monetary. This can simply be what is exchanged for the product or services, e.g. time, energy, or attention. Those delegates are participants and target groups can afford to pay without any discomfort.

10.3.3.3 Place

The series of event that makes up the overall MICE segment embrace numerous elements of festivals, conferences, sport arenas, exhibition, convention all of vividly demands a venue to successfully take place. Historically staging of events were done at open space landmarks. However, the establishment of conference and exhibition hall today has fadeout those practices only remain on doing the traditional festivals in Africa. Hence, a place refers to how the MICE objective can be achieved, the channel of a product or service is sold (e.g. online or by wholesale and retail). The geographic region or industry to which segment for the young adults, families, business, people, etc. also referred to how the environment can affect sales.

These four elements are often referred to as the marketing mixture. Today has been so instrumental to all sectors of travel, tourism, and event industry all of which interrelated and intermarriage, which a marketer can use to craft marketing plan. Similarly, MICE events are done on bidding system on. The city or destination to host local, national and international event will depend on the qualities of available venues. Few examples have demonstrated rapid growing. The question remains to Africa? Africa has few or none up-to-standard venues to host multi-national events. The motivation international planners are to choose the African nations, with the growing number of student undertaking MICE as a career, soon it will be a thing of the past.

10.3.4 MICE Promotion

Promotion includes advertising, sales, publicity, and personal selling, and branding. This is referred to various methods of promoting the product, brand, company or the event organizers. They should motivate and attract sponsorship at high turnout of various delegates from all over the world. The media fraternity has been the most popular form of promoting event through the use of internet and its related technological facilities has aided in the process of message transmission to the target groups. Today, almost all event planners have online information center. They were people all over the globe logging on to access the information to regard the various events, calendar, registration, and other procedures. These are made available on the search engine and social networks. The principles of promotion meet the marketing objectives as follows.

10.3.4.1 Increase Profit

The principles of promotion can meet the market objectives by customers having the right product for the right price. This should be offered as affordable for customers. The business integration at the right place and use effective means of product promotion for the customers must have the product knowledge usually done in exhibition forums. This can be done through the use of Internet, Radio, Television, Sunday Newspapers, and so on. The customer to customer information is always effective in getting the right information from the media. Customers know the products and the prices, thus choose what to buy. The product of their interest will motivate them to purchase through organization or by him or herself. If buying goods through organization gets more benefits or discounts, the customers will purchase their products and service it increase their profit.

10.3.4.2 Increase Awareness

Increasing awareness on product and service must be promoted for customers very frequent. They become aware of it the promotional activities in travel, tourism, and event industry can be done by Internet, Television, Radio, and public relation. This will make customers to have the first hand information about the products. These promotions increase organizations exposure to customers about the products through their organization. Therefore, it will enhance their decrease during weekend. The lower the products by weekend, the more customers will come to purchase them. However, if the price of certain products increase there will be lesser customers to purchase, unless the products are in the need of individual, family, or else.

Business to business integration and partnership opportunities are some of the successful contact that can be established by companies. While business organization continues to build trust worthiness with their consumers, trade shows, exhibition and expositions are good examples of business to business integrations, and consumer to business.

10.3.4.3 Increase Sustainability

Since 2012, ISO 20121 is developed for an international standard for sustainable event management. This code aims to help events to improve their sustainability throughout event management cycle. All managerial approaches are including self-declaration, confirmation of conformance by all parties regarding to clients and stakeholders. Finally, an independent third party, a certification body, can certify their pro-environmental behaviors, i.e., reducing electricity consumption, water, and paper. This saving can be covered reducing carbon footprints, printer ink consumptions, and plastic bottle uses.

Box 10.4: Case Study on One of the Most Popular Events: Eco-festival in West African State of Gambia

- **Why Eco-Festival?**

Eco-Festival will promote the conservation of traditional cultural patterns, their artistic expressions and place of historic and cultural importance. This will represent the heritage of the Gambia; and minimizing any adverse impact of tourism on the traditional cultures; thus maintaining a sense of cultural heritage and enriching national cultures diversity. This kind of festival will introduce a new type of tourism product to the destination. It will result in a greater overall tourist flow. The eco-part of the festival is minimizing the noise, carbon pollution plus waste management in providing a funding towards environmental conversation and preservation by an annual tree planting exercises. The aims and objectives of eco-festivals are:

- To promote the cultural activities of destinations in creating a unique market in the generating countries.
- Aiding the destination to grow edge over the competitors
- Retain and maintain the authenticity of the local culture
- Facilitate the decentralizing of tourism benefit to regions across the country in promoting the social livelihood of the locals.
- Aiding the destination for cultural festivals.

- **The Essence of Eco-Festival**

Event Administration: The team should include some members of Gambia Tourism Authority working in close collaboration with Future Tourism Foundation. Responsibility of the planning and staging of the event, including the administrative logistics and finance will be handling by the two bodies.

Event Sustenance: After a successful funding of the event, all the subsequent responsibilities should be shouldered by the members of the team in support of the Gambia Tourism Authority, Ministry of Tourism and Culture, National Center for Arts and Culture, Association of Small Scale Enterprise in Tourist. It is going to be initiated to be an added product to the existing once of destination Gambia.

Event Justification: The acceptable reasons for initiating these noble events are to provide cultural experience and make the destination popular for its diverse cultures. It is most of our events, viewed did not contain much of our culture that can easily create market in the generating countries and serve as a panacea in retaining and maintaining them for the future generation. Putting the trend of tourism into consideration, the highest percentage of tourist traveling is more in for cultural experience that any other attraction.

Event Contribution: The event will contribute in upgrading the social economic status of the indigenous people and promote their culture and create awareness for the destination. These will aid in drawing edge over competitors by grapping more tourists from the generating countries. For example the responsible travelers, Eco-tourist etc., if maintained, will make the destination known for its uniqueness in culture (Fig. 10.3).

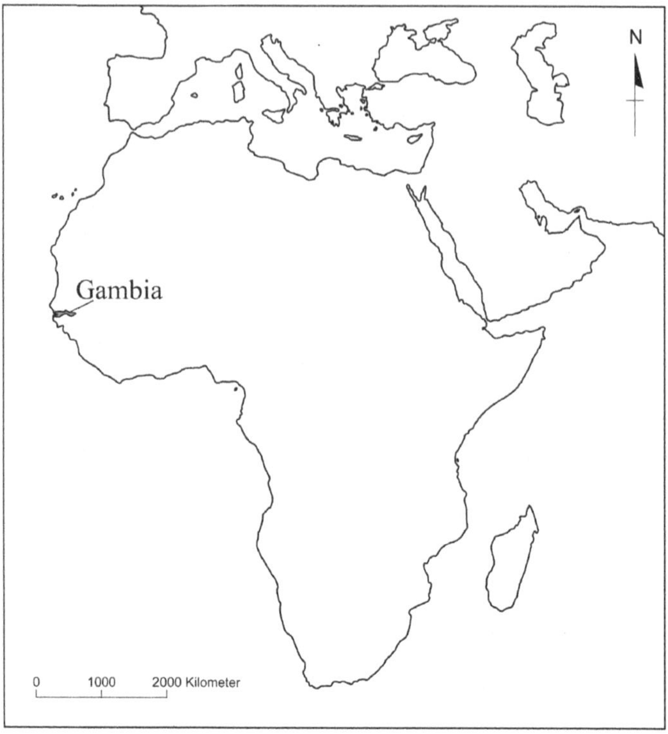

Fig. 10.3 Location of the Gambia (Illustrated by Yi-Te Chiang)

- **Toward Cultural Life**

Eco-Festival is geared towards portraying the real cultural image of various tribes. In all aspects, their culture is being considered as eco-form. This includes its significance in allowing audience to have clear picture and understanding. This is to serve as a panacea in minimizing the talent shows. The show lies in our festivals providing real cultural experience for both the local communities and guests. In promoting the culture of the indigenous, some people retain and sustain prestige culture of destination in Gambia.

The trend on tourism today is towards cultural and responsible experience. This is an experiencing of different cultural norms and values in effecting smooth communication. The festival will consist of both the display of cultural dresses in several cultural life, cultural events and the preparation of cultural foods and beverages in their values.

Tourism is based primarily on beach resorts and urban life with organized excursion taken into countryside. This type of tourism does not offer tourist the opportunity to experience more Gambian way of life. They meet Gambian on spontaneous basis. Some residents are being socially degrading seeing African culture. Local people feel they must meet the expectation for tourists. We are worried about the demonstration effecting tourists especially to young people. They loss traditional values. This sense is based on which we think of an approach in developing a Niche product that directly benefits their communities. They bring more cultural understanding and interesting experience to tourists. The niche product will be an Eco-Festival where it comprises of a three-day event. Eco-Festival is geared towards promoting the preservation and conservation of both the environment and the local communities.

The development concept of the plan is focused on the coordinated realization of the tourism potential of the cultural, historical, and environmental resource—special to Gambia. Ways to give tourists greater exposure to Gambian culture is suggested through organized presentation of song and dance, exhibition of traditional cultural foods, tea, and beverages.

Cultures are norms and values for the future generation. The festival will showcase the different facets of the Gambian culture in its authentic form. This is a way will revitalize and enriched the culture and as well as promote social livelihood of the locals to venture into such as activities which can create job opportunities for the locals. The starting of eco-festival will be held in Senegambia Resort or Independence Stadium. The decision is simply because of its pristine and natural environment, it is accessible and comprised of a preserved culture.

- **Day Plans**

The plans are expectedly achievable by confining to specific, measurable, achievable, realistic and timed activities. They will be conducted in three days. It will have strong support of both national and regional government, very important.

The technical and financial assistance, as well as dedicated sharing of roles and responsibilities, will be upheld by members of the foundation, Gambia Tourism Authority, National Centre for Arts and Culture. These are all valuable. Education of the villagers and local participants are necessary aspects of implementing the event, which will be done by the members of the foundation. All activities will be dramatized in sequence providing audience with clear picture and understanding to what is been portray:

Day one (1) Cultural ride: Each and every ethnic group would showcase their different cultural talents, wear in different stages of their being, which should be back by pure traditional musical instruments, like "Kora", "Njembe", "Geo-Tantang'ho" etc.

Day two (2) Traditional Food and Beverage Sampling and Tasting: Each and every ethnic group will showcase their pattern of food preparation in compliance with health and safety regulation. These will lead them to portray the nutritional values of these food and beverages, including "Benechin", "Domoda", Fufu, etc.

Day three (3) Eco-wedding: This is the service that will showcase the cultural wedding style of a chosen tribe, by the brides from how wives are being chosen, the marriage, the cancelling of the brides by the elders, the wedding ceremony in an eco-form.

The activities of the said services will be dramatize in sequences to visualize all the processes in open place allowing the brides to understand the cultural aspects of our wedding that will promote cross cultural exchange, as well as make the Gambia known for traditional weddings.

10.4 Conclusion

The MICE Industry is a young industry. However, it is rapidly growing at most parts of the world. The word MICE means meeting, incentive, conference and exhibition, otherwise known as event industry in America and Europe. However, in Australia and Asia, the term is widely known as MICE. It has become a major economic sector for the nation. MICE industry is a newborn emerging development in overall tourism industry. It is growing rapidly in most destinations, especially rapidly expanding and growing in continent. It is leading to the construction of high quality conference halls and exhibition centers, which captures the attention of international event organizers. It is a fascinating industry that leads to the certification programs and the formation of academic curriculums in universities and colleges. It is also allowing various professional career development programs for everyone with desire for planning, management, organization, and launching of occasions to undertake courses at undergraduate and graduate schools of education.

The growing events for more venues have led to the establishment and construction of conference, exhibition halls, and centers with higher qualities and facilities to meet the growing demand of planners. Development plans and government department aim to develop, plan, promote and implement the MICE industry and its related activities drafts agendas. Statistics has shown the success registered in Asia, America, Australia, and Europe. In contrast, the academic brain developments of the MICE experts in Africa are still on its infant stage. However, the prediction shows that it will become a top history in the future.

References

Chen HC, Chiou CY, Yeh CY, Lai HL (2012) A study of the enhancement of service quality and satisfaction by Taiwan MICE service project. Procedia-Soc Behav Sci 40:382–388

Goldblatt J (2002) Special events: global event management in the 21st century. John Wiley & Sons, New York, USA

ICCA (2011) 2011 ICCA Statistics Report. International Congress and Convention Association, Amsterdam, Netherlands

ICCA (2017) 2017 ICCA statistics report country & city rankings. International Congress and Convention Association, Amsterdam, Netherlands

OECD (2018) Economic outlook for Southeast Asia, China and India 2018: fostering growth through digitalisation. Organization for Economic Co-operation and Development, Paris, France

Publishers TN, Packer JI, Tenney MC, White W (1995) Nelson's Illustrated Encyclopedia of Bible Facts. Thomas Nelson, Nashville, TN, USA

Sable K (2019) Asia-Pacific MICE industry: opportunity and Forecast, 2018–2025. Allied Market Research, London, UK

Schlentrich UA (2008) The MICE industry: meetings, incentives conventions and exhibitions. In: Brotherton B, Wood RC (eds) The Sage handbook of hospitality management. Sage, London, UK, pp 400–420

Taiwan Business Topics (2016) 2016 Taiwan White Paper 46, 6. https://amcham.com.tw/

Tang H-W (2014) Constructing a competence model for international professionals in the MICE industry: an analytic hierarchy process approach. J Hosp, Leis, Sport Tour Educ 15:34–49

UNWTO (2012) MICE industry—an asia-pacific perspective. World Tourism Organization, Madrid, Spain

UNWTO (2014) 8th UNWTO Asia/Pacific executive training program on tourism policy and strategy, theme: MICE—a new paradigm for tourism. Technical Note. 28 April to 1 May 2014, Bali, Indonesia. World Tourism Organization, Madrid, Spain

Zheng X (2018) Research on the marketing strategy of MICE cities, taking Asia as an example. Adv Soc Sci, Educ Humlties Res 205:945–956

Chapter 11
Future Tourism

11.1 Introduction

Tourism industry has been world fastest growing industry globally with lots of economic, social, and environmental benefits. It is not only deterring to a peripheral scope of human imagination but also proven being doubts from super structural, infrastructural development, and higher economic growth to cultural empowerment. Despite the world economic and financial crises, the future of tourism is determined to be successful with more travelers, hotel development, emerging destination, leisure, and recreation centers. There will be more rooms for environmental and cultural exploration, all of which will face a tremendous evolution at present and few years to follow later. In the future prospect of the tourism industry, it is promising with more and more investors, new airlines, hotel accommodation, and employees are having disposable income with long breaks. This will become a motivating and appealing factor for more tourists to travel and discover various destinations for their choices.

Destination across Asia, America, Africa, Europe, and Australia, are expected to have a higher growing and booming future of this tourism industry. Most development plans and agendas are laid on a vision to be achieved in 2030. Destination marketing managers with the so-called tourism 2030 vision is aimed at the global development and promotion of tourism worldwide, forecasting of which has shown a positive achievement of the dreams toward Sustainable Development Goals (SDGs) (United Nations 2015).

The future of tourism is positive for many destinations but more for the Asian continent, more especially China, Taiwan, India, Singapore, Malaysia, Philippines, and Thailand. They are simply backed by the higher successful economic growth. For the next 10 years to come, where most of the destinations are targeted, the international tourist arrivals are expected to double and increased at a good percentage. The Asia and Pacific are expected to gain more of the new international tourist arrivals coupled by southern countries like Brazil, simply due to their higher economic growth and higher population.

W.-T. Fang, *Tourism in Emerging Economies*,
https://doi.org/10.1007/978-981-15-2463-9_11

Researches have shown that all aspects of future tourisms will have some elements of "ECO" being built in it. To see that tourism contributes immensely and meaningfully in destinations economically, culturally, and organizationally, we have to involve in the agendas. Statistics from the World Travel Market speech by "*Marclo Favillo—Executive Director In-Charge of Competitiveness, External Relations and Partnership at UNWTO*" in 2011 revealed that "Global growth in international tourist arrivals is expected to continue, but at a more moderate pace, from 4.2% each year" (1980–2020) to 3.3% (2010–2030). The results involve four factors:

- The base volumes are higher, so the smaller increases still add substantial numbers;
- Lower GDP growth, as economies mature;
- A lower elasticity of travel to GDP; and
- A shift from falling transport cost increasing ones.

Tourism towards 2030 shows that there is still a substantial potential for further expansion in coming decades, established as well as new destinations. It can benefit from this trend and opportunity, provided they do shape the adequate conditions and policies with regards to business environment, infrastructure, facilitations, marketing, and human resources. Along with opportunities, challenges also arise in maximizing social and economic benefit, and minimizing negative impacts. All of these are strong evidences showcasing the positive growth and expansion of the tourism.

Definitely, this moment gives impression to be a suitable point in time to be reflecting on the possibilities of a system transformations or paradigm change of the global tourism industry. It is a feature pride of the contemporary period to emphasize that the future of tourism is generally an evolution. The present is more a sharp transitional era which showcase echo calling into brighter future. With successful vision of the industry authentically supported by diverse cultures, liberal government policies, and new scientific inventions emerge in this third generation. It is interacted so quickly into the tourism system. For example, the application of smartphones is important that enable tourist to move around independently.

11.1.1 Motivation and Thoughts

Tourism is so dynamic. It has portrayed a higher sense of evolution from the golden mass packed holidays to present day as an independent traveler. Tailor-made tours focus on sustainability and responsible travel to emerging destinations of abundant environmental, cultural, recreational, and event activities. Tourism looks at the tourism growth and development in the future years to come. It is basically about prospecting tourism at all dimensions—all about connecting, enlightening, and aspiring the tourism industry for years to come. The revolution was from the technological development of tourism facilities in Asia, to the responsible eco-resorts in Africa, to the more independent and specialized travelers in Europe, coupled by the higher increasing staging of events in America and Australia.

Looking at tourism from holistic and academic point of view, we can narrow down our perspective into three dimensions. We use the experience about the (1) PAST numbers of both domestic and international arrivals, (2) PRESENT tourism development, growth and perspective, (3) TOMORROW or FUTURE potential growth of the industry. They are geared towards strategically measuring the emerging trends of the overall tourism business in destinations globally, as well assessing challenges and opportunities likely to occur over the next ten years. It is then identified necessary action and key priority areas for the future.

When we look in detail, we can see that the prospect of the tourism industry is bright. Despites facing problems by the world, most trouble in financial crises, people are logging for more travel destinations. The new forms of tourism activities and concern are over the well-being that relate to the host communities. Internationally, people are earning their incomes through progressively increasing and continuing the process of financial boost. This increases the affluence of disposable income which will further enhance the tourism industry to continuously prosper. The potential travelers will seek for destinations, growth in transportation which facilitates the ease of movement. On the other side, the emerging destination and quality products will serve as motivating factors.

Today, there is a wider number of tourism curricula offered at undergraduate and postgraduate levels. These curricula with international tourism conference and international marketing tours are collaborated with the media of British Broadcasting Cooperation (BBC) of UK and Cable News Network (CNN) in the U.S.A. All of the features travel documentaries about destinations are characterized by incredible development and growing integration. There is a growing demand that tourism covers areas of environment, economy, and social areas. Tourism has incurred higher involvement of community's stakeholders. The relations are increasingly seen as central interest to achieving desired outcomes of future tourism development. They all lie on three major influences as affirm mentions to products in demarcation, dynamic packaging, and new emerging destinations. The future tourism has the elements of the present and past. Thus, tourism has been developing and utilizing to attract potential tourists, the future of which is also more reliant in such initiatives.

11.1.1.1 Destinations Aspirations

Future tourism has been inspired and forecasted by most destinations. The current tourism operators are heavily depending on tourism as their number one economic earner to their GDP. They hope to bring in lots of foreign earnings, reducing unemployment rate, creating social awareness, and all types of creativity. All of these bring the future tourism beyond the normal sun, sand, and sea package called the "triple S" toward sustainability (Matzarakis 2014). In the few years to come, the tourism industry diversification will give room to more adventurous tourism categories: ecotourism, responsible tourism, ethical tourism, space tourism, volunteer tourism, educational, and event tourism. There will be more privileges growing interest in such areas than that of the conventional mass packages holidays.

Conceivably, we cannot predict about the future of any given business. In so saying, tourism is an exception in both evidences. The future of tourism is the experience of today's tourism. The more interested travelers, the more demanded tourism employees will be, required more tourism infrastructural development, such as the diversity of accommodations, facilities, attractions, destinations, practices, and forms. These are the trends in the industry. The positive will image picture about the future development of this tourism industry. This is the liberal initiative of ecotourism and responsible tourism. When there is an increase in demanding of employees, the travel, tourism, leisure, and recreation industries, will increase. Parallel with very good tools have been detected in social networks like Facebook, Twitter, and LinkedIn, all of which are used to bring the world into a global village, facilitating marketing, and promotion for the tourism industry products and services online.

These had increased awareness about destination images through peer to peer interactions. The social media influence destinations' images, most of them have their fan on pages on social networks. People in free marketing will do a gorilla format. Being so effective spreading the message and cost cheap across all dimensions, they look at things beyond all angles with deep focus on tourism—domestic and international stage, economic, and environmental opportunities.

Figure 11.1 shows evolution graph on how tourism developed globally from the 20th to the 21th century. There has being a considerable incensement in global growth of tourism industry which is expected to continue to the future with more international tourists (Figs. 11.2 and 11.3).

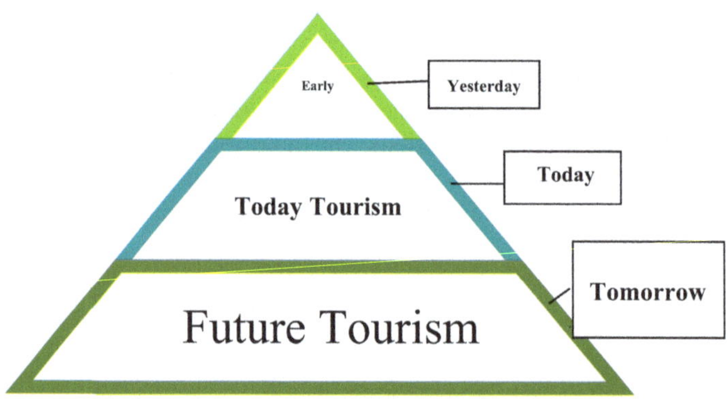

Fig. 11.1 The diagram portrays the evolution of travel and tourism industry, from the commencement of mass-packages tours by Thomas Cook (UK), to today's diversifications of products and services of destinations across the globe, powered by a brighter future of space tourism, ethical tours, virtual realities, 7+ star hotel, helicopters, and under water hotels, all of which show a promising future of the tourism industry

Fig. 11.2 After laying the sustainable development goals (SDGs) by the United Nations, tourism upholds it future development by uniting, educating, and inspiring the industry on four major areas. They are basically having concentration on environmental protection in sustaining our planet earth, climate change, poverty reduction, and product diversifications and destination management for greenery tourism (Photos by Wei-Ta Fang)

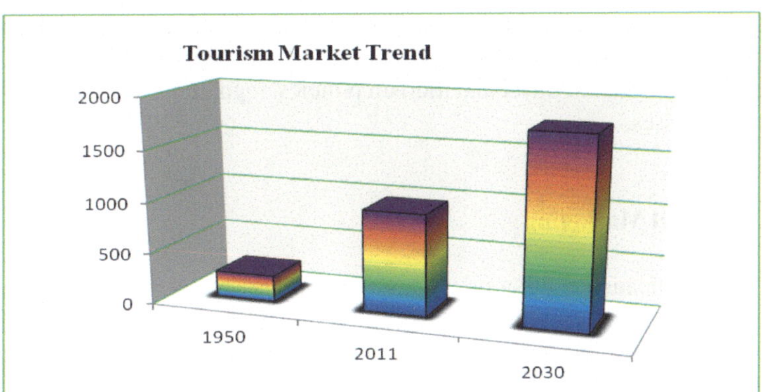

Fig. 11.3 The movement shows the progressive growth of tourism in two eras, past and the future, through the investigations of tourism development and growth over the years in different destinations to date

11.1.1.2 Product Demarcation

Product Demarcation was started from the origin of tourism by Thomas Cook (Brendon 1991). Mass package tour, in addition, was from temperance to Sand, Sun, and Sea package of Sub-Saharan Africa to eco-adventure tourism in East Africa, to event tourism in Asia. The results of future tourism cannot be over-emphasized as it not shown the diversification of hotel accommodation from 1-star to 7th star. With diversifying the products, services, and customer experiences, tour operators and travel agencies are shifting from package tours to more independent.

A special interest, tailor-made tours, and holidays, all are growing in higher interest. Destinations, tour operators, and ancillary service providers are all trying to offer different products and services that differ in nature and context, from the interactive perspective. This will become visible in the practice and discussion of community-based tourism planning in Africa to the development of MICE industry in Asian Pacific to the cruise holidays of United States.

11.1.1.3 Dynamic Packaging

Dynamic packaging is self-motivated that offers more optimistic perspective of tourism development across all sectors of the industry. It has visitor attractions, leisure centers, travel agencies, tour operators, and destinations which is becoming much cheaper. With the creation of online applications and search engines for airfares and excursion, tourists can book their tickets online at one's door step. The internet packaging, which is approaching 90% of the world's business industry, has attracted a wider audience as everyone gets a personal computer. Dynamic packaging of tourism products has made the future brighter with vibrant products and services ready to offer in market. This package is backed by the support of local and expert knowledge about general travel and tourism policies, regulation, and trend about the tourism businesses.

11.1.1.4 Novel Marketing

With global unification about tourism development in—both urban settings to rural peripheral communities, there is an increasing trend of emerging destination and products. The utilizations of internet has become the major influence on the planning and branding of destinations. In the few years to come, there will be new hidden destinations to be discovered with wonderful products that will be beyond expectation. Imagine of space tourism first done by NASA to the under-water hotel of United Arab Emirate to the 7th Star Hotel, much of them are too revolutionized in the future. These tourism packages will become new international attentions, The UNWTO (2011) predicted that this international tourist arrivals would be expected to increase to 100 million in the future. During 2018, it increased to 1.4 billion. This is especially to be a shift in consumer behaviors towards tourism product and services.

11.1.2 Diversification of Tourism

The international and domestic tourism industries are tremendously going on a rapid development and evolution. This industry has witnessed some major diversification over the years, all of which has compliment the progressive growth of the industry. The tourism brings many economic, social, and environmental benefits in the destinations. Tourism industry is bright. Bright in the sense that since its inception for educational, sporting, religious, and social purpose from pre-history of package tours from temperance meetings to religious trip to Mecca, Rome, and Canterbury to grand tours around Europe, to the present tour attending of Olympics in Europe, and the MICE events in Taiwan, Hong Kong, Singapore, and Australia.

Travel motivations are placed on various reasons back by ability and time. Today the tourism industry is vast that the trend moves from conventional mass-package to special interest discovery and educational form of vacations. They are tourists that do not only enjoy the travelling but also the rewards at host destinations. Future tourism will observe diverse tourism principles and practices all of which will be geared towards sustainability and responsible travel. The following are highlights of prospect tourism:

- **Ethical/Eco-Travel**
- **Museum**
- **Theme Park**
- **Sun, Sand, and Sea**
- **Cruising**
- **1–5 Star Hotels**
- **Specialist Holidays**
- **Safari Parks**.

11.1.2.1 Ethical/Eco-Travel

It is also recognized as an conservational tourism. This is a type of tourism experience that appeals to ecologically and socially cognizant folks, who explore places of natural beauty with sensitivity to payback from conservation to biodiversity. Generally speaking, ecotourism focuses on volunteering, personal growth, and learning new ways to live on the planet. Ecotourism has, today, attracted various disciplines of life. It has becoming a theoretical experience for academicians, travel consultants, and destination managers. They are enriching those who dare into researching and understanding the environment around them. It gives us insight into our impacts as human beings, and also a greater appreciation of our own natural habitats (Fig. 11.4).

Fig. 11.4 Eco-travel is classically engages journeys (tours) to destinations, where flora, fauna, and cultural heritage are the primary attractions for visitation (Photos by Wei-Ta Fang)

The international organizations like United Nation World Tourism Organization (UNWTO), Ethical Travel and youth groups in destinations, namely Taiwan Youth Climate Coalition, South Gambia Foundation, Gambia Tourism Support, Gambia Is Good, Sandele Eco-Retreat, all are engaged in activities for keeping-up the tours. The tourism includes programs that curtail the unsupportive aspects of conventional tourism on the environment. It enhances cultural integrity of local people. Therefore, in addition to evaluating environmental and cultural factors, an integral part of ecotourism is the promotion of recycling, energy efficiency, water conservation, and creation of economic opportunities for the local communities. In the future, environmental issues on climate change and its awareness for conservation come into play even more as educated tourists to choose accommodations that are certified green. Environmental responsibility is now a selling point and hospitality association worldwide. It is advocating the implementation of eco-friendliness in hotels, inns, and resorts.

11.1.2.2 Museums

The trend of tourism is now moving from traditional and old-fashioned of architectural houses to popular modern-typed of tourism. They are the prominent institutions in services to society and development opened to the public to acquire, conserve, researches, communicate, and exhibit the tangible and intangible heritage of humanities. Museums are enable tourists to explore collections for inspiration, learning, and enjoyment. They are institutions that collect, safeguard, and make accessible artifacts and specimen for publics. They also hold in trust for the society. The tourism industry relies on these archives to capture the history of local people, whilst on the other hand, giving tourist a wonderful unique experience.

Sightseeing of historical heritage sights are gaining popular with more travelers going local. They will be experiencing the authenticity of local culture. Following the unexpected growth in museums during the last two centuries, various travelers want to experience museums in destinations. These ideas are growing in higher interest, for example, the Hsinchu City (Taiwan) Glass Museum has been an ideal place for domestic and international tourists. In addition, the Kachikally Crocodile Pool (Gambia) is not only famous for its crocodile and folklore, but it is a museum that attracts various tourists—locally and internationally. The expectations are growing far above the ground that this segment of the industry will continue to flourish in years to come. There were international tourists who go local. This makes local people feel more authentic. Few remaining museums in Europe and Asia are gaining popularity. Some examples are like the Museum of Modern Art in New York (U.S.A.), Aberdeen Art Gallery and Museum in Aberdeen (UK), Victoria & Albert Museum in London (UK), National Gallery in Prague (Czech Republic), and Weissenhofsiedlung in in Stuttgart (Germany), and National Palace Museum in Taipei (Taiwan).

11.1.2.3 Theme Park

Leisure and recreation constitute of the travel motivations in planning and undertaking of both—domestic and international tourism. Theme parks are on the move on development in various destinations. Basically, they are collections of rides and other entertainment attractions assembled for the purpose of entertaining a large group of people, and individuals alike. Theme parks are facilities of the third generation. They are more complex than an easy city park or recreational area, usually providing attractions meant to cater to adults, teenagers, and small children. A theme park is a type of amusement park. It has been built around one or more themes, like an American West theme, Orlando Theme Park of Florida, Paris, Atlantis, or else. Today, the terms theme parks are often used widely and interchangeably.

Often, a theme park will have various lands (sections) of the park devoted to tell a particular story for entertaining audience. Non-theme amusement park rides will usually have little of teeming or additional design elements. On the other hand, a theme park has all the rides, like in Magic Kingdom in Walt Disney World. The future of tourism predicts an increasing numbers of theme parks globally established. All of them will be a save heaven for both local and international tourists with lots of sporting, leisure, and entertainment activities.

Fig. 11.5 Sun, Sea, and Sand Concept with Shoal Bay Resort & SPA and the sandy beach in Shoal Bay, Australia (Photos by Arba'at Hassan)

11.1.2.4 Sun, Sand and Sea

This product is the most popular and common form of product in tourism. Currently, it has been used worldwide. The entire destination market is to attract tourists. Tourists travel and enjoy the sand, sun, and the sea (3S), to escape cold weather of the winter in Europe, America, Asia, and Australia. Tourist now travels from normal home for holidays to enjoy the product. They will relax under the sun beds at the beach, and later to get into the sea. When they go up to the beach, they feel the sea breeze and wonderful sunshine. Their spending time of stay, the trend showcase, the increasing development in the pattern of the triple S, and the added value of style notably meaning of life style of locals. Increasing diversification on the continually trend of tourism makes it complex for many uprising destinations.

They discover and build-up new products and services which demanded by customers. All are well to have edge over their competitors. Meanwhile, the sun, sea, and sand concept is still remained highly competitive. Many people focus on the market price with access costs as a package sold by most inbound and outbound tour operators. In more years to come, hotel allocations will continue to flourish on the sandy beaches of Gambia, United Arab Emirates, Taiwan, Thailand, America, and Australia (Fig. 11.5).

11.1.2.5 Cruising

Cruising has become a major part of the tourism industry. With millions of passengers each year mainly the aging population undertakes a journey by sea, are visiting coastal cities and shipping post for a holiday or vacation. Cruise is a lifestyle that involves living for extended time on a boat, while traveling from a place to another place. It is for pleasure, plus visiting cities and shopping handicrafts, at various destinations.

Cruising is generally referred to trips or undertaken journey for days: few days or more. However, this can be extended to round-the-world voyages. Tourists now travel to experience and enjoy their life in the water by cruising as holidays. Egypt is highly known of its Nile Cruising, also couples explore cruising the Bahamas. Cruising is also vibrant through its good reputation and successful experience (Fig. 11.6).

Fig. 11.6 Cruising is among the fastest growing facets of the industry, more especially for the economy of the United States (top) (Photos by Wei-Ta Fang). Their highly furnished cruises are operating at random, voyaging across the globe from swallow seas to deep Atlantic Ocean. Photos (bottom) of Nile River Cruising, Cairo, Egypt (Photos by Arba'at Hassan)

11.1.2.6 1–5 Star Hotels (Star Classification)

Internationally, all hotels are backed by a system of grading which rated hotels according to the quality service and standard. The star classification system is a common one for rating hotels. Higher the number of the star ratings, for example number 7, indicates more luxury, and the standard of the facilities in the hotel is. Hotels are independently assessed in traditional systems and rest heavily on the facilities provided. Smaller hotels, which quality of accommodation could fall into one class but the lack of an item such as an elevator and other necessary facilities and appliances. It would prevent it from reaching a higher categorization, in the quality of the accommodation, and the food within of hotels.

The conventional system of tourism was purely relying on such so-called reputation. This is still continued to flourish to future of the industry. The future will begin with the home news of higher quality of international chain hotels like Hilton, Hyatt, Sheraton, Shangri-La, Marriott, Corinthia Atlantic, and Novotel. Among other hotels are few international hotels known for first rated quality services classified as 5-stars. Good numbers of customers, staying in these accommodations, feel much convenient to be satisfied with their services. The trend is shifting attracting both domestic and international customers to experience luxuries fancy hotels.

11.1.2.7 Specialist Holidays

Tourism is being a dynamic and evolving industry, from contemporary mass tourist to today special interest holidays. These are holidays in which people come together to share common interests, or to take the opportunity to learn something completely new, as part of their holiday in destination. It can be in the form of academic meetings, ornithologist fair, learning about the aboriginal dance of Taiwanese, or learning how to use a chopstick. There are more than 150 special holidays on offer by various tour operators like Thomas Cook, MyTravel, Gambia Experience, and Ethical Portal Travel. They are known to add pleasure to the tourists who are looking for more than the sun, sand, and sea's experiences. Some of the examples of special holidays are antique, sports, academic travel, cultural encounters, and bird watching.

11.1.2.8 Safari Parks

Wildlife watching has been very popular form of travel interest among tourists. Game reserves and zoos are also become major economic contributors for destination. Safaris are currently popular in the tourism industry. Commercial tourist attractions are where visitors can drive in their own vehicles and/or observe the wildlife in nature or wild rather than viewing animals in cages or small enclosures. These Safari tours are mostly conducted in Africa even some can be found elsewhere. The main attraction is large animals from Sub-Saharan Africa like giraffes, lions, rhinoceros, elephants, zebras, antelopes, and so on can be watched or ridden at the site.

Fig. 11.7 Taipei Zoo is remained home to be one of the Safari parks where still remains icons for the future travel and tourism market with more and more tourists are venturing to unique life time experience by having a close contact the wildlife (Photos Wei-Ta Fang)

Giraffes being fed by visitors in the West Midland Safari Park, generally are larger in size than a zoo. Safaris in other countries than Africa offer watching of wider varieties of animals, in small area compared to game reserves in its original habitat in Africa. The concept of ecotourism has increased the popularity of safari camps and game drives in most African countries. It becomes major markets for European and Scandinavian tourists (Fig. 11.7).

11.1.3 Future Trends in Tourism

Tourism, like any other businesses around the world, has been jeopardized by the reverse impact of recession for many years. As a result, tourists are being reluctant to spend their disposable income to go on vacations, exception for those retired, students, and semi-retired segments. The economic recovery of nations has shown a giant boom of the industry on destinations like France, U.S.A., China, and other developed countries. Surprisingly, tourism as the world-admired industry has less or minimal repercussion has destinations continues and will continue to show potential growth in sales increment on tourist arrivals as a potential tourist destination.

Despite the economic crises of the world today, tourism still flourishes with bunch of people going on vacation. The United States are opening up their borders for easy access for tourists. The year 2012 witnessed President Obama spoke on ease of visa access for tourists of developed economies like China, India, Taiwan, allowing citizen of these nations easy entry to America to experience and help boost their economy through visit to Orlando (Florida). That year showcased the free visa application for Taiwanese citizens to travel to Hong Kong on holidays. Whilst majority of the African destination are opened for visit less visa procedures. These development and advancement of tourists witnessed the immense novel experience being brought to meet the demand and exceeded the expectations of the customers, precisely;

- **Space Tourism**
- **Virtual Tour or Panoramic Tour**
- **Hotel Room Design**
- **Super Long Breaks**
- **Underwater Hotels**
- **Ethical Working Holidays**
- **Sport Tourism.**

11.1.3.1 Space Tourism

Space Tourism is the modern observable fact of tourists paying for journeys into the space, even though orbital space sightseeing opportunities are limited and luxurious. This is happened with only the Russian Space Agency providing transport to the space. Predictions are expected to expand in years to come, with more tourists wanting to experience the space tourism. Among the primary attractions of space tourism are the uniqueness of the experience, the excitement and admiration of looking at Earth from space. The experience's notion is an exclusive status, and various advantages of weightlessness. The space tourism industry is being targeted by spaceports in numerous locations, including California, Oklahoma, New Mexico, Florida, Virginia, Alaska, Wisconsin, Estrange in Sweden, as well as the United Arab Emirates.

11.1.3.2 Virtual Tour or Panoramic Tour

This type or tourism is a simulation of an existing site, generally self-possessed of panoramic descriptions, as opposed to actual tourism. A virtual tour is typically accessed on a personal computer, basically it is a systems of animation design with real images of an attraction or destination on a 3-dimensional (3D) drawing with a computerize background. However, it is often used to describe a variety of video and photographic-based media.

The expression of saying panorama (scenery view) is to indicate a continuous sight, since a virtual can be either a series of photographs or panning video footage. However, the phrases "virtual tour" has mostly been associated with virtual tours created using still cameras. Such virtual tours are made up of a number of shots taken from a single vantage point. The camera and lens are rotated around what is referred to as a nodal point (the exact point at the back of the lens where the light converges). The internet will play a major role in people's itineraries. Three-dimensional imaging of places will be important in planning one's trip. Also, travelers are seeking economical alternatives, deals, and packages—thus, the emergence of more travel agencies on websites.

11.1.3.3 Hotel Room Design

Tourism industry is undergoing rapid development across all facets of the industry. In so doing, technology is not an exception. In the future, hotel rooms will be designed to meet the expectations and requirements of hotel guests using tomorrow's technology. Future hotel rooms will feature the latest technology in the fields of media and communication of emails, phone calls, internet connections, liquid crystal display (LCD), and televisions with multi-channels. These are in mixture with archetypal products fashioned by renowned manufactures. Client comfort will be the rule while the technology functions make available the possibility for individual control of media, light, and climate.

Some of the innovations are predicted to include anti-jet-lag lights, an active comfort bed, a personal spa area, an intelligent mirror, and a large media display window, all of which are geared to go extra-miles beyond the expectations of the customers. Industrialized destinations like Taiwan, United Arab Emirates, United Kingdom, U.S.A., and Singapore. They are on the move with the technological trend, whilst less developed countries of Africa are on-going more eco-friendly to conserve the energy and more sustainability.

11.1.3.4 Super Long Break

These are the current developments in the tourism industry. People are jetting off on long weekends of two months or more to travel. They enjoy themselves with their families to destinations like Dubai, the U.S.A.'s cities, and Hong Kong! These demands for such frantic long and mid-haul trips are responses to workers being too overworked to take their full holiday entitlement and have a proper holiday. Employees work hours and months making savings to undertake a vacation to tour around the world. Organizations are giving study leaves and maternal leaves to staff on which they are entitled. They are to stay away from work for a longer period of time during which discoveries, self-upgrading, and vacation are taken. In America, it is the undertaking of cruise journeys by ageing population of retired and semi-retired citizens.

11.1.3.5 Underwater Hotels

The advancement of architectural skills and innovation in building design, technology and engineering will further enhance evolution of the industry. The future developments of tourism accommodation are expected to be built—both on land and under the water, where tourists can have the opportunity to see and experience the real life underwater. These hotels are being built underwater to create the first ever facilities in the marine as it is believed that the future rests in the sea. It is certain that one day in the future, a whole city will be built in the sea. Therefore, this is why plans are to slot in many different elements connected with the ocean. This is also why underwater resort hotels are now built under the sea.

The eye shadow will be ocean-based, the cinemas will screen films that focus on aquatic themes, and a children's sea world will educate, as well as entertain them. These visions the formation of facilities and opportunities are to those who do not dive or do not even swim. They can experience the tranquility and inspiration of underwater world. The aim is to inspire people to develop new awareness of the sea. At the same time, they are well emphasizing the positive aspects of water. Few examples are already established underwater hotels. They are Hydropolis (United Arab Emirates), Jules Undersea Lodge (Florida, U.S.A.), Utter Inn (Sweden), and Poseidon Undersea Resort (Fiji).

11.1.3.6 Ethical Working Holidays

Ethical tourism is gaining popularity among future travelers. It is predicted that over half of future travelers would be willing to pay more for their overseas holiday if they were guaranteed that the "money goes towards preservation of the local environment". Workers in the destinations are guaranteed good wages and working conditions, or the money goes to support local charities, environmental conservation projects, energy saving plans, poverty reductions, and clean water provisions. Travelers are becoming more aware of ethical issues and therefore chose holidays that are more ethical than the tradition leisure type of holidays. Organizations are also giving ethical working holidays to their staffs who wish to travel and take part in helping their host destinations by volunteering or even through philanthropic gestures.

11.1.3.7 Sport Tourism

The international travel and tourism are well renowned for sporting events from the prehistory charioting race to the present Olympics. Sport tourism is the travel of tourists to attend and take part in leisure, recreation, sporting events around the world from national MBA, Soccer, and Rugby, and volleyball tournament to International football leagues. Today, the most widely sports are the Champions league of UK, Premier league, UFA and Olympics, all of which draw participants from far and wide. At the same time, the Olympics and FIFA World Cup are the talk of town, seizing the headlines for the enormous main-stream of destinations. These events have been the key economic driver of destination who hosted them.

Based on those, there are biddings for countries to host the Olympics whilst FIFA remains in Europe. Good numbers of destinations have high-quality reputation created by sport notable Brazil, Caribbean, and Thailand as renowned for their winter training, golf, water sports, or other specialist destinations. Leading analysts predict that sport tourism is set to take a central position at the vanguard of future growth of tourism industries. With participation vacations (holidays with an activity attached like golf, tennis, and adventure) will lead for consumer purchasing trends. Sport Tourism has contributed an astonishing 14% of overall travel and tourism receipts and is set to grow exponentially over the next decade (Harris 2012).

11.2 Future Tourism Strategy and Market

11.2.1 Objective

Future of the travel and tourism business will be on evolution for destinations, as well as an amalgamation business. Here, tourism organizations around the world, specifically in destinations, will be working together in partnership or integrating as one large business entity. They will make themselves to achieve their fundamental goals by having ages over their competitors in the market. In a business of revolution, travel and tourism will register success in several areas. They are the guiding principle, legislation, convention, development, and human resource development, official recognition in both peripheral and urban settings, national and international advertising to promotion.

The industry has over the year witness the successful operation of international chain hotels. They get the following benefits of blueprint and beyond. Destinations are relying vividly on tourism. They will endeavor to ensure the good brand name for gratitude and quality control on sightseeing possessions. While creating loyalty amongst customers, they will enrich educationally for service proficiency of the tourism workforce. Economically and business wisely, the industry will go beyond human imagination to empower the status of tourism business in the destinations. Therefore, tourism will be assets for many of the following key areas:

- **Economy Weighing Machines**
- **Expansion of Industry Horizons**
- **Exclusive Production**
- **Diversification of the Industry**

11.2.1.1 Economy Weighing Machines

It has been the cost advantage that travel and tourism organization will obtain suitable benefit to their integration with other organization, either locally or/and internationally. The common advantage of which will be future purchasing bulk materials through long-term contracts. This can be a tour operator purchasing bulk seats and rooms at a hotel. While decision-making growing of specialization of managers, this could be financial obtaining lower-interest charges when borrowing from banks, having access to a greater range of financial instruments. All marketing spreading the cost of advertising will be over a greater range of output in media markets. Each of these factors reduces the long run average costs of production by shifting the short-run average total cost curve down and to the right.

11.2.1.2 Expansion of Industry Horizons

Following the integration of travel, hospitality, event, and tourism organizations are working together as one business entity. The overall industry will continue to grow bigger and superior. There will be an intensification of an organization through the acquisition of comparable corporation. This can be either in the same level of business or different level on the chain of the distribution in the anticipation of achieving financial system of balance and market competency. Unrestrained imaginative spreading out business across most regions, this can show the way to powerful domination of tourism remaining, as the world most viable industry across all level of economics, socials, and environmental developments.

Associated with future is perpendicular development of tourism business enterprises. They are in the course of the acquirement of companies (agencies) that construct the intermediate travel and tourism products and services. They are in the form of hotel accommodation, flight seat, package tour and excursion, needed by the clients. Any such value chain activities can be acquired through vertical expansion/horizontal incorporation. The industry is experience as the Gambia Experience about European tour operator which has their own flight, travel agency, and hotel. This type of trend is expected to become globalized.

11.2.1.3 Exclusive Production

Desirable increase of tourist demands will further result in as elite. This is to refer to a system when an organization in the industry; been a seller, the travel agent or wholesaler is the tour operator, will try to procure from the contractors the airline and the hotels. The accommodation provider and airline on the considerate of assimilation, that no other distributor, will be allotted or taken delivery of supplies in a given area accelerating the rate of completion for qualitative and quantitative production that suits customers' needs. Whilst the sales outlets are owned by the supplier, airline, and hotel motivate vertical integration to the outlets. The self-governing productions can be bought or merged by larger companies to expand their scale of operation of productivity. Exclusive production of the industry will be an obstacle to access. It should be capable of the safeguarding business integrity on the justification that it is an advantage to clients as it can tolerate after sale service.

11.2.1.4 Diversification of the Industry

Tourist industry emerging is facing evolution, simply due to the diversification of production and packaging of travel and tourism products. The diversification involves a variety of development from product development, product launching, and advertising approaches in destinations. All of those are shifting from a traditional format to more complex technological strategized approaches. Such developments and packaging give room to increase profitability of destinations. The trade organizations will integrate through their amalgamation with agencies. On the same scale and level of production, mutual acquire to gain sales and quality product, are done by obtaining new products to new markets. There will be a share of experience in the production of products. Today the industry is experiencing such transition from the conventional mass-package tourism to eco, ethical, sustainable, responsible, pro-poor, and event tourism. They are all resulting the diversification of tourism and tourist experience the business for the future. It is expected to continue to flourish with advance researchers and professionals undertaking managerial positions in the industry.

11.2.2 Trends and Paradigm Shift—Road to 2030

Flourishing of the world tourism is optimistic and consistent in the sense that, tourism will be one of the major economic superpower for nations in the future. Destinations will be capitalizing on trends of marketing strategies, quality supply, sustainability of products and services offered. The rising affluence of diversification on trend and customer needs will have strong authority on destination management at regional and national level. Tourism will showcase a tremendous demand for package and special interest holidays in bulk qualities, the future of the world, despite the financial credit crises. There will be peace, prosperity, and experience travelers who will be adhering to discover new sites and destinations. They are from traditional weekend breaks off work to the more sophisticate holidays in simple terms bank holidays which shall last for a month to travel on vacation.

Tourism is being the world fastest growing industries. It has prescribed by the World Tourism Organization (WTO). The rapidly economic development and social prosperity of destinations and global diversity in culture has made it a worthwhile venture for tourism discovery. It continues to make it valuable for an experiencing far more than just mere sun, sand, and sea package. The aviation industry is gradually expanding with more flights operating domestically, nationally, and internationally. The intercontinental carriers are making travels easier, accessible and affordable. Lately the services are supplemented low-cost carriers making traveling more positive and possible. Therefore, the tourism by air is becoming more popular as people who need to travel are facilitated by air flights. On other means of transportations, people use the facility of online booking and reservation systems. These become cheaper and convenience as people who travel by air can purchase their tickets and make hotel reservation using online. The search engines by virtue of economic development and social development are emerging markets in Asian Pacific, Africa, and Latin America, precisely the processing of visa and travel particulars are becoming more convenient and easy than ever before.

The global concern for climate change and carbon emission control will, as well at long run, resulted in the rise on fuel prices. This is on contrary impacts of rise on fares. However, this will not deter natives of developed and developing economies from traveling as they will have the ability and power to still go on vacations. As the expenditure of flying rises, they will exceptional choose an international travel. Tourists will no longer consider exclusively to be relaxed boldness regarding departures out of their normal place of residence to go on vacations. Surely, holidays will become more significant as time and experience dictate the future holidays. The future of travel and tourism experience will surely depend on motives and treasure. It will be worthy more of rationales, with not only our own needs in mind, but also those of the destination to go for vacation. In so doing, we will see tourists designing up for holidays with more proposals on how destinations benefit them. In summary, vacations have become a necessity to people to take a break from their daily works.

- **Future Tourism Forecast on China (Asia)**

Mainland China is located in the eastern Asia and on the western coast of the Pacific Ocean. It has an approximate territory spot of 9.6 million square kilometers, with an inland and coastal water spot of more than 4.7 million square kilometers. On the eastern and southern continental are coastlines extending for about 18,000 km. Its vast maritime territory is studded with 7,600 islands. China shares common borders with 14 countries among them include India, Vietnam, North Korea, Kazakhstan, Mongolia, Russia, Laos, Bhutan, Nepal, Pakistan, and Afghanistan, and is adjacent to eight (8) nations on the ocean. There are four (4) municipalities directly under central government administration, 23 provinces, five (5) autonomous regions, and two (2) special administrative regions. The China capital city is Beijing.

This country has a higher growing tourism prospects for the future, simply due to its booming economy and forecasts on. Its higher population and infrastructural development are powered by numerous attractions. Sites of significance to tourist, both in urban settings of Beijing and Shanghai and rural areas are rich in tradition and fine arts. In fact, China is highly recommended and recognized over its hi-tech products of electronic and electrical appliances. China is taking the lead for international and domestic tourism level. The prediction has a good future to be on the itinerary of interest to be visited by international, sub-regional, and regional tourist. China is endowed with vast attractions which make them appealed to tourists. This pristine destination gets world cultural heritage site like Mountain Wuyi, Buddha scenic area, Mountain Taishan, and temples. They are gaining popularity off religious significance in Asian continent. The destinations have many to offer to public. The main tourist attractions in Beijing are related to evening entertainments, one-day tour of Beijing, and a half-day tour of the famous historic in Beijing.

Fig. 11.8 In recent year, China gets good reputations for hosting of expositions, exhibitions, conventions, festivals, and fairs (Photos by Wei-Ta Fang)

China tourism industry will witness tremendous experience during their tour. Over the years the possession of attraction are appealing to most domestic and international tourists, either for leisure, religious, educational, or business. However, some pulling factors to China are the tourism cities, scenic spots, foods, shopping, and the fine arts world heritage sites. It has become a venue for world most renowned events of the days (Fig. 11.8).

Mainland China has the oldest historical legacy than any other nation of the world. The unique spectacular destination features many places of interest to international world. Among them is its great wall, well known as The Great Wall of China, to symbolize the old China. In addition, the summer palace is portraying special blueprint of the Chinese landscape design. It is specifically providing a monument that stands from the Emperor to his marble. There are ancient towns and cities like Zhouzhuang, Wuzhen, Fenghuang Ancient Town, Wuyuan, and Shangri-La. They are the remains of the original sense of authentic traditional China. They are still flourishing showcasing architectures, handicraft, and technology dating back to the old years. China has a tourist destination still possess wider scenic spots of waterfall, lakes, and magnificent seashores. All of them are well-known of leisure, scenery view and health purposes. China is the land of dramatic venue for future tourism, let's go and experience for ourselves.

- **Future Tourism Forecast in the U.S.A. (North America)**

The United States of America (U.S.A.) has been on the forefront of both international and domestic tourism. It has been ranked as the second world and most popular visited destination. It has affirmed by UNWTO Report (UNWTO 2011; 2017) of world top tourist destinations. America is sanctified with numerous manmade and natural attractions. They appeal to the local and international communities. This land's tourism industry is very positive on the opportunities of filling the demand capacity of the market. The former President Barack Obama (1961–) signed the bill with law called Travel Promotion Act (TPA) in 2010. It had settle-up a new non-profit organization called Brand U.S.A. All of them are geared to promote the U.S.A. as a world class tourist destination. It is added with primary focus to advertise and promote America. It has travel destinations to international communities. Based on these recent developments, America is determined to be one of the most reliable future tourism global destinations.

Recently, America has visa free for 38 countries in the world. Citizen visiting the U.S.A. will be permitted to enter without undergoing the process of obtaining a tourist visa. The former Obama administration was working on strategies to increase the number of countries to visit America without a visa. This showed the promising future of tourism to continue in the U.S.A. with more people having the access to enter for vacation and business purposes. America's doors are opened for countries with rapidly growing economies, huge population, and emerging middle-class population examples. They are China, India, and Brazil. Tourists from those countries will continue to double. In so doing, it will accelerate at an approximate 40% visa processing for visitors from China and Brazil. Future showcases are steady status of America being a renowned popular top destination of the world with increased manmade attractions appealing to the international community.

Box 11.1: Future Tourism Forecast on Orlando, Florida (U.S.A.)

Every year ten millions of tourists all over the planet go for vacation to America. Orlando has been so instrumental to American's economy with the reality of being the tourism wonders of the day. In 2018, 80 million international visitors visited America. That helped the tourism industry generated over 1860 billion from International Travel Imports in 2018 (US Travel Association 2019). Tourism is the number one service that America sells to other countries of the world. America is opened for commerce. The existing steps to boost country's tourism business can increase the financial system and generate further employment. America gets the best producers to sell to the world today. The most entertaining destination of the world (Disney World) is the land of extraordinary national wonders. These are added with Robert Gate Bridge (Empire State Building), Gateway Arch, the land of iconic cities and many more.

Attractions constitute one of the main components of tourist itinerary. It builds the U.S.A. appealing from the possession of its theme park of Disney World in Orlando, Florida. The international arrivals of all tourists enter on different purposes ranging from leisure, academic, joint-family businesses, and sporting events. Tourism in America will continue to flourish with more people entering the states on tourism.

Seaside in northwest Florida are graded as one of the finest beaches in the world. In addition, other finest beaches with quantities of extravagant in the world seem also to be found and located in the U.S.A. Beaches in Walton, Bay, and Gulf Counties have been preferred for the number one ranking by Dr. Stephen P. Leatherman (1947–), an expert on beach quality and author of America's Best Beaches (Fig. 11.9).

Fig. 11.9 Opening of entry doors of fast and efficient visa service into the U.S.A. has boosted the America's tourism industry, as well as creating employment (Photos by Wei-Ta Fang)

The number one top amusement center serves as the main attributing to boost the future of America. It is the Walt Disney World in Orlando being the Magic Kingdom has been magnificent venue which continues to appeal tourists of diverse interest in the future.

- **Future Tourism Forecast on Gambia (Africa)**

The new product and service that could be introduced in Gambia to help improve the tourism industry in the near future is the River Gambia. Tourists cruising are using medium size ships than ocean-going with cruise ships are typically targeting passengers internally (domestic) and oversea (international). Tourists in sub-region and beyond are tourists at borders to Europe and Scandinavia. River Gambia cruising will have tourists from inside into the countryside. They are usually tourists relaxing in the cruise and interesting seeing the land nearby. Gambia River cruises will usually last from 2 to 3 days, although some can last until 5 days or longer.

Facilities in cruises are the sun decks, dining rooms, lounges, fitness facilities, swimming pools, casinos, and other entertainment. Accommodation, meals onboard, entertainment, special events holidays, festivals, contests, and concerts are usually included in the cruise price, while bar expenses, massage, laundry and cleaning, and phone calls are not. Those services are free of charge to any tourists who are on the cruise.

The last day of tourism will be guided tour to historical and cultural sites, visiting local attractions, museums, and galleries, and other points of interest to attraction. Some of the places are like Janjabureh, Mungo Park Memorial, and Juffereh the home of Kunta Kinteh, and Gorilla Island. During the tour, a guide will inform the tourist about the area they are visiting.

River Gambia cruise is different from a river excursion. It will have a break for the sightseeing of the land so tourists can have a bird watching and fishing activities. They can spend as the time specified by the tour guide. Tourist of all types will be attracted to the views. They will be catered certain activities like ocean enjoy cruising. They can relax like at home. They can also enjoy the rhythm of life floating while in the cruise. Currently in Gambia, there is only one exciting in the country. It is about Dentong Bridge. It is used when a visitor takes trips to Juffereh alone; while the River Gambia Cruise going around the entire program during visitation of interesting attractions and historical sites along the Gambia River. All of those, become a part of the voyage along the Gambia River—one of the fascinating future products of the Gambia.

11.2.3 Empowerment and Local Wisdom

Empowerment in it simple term means to authorize or give power to a person. Concurrently, empowerment has the world tourism development faces diversification and higher level of competition in the market. It is similar trend that witnesses a rapid growth of human demography. The international population is expected to double in the future particularly in nations of developed economy. Countries which are more visited have many destinations, like China and India, are booming their economies. Tourism still remains evident of being one step ahead of other sectors, providing more features and greater values to consumers.

The future of tourism will witness more bona fide and authentic encounters culturally with local cultures than the ordinary mass sun, sand, and sea package. Over here, tourists can only stay on their sun beds along the coastal site to be sunburn. However, comes the future, tourism will further go beyond that practice and philosophy. It will be developed to provide the highest products and services which meet both tourists: locally and internationally. The international standards must be certified by observing organizations. It will be about the positive reception of local people and their cultural diversities, their norms, and values. These will further strike to maximize cross-cultural communication and broader understandings by making sure those tourists get what they want. At the same time, the mutual local people also benefits from tourism.

11.3 Future Tourism Planning and Management

Future tourism planning and management are essential. It is the development and administration of tourism operations in destination, concerning with the overall tourism business at different levels of the scale of operation. They must provide inclusive tours, package tours, travel agents, corporate travels, vacation planning, online booking systems, and social networks media for telemarketing. These planning and management will be concerned with the diversification of tourism. The unifying tourism products and services, as well as coming up with added services to win the heart of its potential customers, are indeed important. Actually destinations are becoming innovators in designing, marketing, management, and development of destinations.

11.3.1 Domestic Travel

As the concern for future tourism creation and designs continues, there will be more local experiences. At a particular time, destinations will better still have something to offer for both domestic and international tourists. Such experience will be the discovery of local cultural diversity and appreciation of local people and their norms and values. Many local people in the Philippines travel to the province along to the seaside with leisure and weekend stays. Meanwhile, the African people travel on weekend breaks to attend the cultural festivals and ceremonies, where as in Taiwan, people travel to "Houli" to go biking trips, as well as attending festivals like the dragon festivals. These festivals found to be mostly occurred and diversified in the future. It is, therefore, more calling for more appreciation for the local cultures, so will keep the traditional bodies preserved until the future.

11.3.2 Searching for Authentic Experience (KISS = Keep It Short and Simple)

Future tourism is focusing more on local authentic experience with a frame-work ongoing profoundly in the local communities, spending holidays with local people. This will be admiring something specials in safari camps, tents, temples, and home-stays. Ecotourism is soundly environmental. The community travel is with certainty to flourish with creativity of broadening ideas of horizon and promotion of cultural understanding. One of the factors supporting prosperity of tourism will be the culture. It is by raising awareness and informing tourist about community based tourism. It is, through the responsible travel and development of products and services, will highly appreciate the local culture economically, socially and environmentally. Domestic and international travel will continue increasingly on discovery of interesting and mysterious attractions. In order for the destination to maintain a balance of local communal involvement in tourism, to sustain domestic market share of domestic tourist, they will requirement to supply new to a home customer by giving them a unique experience about their own locality.

11.3.3 Diversification in Mode of Transportations

The rising concern over major troubles facing our earth today is the global warming. It is blamed on international travel, through the emission of smoke from flights. The affirmed concern has awakened the attention on international community. The customers are cross-checking on carriers with green certification and those responsibilities with contribution to the maintenance of such effects. However, the alarming issues over the rapidly incensement of fuel price has become an international dilemma with falls in many tourism operations going down. The international aviation industry is no exception.

Diversity means transportation despite the higher increase in fuel prices. This will not stop people from traveling. The question remains on how and through what? From the early inception of travel, it is renewed by using of airplanes on international travels. This is expected to change in the future due to the higher charges on flight tickets. People will switch to other mode of transportation, simply we progressively witness a positive reception of other means of transport. This is the undertaking of journeys through the uses of sea or/and land carriers. They are like the boat, train, and bicycle all of which are in advancing in attractiveness.

Distance travel possibly will be turn down as economic and environmental essentials stipulate the use of trains. Mass Rapid Transit (MRT) is becoming popular than ever before in places such as America, Asia, and Europe. There is increasing number of schedule and charter bus services operating inter-county, within towns, and cities whilst fascinating enough families are taking ride-using bikes to enjoy early summer morning blue skies in the coastal and rural settings.

Further, interestingly in the face of the use of local slow mode travel, in future tourism will distinguish airlines increasingly register success in good number of flights. They will be utilizing environmentally pleasant bio-fuels, which are neither that much effective but still congestive than the ordinary fuel and gasoline. With the advancement of technology, Taiwan has come up with new scooters that are environmentally friendly in the year 2011 namely, Gogoro, the electric scooter. This development has resulting great example in the advancement of future tourism which is eco-friendlier mode.

11.3.4 Destinations Evolution

The future call of change is already having a philosophical effect on destinations. It will further continue than ever before, almost immediately since the inception on global travel and tourism industry, but it regionalized and eventually becomes an international business of destinations. This tendency is in progress as destination's tourism development areas and facilities of accommodations. The service providers are centering on becoming responsible in their operation of meeting the needs of present customers, while cooperating benefits for the future. They are through actions demonstrated the movement of sustainable tourism initiatives of maintaining a sense of attraction. They are also more authentic and unique in experiences. The destination has further become extra perceptive in their advance application of information technology for marketing and promotion offers to destinations, from the white sandy beaches of Kenting to the eco-friendly tours of Sun Moon Lake of Taiwan.

11.3.5 Concept Industry Experts in Destinations

Future wise, the tourism industry will be about uniting human concepts. Educationally and inspiring those thoughts will insist them in the industry to face the challenges of climate change, poverty reduction, and sustainability of the planet earth. Such progression realizing the dreams of the future led to the cluster development of attractions, accommodation, facilities, accessibility, and awareness programs of destinations. By attracting special types of customers, they are responsible in their reactions during holidays and at the same time are good spenders. These innovative minds at work is powered by knowledge based researches and implantation of skills that follow the market trends leading to a brighter future in apprehending emerging fashions in travel and tourism industry.

Box 11.2: Case Studies: Top 10 Most Future Popular Destinations 2017

Tourism industry in its purest form of prosperity for years to come will be flourishing and destinations will be competing to be ranked among the world most popular destinations. This is depending on the higher magnitude of international arrivals. International tourism will continue to prosper further in the future with more domestic and international travels. According to the World Tourism Organization (WTO) the international tourist arrivals in destinations will continue to grow in the year 2010–2030. However, at further reasonable rate than the past decades, the number of international tourist arrivals worldwide will increase by an average of 3.3% a year. As a result, an average of 43 million additional international tourists will join the tourism marketplace every year. Statistically, the following destinations are the world most popular destinations of the year 2017. This was prescribed according to the number and percent of tourist arrivals.

Top 10 Most Popular Tourist Destinations in the World Year 2017

Here is the list of top 10 Most Popular Tourist Destinations in the World for the Year 2017 as indicated by UNWTO Market Trends:

1. France (82.6 million visitors)
2. U.S.A. (75.6 million visitors)
3. Spain (75.6 million visitors)
4. China (59.3 million visitors)
5. Italy (52.4 million visitors)
6. UK (35.8 million visitors)
7. Germany (35.6 million visitors)
8. Mexico (35.0 million visitors)
9. Thailand (32.6 million visitors)
10. Turkey (30.0 million visitors)

Globally, as an expert category, tourism ranks fourth after fuels, chemicals, as automotive products. Tourists, however, will enter marketplace every year. According to UNWTO forecast for 2010–2030, the international tourism will increase by 43 million international marketplaces every year (Kester 2011; UNWTO 2011). International tourist arrivals grew 6% totaling 1.4 billion during 2018 (UNWTO 2019).

11.4 Conclusion

Tourism being the rapidly expanding industry globally has a strong positive future. This industry will continue to grow and flourish beyond reasonable boundaries with more innovations, technological, and responsible development. It should appeal to the demand of the tourists. The global diversification of product development, innovation, and service delivery indicates a success story of the sector to forge ahead. Recently, there has been a rapid in technological facilities and aviation industries, all are supporting the future advancement of the industry. On the other hand, human demography of having more birth rate than death in the growth of both domestic and international tourism.

People are still traveling within their home countries and across international borders. Destinations are changing their style and pattern operation from the conventional sand, sun, and sea package, change to a more sophisticated, ethical form of travel and tourism industry. The trend is, however, goes beyond human imagination about going to extra miles. People want to serve the taste and flavor of client through the establishment of underwater hotel and space tourism. All those in the long run will attract more tourists. On the other hand, destination will discover and establish more manmade attractions, and then exploit the marine habitat. Future tourism will encompass everything with diversity of products, services, and tourism experience. Vacations will become unique and go to better ways in some destinations. These are to support local people, local livelihood, environmental conservation, infrastructural, and super-structural development in destinations, with value added services (VAS).

In a nutshell industry trends are going greener, using sustainable energy, alternative transports, and advance technology evolving the operation modes. The operation will grow bigger, while attraction becomes more authentic in package. Finally, tourism will move toward economically, socially, and environmentally. The business will meet the demands of tourists in tourism business.

References

Brendon P (1991) Thomas Cook: 150 years of popular tourism. Secker & Warburg, London, UK

Harris A (2012) Sports tourism set to take over travel industry. https://businessbarbados.com/industries/tourism/sports-tourism-set-travel-industry/

Kester JGC (2011) 2010 international tourism results and prospects for 2011. UNWTO News Conference HQ, Madrid, Spain

Matzarakis A (2014) Transfer of climate data for tourism applications—the climate-tourism/transfer-information-scheme. Sustain Environ Res 24:273–280

United Nations (2015) Transforming our world: the 2030 agenda for sustainable development. The General Assembly, United Nations, New York, NY, U.S.A.

UNWTO (2011) Tourism market trends. World Tourism Organization, Madrid, Spain

UNWTO (2017) UNWTO tourism highlights. World Tourism Organization, Madrid, Spain

UNWTO (2019) International tourist arrivals reach 1.4 billion two years ahead of forecasts, Press Release. PR No.: PR 19003, 21 January 2019. World Tourism Organization, Madrid, Spain

US Travel Association (2019) U.S. travel answer sheet. US Travel Association. https://www.ustravel.org/answersheet

Chapter 12
Epilogue

International business of travel, tourism, and event industry is very instrumental economically, socially, and environmentally. Today nations around the world compete to meet the demand of both domestic and international travelers. Tourism has being a successful industry in many destinations; the industry provides employment benefits, cross-cultural awareness, super-structural, and infrastructural development. This noble economic service sector has served as an economic backbone of developed and developing countries. Tourism operations and businesses ventures are diverse in nature from the small-scale peripheral operation in local communities of Africa to multi-national cooperate entities in mega cities of Europe, America, and Asia. International tourism business has being the world most popular and vibrant sector, providing lot benefits economically, socially, and environmentally. The industry serves as a weapon to fight against unemployment, environmental degradation, and cultural decline.

This chapter has been wrapped up each chapter in this book. The content chapters of this book could be discussed crucial issues in previous chapter as detected as follows from tourism education, planning, design, management and, the final impacts, positively and/or negatively.

12.1 Tourism Education

Tourism education is tourism course evolves into various approaches. Among the essentials one is the up-coming trend of SAVE program involving Students, Academic, Volunteer, and Educational (SAVE). This is a new growing segment of the international tourism, with this recent development companies, schools, and agencies are supporting students for international outreach programs on various events. The events are like internship programs, field trips, international student exchange programs, study visits, workshops, and volunteering (see Fig. 12.1). It has facilitated tourism education by providing students face-to-face trainings and firsthand

W.-T. Fang, *Tourism in Emerging Economies*,
https://doi.org/10.1007/978-981-15-2463-9_12

Fig. 12.1 Educational tourism is tourism course evolves into various approaches. Among the essentials one is the up-coming trend of SAVE program which involve Students, Academic, Volunteer, and Educational (SAVE) at Guandu Nature Park, Taipei, Taiwan (Photo by Wei-Ta Fang)

experiences which equipped them practical with skills to meet the job markets and human-resource needs of the industries.

Furthermore, tourism, travel, and hospitality are like any other academic fields of studies. It requires up-to-date studies, researches, and knowledge acquisition. The structure of the tourism industry is diverse which constitutes various careers and job outlooks, all combining to form the overall industrial workforce. The most evident point today is the fact that tourism cannot be left in isolation without one being equipped with the relevant academic know-how, knowledge, and professional skills. This is very important when dealing with the industry trends and diverse customers with the goal to satisfy their unlimited needs. In so doing, one of the best weapons is to major your career in tourism academic disciplines on courses offered on tourism, travel, and hospitality. This is because it relates disciplines at schools, professional training institutes, colleges, universities offering structure trainings, teaching, and courses on tourism management and services. Tourism education career courses are abundant in institutes, colleges, and universities for individuals wanting to work in hotels, tours guides, resorts, cruise ships, entertainment parks, event planners, casinos, travel agencies, convention and exhibition centers, and many more. Have you ever gone on a vacation or planning your trip! Think about it! There are professional training personnel who combine and guide your travel itinerary from your point of departure to your destination of arrival, taking care of your transportation, accommodation, food and beverages, and entertainments. They are peoples who have acquired basic or advance training in tourism.

Today, there are various courses offered on tourism, ranging from certificate, diploma, advance diploma, undergraduate, graduate, and postgraduate degree courses. One of the key essential scholastic objectives for tourism education is to construct a curriculum of study at colleges and universities for practical researches of cultivating human resources in the industry. Additionally, there should be theory programs provided to build up professional management team with frameworks to facilitate them with analytical skills, operational skills, and managerial skills of the global tourism industries. These programs will have potential contributions to further strengthen the growth and development from environmental planning.

12.2 Environmental Planning and Heuristic Tour Guide

Environmental planning deals with the principles, philosophies, practice of ecological managements of the biodiversity. It embraces a wider collection of the fauna and flora. In its simple term, environment deals with the surroundings where we live and some other things that constitute our surroundings within our vicinity, like plants and animals. These are instrumental towards a sustainable meaningful live. Today environmental planning has become a global issue with involvement of the world renowned organizations like United Nations, International Union for Conservation of Nature (IUCN), RAMSAR Convention, and World Wildlife Fund (WWF). They are all advocating for environmental protection and development. It is also to reduce the impact of carbon emission, as well as to protect and preserve endangered species and wetlands. All wetlands serve as breeding grounds for living organisms. These can also be used for educational and research purposes for students majoring in environmental education. Today, the international communities are conceptualizing the vision of green environment. Statistically, report has shown that approximately 75 percent of the international community's live in regions that are liable to natural disasters. Among human negligence is a key factor, some of this places have already being affected by earthquake, tsunami, tropical cyclones, flood, or drought at least once between the year 1980–2017 (Alexander 2017).

12.2.1 Environmental Planning in Global Perspective

Environmental Planning in countries across the world, has being a global issue and concern for all governments, international organizations, schools, philanthropies, and communities based organizations. All of these are advocating and collaborating to make this world a better place for the human race. Today, our environment is threatened by many factors. Among of the factors are the effects of carbon emission of smoke resulting from the high-tech industries, factories, cars, flights, coupled by noise pollution, congestion and over population in urban areas. These factors are resulting in degradation of reserve lands for establishment of new housing and real estates.

On the other parts of world, precisely in Africa, there is higher dependency on environmental assets. These have threatened the productivity of the wetlands which serve as habitats for birds, mangroves, and oysters. All of them contribute significantly for the economic growth of the nations. The efforts are from environmental engineers and some concepts of environmental planning are from their suggestions of strategic plans and formulated policies. They are all preserved the integrity of natural environment, at some element, setting up of capacities of residential developments with restriction of settlement at certain geographical locations.

12.2.2 Environmental Planning Considers Human Development

Environmental planning and development are the procedure and methods of considering the rate of human development. This is in relation to responsibly considering the impacts and effects on natural environment. The introduction ideas of sustainable developments to governments and local communities enable assessment making to convey sustainable development with outstanding thoughts and concerns of sustainability. They are geared toward the enhancement of local community livelihood, ecotourism, and greenery development for sustainable use of natural resources. The concept revolves in some parts of the world ensuring to establishment of community forestry schemes, bookkeeping projects, agro-forestry, women gardening, and rural micro-finance sub-projects. These are all geared to support the preservation and conservation of natural environment, in such a way that appeals to the local residents economically and socially. The integration of marine and coastal environment into the practices and philosophies of environmental planning has further strengthened the involvement of central government. They actively work in partnership with local communities and funding donors with utilizing universal frameworks to attain the justifiable human reactions towards the natural environment, and its endangered species, and resources.

12.2.3 Tourism Dependency on Environment

Tourism industry has heavy dependency on the environment. It features has many travelers to experience the beauty of scenery of destinations. Based on such affirmed contribution on tourism and tourist experiences have paved sound policies, development agendas, and regulation concern with the welfare of environmental sustainability. Through balancing the human interaction towards mitigating environmental degradation between human reaction and ecosystem, it has successful resulted establishment of parks and wildlife management schemes. These schemes have preserved endangered species of the biodiversity.

The ecological and environmental activities today are planned and dedicated to administrators. The global concern over greenhouse effects are increasing thus slowing the rate of development of biological features. These process is successful, timely, and effectively with the largely involvement of international organizations. Local residents answer the call by embarking on green roofing. This has been clear as crystal in Asian continent. There are many accommodations in Taiwan ornamentally designing on the terrace of the building. People need good life going with better environmental views detecting as an impartial approach of living symbiosis. This approach meets with the nature for preserving better quality of lives requested people's needs for the present as well as for the future generations.

12.2.4 Heuristic Tour-Guide

Tourists are people living in their normal environments for temporary short-term visits. The visit can be for leisure, business, academic, religious, and/or other purposes. With an aim of familiarizing them with local culture, they are fulfilling their aspired dreams of visiting heritage sites. The tasting of delicacy of local cuisines will be spent with a local family. These people in either traveling individually or in a group are accompanied by well-trained persons (tour guides). They are their contact persons, giving them factual relevant information about their point of interest.

Heuristic tour guides are those persons with experience based, hardworking, problem solving, and inspirational tourist guiding. They are people who always work extra miles to meet tourist needs and expectations. They provide tourists satisfactory rewarding trips, as they are helpful, supportive as required by their jobs, and modest by nature. Heuristic tourist guide is qualified trained personnel. They work for tour company license or on freelance based guiding the tourists on conducted organizer. During outing, they give them information to make sure tourists' health, safety, and security in a diplomacy and professional conduct. Heuristic guide tourism on excursions and other programs are operated locally on domestic bases and international occasion (outbound).

Domestic tour guides are the tour guides taking care of excursions and tours within receiving destinations. They collaborate with ground operators. The ground handlers who are agents for oversea tour operators will be in contract for a specific period time to render service to the clients. Outbound tour guides are licenses trained professional guide operating on international bases, travelling with groups of tourists from their originating destination to receiving destination, taking of their needs, and helping them in all their reservations and other necessities. They work with local guides upon arrival at their oversea destinations. The example of outbound tour guide is John, travelling with a group from Singapore to Leofoo Resort Guanshi in Taiwan, where he handles their reservations and sort out their rooms and excursions.

12.3 Landscape and Green Accommodation Design

Landscape and green accommodation are the technical and development oriented approaches used in land uses and modern state of the earth. Land development of the scare noble resources of land and its features include the hills, mountains, and wetlands (like swamps, lakes, and ponds). They are habitat of green plants that contribute to the beautification of the land surface. This is shown on how important the land constitute as the fundamental fixed-asset of human survival based on structural development of housing estates and the effective meaningful contribution to confront the emerging danger of higher ultra exposure of sunlight to the earth. Today, landscape and green accommodation design have contributed to cooling the weather and unification of scenic spots. They are characterized by a quality view of the land surface.

This fruitful area of focus is a factor on humanizing the environmental region. They are basic factors for development of the ecological units in a holistic move toward an eco-friendly world which embraces both the aquatic and terrestrial habitats. Landscape and green accommodation are design pattern and architectural of formations of infrastructure. The structural development is considered as spacing and beautification of residential and reserved areas. These disciplines focus on human habitat of lands integrating with modern residential development projects. They are central mainstream standards of the land uses. They offer optimistic improvements on the responsible and sustainable uses.

The usage of low carbon emissions and energy consumption is important. Houses are constructed in such a way to facilitate conservation of sunlight using solar energy. The collected solar energy is used for the lighting and other necessary appliances. As a whole, the energy produced from solar appeals to be environmentally friendly as it does not emit high carbon that pollutes the environment. The solar energy collected by wind turbines has helped generate more friendly electrical energy to supplement electrical power to citizens. The invention of wind powers and turbines are used in coastal area which help in generating the electricity because wind is a natural asset contributes to the production of cost effective renewable energy.

12.4 Ecology and Infrastructure Management

Ecology and infrastructure management has recently intermarriage through wider support in maintaining the species of fauna and flora. It focuses on preventing land use of not being catastrophic to natural resources of endangered species. The central point of preserving and conserving the authenticity natural homes of living organism are rapidly effected the growing over of land uses. It has eventually resulted to degradation of reserve lands used for residential and commercial purposes.

Today, this desire to protect the environment has resulted in tree planting exercises in some regions of the world. The formation of parks has resulted to establishment of national parks and wildlife management. Ecological and infrastructural management has helped to promote to ecological conservation method of land uses in establishing forestry projects. It has been an instrumental approach that eco-friendly lodging and camping facilities for tourist uses are helping the sustainable environment. The friendly in energy consumption has responsible approaches in reusing and recycling many resources.

At present, infrastructural development is constructed in a way that the ecosystem must be continuously protected. The involvement of the environmental expatriates in assessment and audit programs must be tighten for threats to environment—flora and fauna. Our environment is at risk and to find possible solutions to face the challenges lie ahead. We must protect and preserve the welfare of our natural resources, and rare species of fauna and flora. The international and local conservational groups must be actively advocating for the global concern which is enthusiastically bringing an effective change for a sustainable growth.

Hualian County of Taiwan has recently initiated and introduced new policies to safeguard the environment. Every vehicle waiting around the train station must be turn off. Their engine and air-conditioning system must be set to reduce the effects of emission and reserve energy. In doing so, many people have responded positively. As a result, the train station has become heaven for all. This shows a clear testimony of how infrastructure is contributing to environmental care. Such management is timely and responsible means to green environment.

12.5 Tourism Assessment and Satellite Account

Tourism accounts for most of the world development either optimistically or unenthusiastically. The contemporary situation of ecological development, principles, and practice, are distinct in various perspectives and planned in different agendas. They are diversified but all are aimed for one objective. Social and economic development of the environment has both negative and positive significance. Positive construction for all-purpose facilities in transportation and recreational purposes are like railways, roads, ferry terminals, and airports. On the other hand, facilities for leisure purposes are the theme park resorts, water parks, golf courses, game reserves, and yachting marinas. All these bring in economic proceeding to the nation's economy through tax collections from enterprises. Those operating businesses survive from admission fees while people receive social awareness. Moreover, they can broaden their networks with others at leisure facilities. Tourism has this potential in inculcating the concept of environmental conversation as generating economic is importance to local people.

In contrast, the negative impact of tourism arises when the number of visitors is being greater than the carrying capacity of the parks, zoo, and reserve lands. This will eventually result in over trampling on the social surface resulting to the exposure of soil erosion, as well increasing in pressure of danger on endangered species. Sometimes, the final tune exposure of the forest will initiate to bush fires as a result of people discarding lighters and cigarettes on dry grass.

The logic behind the environmental planning and design is to diminish legal responsibility of danger that our world is facing today. Looking into possible solution, the existing development should benefit to society in economy. These include the programs of social awareness, sponsorship, and funding for environmental campaign conferences which boost and support the economic status of host country and event organizers. Such program has paved prospects for building risk evaluation and environmental audits for land uses on residential and agricultural purposes. Feasible schemes are launched in most parts of government national development plans. The actions of such designation of real estate projects, government reserve lands, tourism development areas, leisure and recreation facilities, all are constituted national and private enterprises. The environmental development plan is around the philosophy of the new contemporary environmental planning techniques.

Destinations around the world, both in developing and developed economies, have propagated the concepts through mechanisms of safely and responsibly reducing the depletion of environmental and natural resources, like the land uses for resort and hotel building. Industry has high pressure on lands and water uses for the establishment of swimming pools and waterfalls, which when wisely utilized can increase the confrontation, sustainability, and safety of environmental resources. The sewage disposal from hotel and other facilities results in sewage pollution, making life unsuitable for coastal and marine environment. Pollution of the aquatic habitat will result to death of many lives. When discarded by the sea, it will make the lands so unattractive for tourism and locals. Among them are plans triggered on the following issues. Some of these issues are addressed herein with focused on three dimensions environmentally, socially, and economically.

12.5.1 Positive Environmental Audits of Tourism

- **Conservation**: Conservation in its simple term, means preservation and maintenance of the quality environment, both natural and manmade. It is crucial to tourism business, as it attracts a wider number of tourists to go on sightseeing and vacation to those destinations with wider collection of wildlife and scenery spots. However, tourists' relationship with the environment are multi-faces, many activities can have adverse environmental effects. Many of these impacts are linked to the construction of general infrastructure. They are the roads and airports, and tourism facilities—including resorts, hotels, restaurants, shops, etc. Tourism has promoted sound effects on environment by contributing to environmental protection and conservation. It is a way to raise awareness on environmental values that can further serve as a tool to finance protection of natural areas. This will increase socio-economic importance of the community when properly planned and managed.

- **Preservation**: The importance of tourism cannot be overestimated, it is essential directly or indirectly, preserving the world heritage sites, local cultures, and natural environment like wetlands, reserve lands, forests, wildlife, natural parks, etc. Tourism industry has great impacts in preservation of natural environment by raising awareness programs on environmental educations, preservation, key values, and financing protection of natural area as tools. This can even create an economic importance.

12.5.2 Positive Social Impact of Tourism

- **Education for Locals and Tourists**: Tourism industry plays an important role in educating local people, communities, and tourists. This brings the training of local people in most local societies to get benefits to be able to preserve the authenticity of local cultural, attraction sites, and products like tomato paste, Garry from cassava, soap making, etc. This is crucial in promotion of sustainable tourism which helps to bring economic proceedings. Tourists also can log in for training programs to acquire skills from the locals to be culturally aware to understand to local communities. Further, it will strengthen the concepts of volunteer tourism and cross-cultural exchange, thus pave ways for sponsorship programs for locals.
- **Linguistic Benefits**: Language as a medium of communication is no longer an issue of being a barrier. Many citizens have the capabilities to understand and speak diverse international language, all of whom are vital in international trade of tourism business. For this reason, there is higher dealing with people of diverse cultural background. Some tourists are not English native speakers, thus require tourist companies or agency to have language translators to inform them the places they visited. The translators most of the time are a multi-lingual spoken persons (male and/or female). Therefore, this kind of language translators have an advantage of speaking several international languages. They can survive in this tourism industry.

12.5.3 Optimistic Economic Impacts of Tourism

- **Maintaining Local Source of Revenue**: Micro benefits are the local opportunities for restoring local handicrafts and arts. Tourists buy their souvenirs so they earn money for living. One of the significance successes of tourism industry in the world is supporting the local livelihood, environmental, and consciousness of the community. At the same time, they help the country to maintain its economy.

For entrepreneurial activity, some women run their own small and medium size enterprises within the progress of tourism. Many tourism organizations will open and provide services needed by tourists. This brings women to contribute management enterprises through the influence of tourism. This contribution is a positive response to gender empowerment.

- **Occupation for Local Residents**: Tourism industry provides countless people. With employment opportunities can be in the cruise industry, hospitality, and aviation industry, all of which are provided with professions to enable them to earn their living. Many resorts are opening accommodations and catering facilities; hotels and lodges are increasing and even restaurants that create job opportunities for many people. Self-catering services are taken places. Game and theme parks, natural resorts, local bars are open, and even businessmen around hotels, all facilities for tourism appear to be sources for job opportunities to reduce unemployment in the country.
- **Taxes and Economic Benefits**: National economies today are heavy replies on tourism business through the collection of taxes and licenses. Lot of taxes are received from tourism industry as every tourists will have to paying taxes individually as aliens. For the organizations within tourism industry like hotel, travel agencies, tour operators, and airlines companies, the government relies on these taxes to keep its economy growing. Tourism is now one of the largest growing industries and engines to drive country's economy. In addition, tourism is the backbone of the developing economies through taxes levied on both domestic and international tourists. Service providers and organizations involved in tourism pay their due monthly and annually.

Box 12.1: Leofoo Village Resort Guanshi, Taiwan

Leofoo Resort in Guanshi is the foremost and solitary ecological resort in the Asian Island of Taiwan. This magnificent tourist paradise is established to meet the taste and expectation of both domestic and international tourists. It comprises of a resort, theme park, game reserve, and water park. Today it is the most popular attraction in the region with lots of ridings and entertainment facilities, thus making the place a heaven on earth.

Leofoo Village Theme Park comprises of four diverse spots namely African Safari, Wild West, Arabian Kingdom, and South Pacific. Looking from tourism education point of view, it provides internships programs, training, and job opportunities for students majoring in tourism, hospitality, and event management. Students can be trained on various aspects and dimensions. Among of the courses are like front-office operations, food and beverage, tour guiding, and entertainment. Ecologically, it is constructed in an eco-friendly way that conserves a sense and pride of ecotourism through the collection of wildlife species. Some of the wildlife are like giraffe, zebra, rhinos, flamingo, and camels. All of these animals are provided with proper accommodations at close proximity and nice animal natural habitation. Leofoo showcase the structural development of modern ecological, landscape, and green infrastructure designs. It serves as a medium economic benefit, identification of ecological features, and approach of new development in tourism business, thus promotes customers in the tourism industry. The general outlook of Leofoo resort illustrates them as a legend for utilizing natural environment to a modern state of the earth tourist home and sporting center.

According to the journal paper (Fang et al. 2017), *Normative beliefs, attitudes, and social norms: people reduce waste as an index of social relationships when spending leisure time*, this study, working at Leofoo Village Resort Guanshi, Taiwan, has adopted and refined Ajzen's theory of planned behavior, theory of reasoned action, and the value–belief–norm theory by Stern et al. to investigate the effects of normative beliefs, attitudes, and social norms on pro-environmental behaviors. A total of 391 valid responses were collected from visitors to a theme park, Leofoo Village Resort. A structure equation analysis indicated that the overall fit of the proposed model was supported. It was also found that both attitudes and social norms had positive and significant influence on waste reduction. While the results did not reveal any direct relation between normative beliefs and behavioral intentions, normative beliefs had positive direct influence on social norms and attitudes, which in turn had an impact on behavioral intentions. The findings provided further insights about pro-environmental behaviors from an Asia perspective and highlighted important implications for environmental policies and education to reduce waste. All contents please see **Appendix 1**.

12.5.4 Unconstructive Impact of Tourism

- **Erosion**: Advance growth of the travel and tourism industry, with visa waiver, has increased the number of tourist arrivals to destinations. This is eventually resulting to erosion, being a negative shock from tourisms when level of visitor influx in a destination is greater than environmental capabilities to cope with occupancy. For an example, unrestrained mass tourisms bring about potential threats to many natural areas. This can put massive pressure on an area and lead to impact such as soil, water, and vegetation wearing away (Fig. 12.2).
- **Greenhouse Gasses**: The movement of tourist from their originating destination to their vacation destination is done by various modes and means of transport. They come by road, water, and air. It is incessantly ever increasing in response to the rising number of tourists and their greater mobility. Tourism, now accounts almost about 60 percent of air travels. These can contribute to the greenhouse effects. The uses of rail, car, bus, coaches are increasing in respond to mass tourism and cruise transport, as well as the results in air pollution through the emission of smoke. All this source of pollution is rapidly growing as tourism is developing. It has a great negative impact in causing inconvenience of irritation, anxiety, and even hearing loss to humans. Moreover, it causes distress to build life and can cause pollution like air emission, noise, soil waste, littering, oil and chemicals, and even visual pollution.

Fig. 12.2 Unconstructive impact of tourism can put massive pressure on an area and lead to impact such as soil and vegetation wearing away, e.g., tea, beetle nut, and vegetation plantations (Photo by Wei-Ta Fang)

12.5.5 Negative Economic Impacts of Tourism

Seasonality Jobs: Tourism industry is seasonal in some destinations, meaning certain period of the year is peak season while other month off-peak season with less arrival of tourists. During off season, most people working in tourism industry will temporary stop working. Sometimes, during this season these people have other employment elsewhere to earn their living. They cannot sustain comfortable living when this situation persists. They cannot maintain their monthly remuneration for all year around due to the higher dependency and expenses of covering housing and other bills. Off season is always a problem for peoples working in tourism industry. The numbers of people who rely on tourism are very high. Most of the young boys quit school to find a job to support their living. Few of them become jobless and go wonder around beaches or hotels begging money from tourists. This has made many young people dropping out from schools.

12.6 Conclusion

Global perspective on tourism is a key fundamental issue. It pinpoints diverse range of concerns of tourism business. The operation in destination is looking at diversity of tourism operations and outlets in various destinations and how do they make their obligations to meet customers' needs. In meeting their business goals and contributing meaningfully to cooperate social responsibility, it is in safeguarding that the welfare of local residents and the environment, be taken seriously. We endeavor to observe numerous mechanisms of tourism with analytical studies on various cases within diverse destinations from Asian perspective to the smiling coast of Africa.

Almost all researchers have positive stand on culture preservation. The indispensable contribution of tourism socially, economically, and environmentally in both positive and negative aspects, promoting cultural understanding and maintaining their authenticity, fostering environmental protection, bringing in economic and creating employment, must be prioritized for local people. In addition, the awareness programs should accompany tourism business so the commercialization of cultures, degradation of ecological features, and availability of jobs for local people are not jeopardized.

This book *Tourism in Emerging Economies* covers all those facets of tourism development and mutually integrating a comprehensive approach for tourism growth into national and regional development. This will further strengthen livelihood to local people and appeal to the environment. Despite the global financial crisis and threats on the world securities, tourism still flourishes and remains the global top business, thus shows growing emerging trends for more international travelers and more income generation for nations.

References

Alexander DC (2017) Natural disasters. Routledge, New York, NY, U.S.A.

Fang W-T, Ng E, Wang C-M, Hsu M-L (2017) Normative beliefs, attitudes, and social norms: people reduce waste as an index of social relationships when spending leisure time. Sustainability 9:1696. https://doi.org/10.3390/su9101696

Appendix 1

Article

Normative Beliefs, Attitudes, and Social Norms: People Reduce Waste as an Index of Social Relationships When Spending Leisure Time

Wei-Ta Fang [1,*], Eric Ng [2], Ching-Ming Wang [1] and Ming-Lin Hsu [1,*]

[1] Graduate Institute of Environmental Education, National Taiwan Normal University, Taipei 11677, Taiwan; ecowindow168@gmail.com

[2] School of Management and Enterprise, University of Southern Queensland, Toowoomba, QLD 4350, Australia; eric.ng@usq.edu.au

* Correspondence: wtfang@ntnu.edu.tw (W.-T.F.); drumonkimo@gmail.com (M.-L.H.); Tel.: +886-2-77346558 (W.-T.F.); Tel: +886-3-3882461 (M.-L.H.)

Published: 22 September 2017 in open access, Sustainability

Abstract: This study has adopted and refined Ajzen's theory of planned behavior, theory of reasoned action, and the value–belief–norm theory by Stern et al. to investigate the effects of normative beliefs, attitudes, and social norms on pro-environmental behavioral intentions. A total of 391 valid responses were collected from visitors to a theme park in Taiwan. A structure equation analysis indicated that the overall fit of the proposed model was supported. It was also found that both attitudes and social norms had positive and significant influence on waste reduction. While the results did not reveal any direct relation between normative beliefs and behavioral intentions, normative beliefs had positive direct influence on social norms and attitudes, which in turn had an impact on behavioral intentions. The findings provided further insights about pro-environmental behavioral intentions from an Asia perspective and highlighted important implications for environmental policies and education to reduce waste.

W.-T. Fang, *Tourism in Emerging Economies*,
https://doi.org/10.1007/978-981-15-2463-9

Keywords: attitude; normative belief; pro-environmental behavioral intentions; social norm; theory of planned behavior; value–belief–norms theory; Taiwan

1. Introduction

Increasing environmental problems have posed a significant threat to environmental sustainability and this has prompted key stakeholders (e.g., policy makers, businesses, and consumers) to invest substantial efforts to enhance environmental behavior, which can be broadly defined as the types of behavior that seek to change the availability of materials or energy from the environment [1]. Accordingly, pro-environmental behavior (PEB), which refers to "behavior that consciously seeks to minimize the negative impact of one's actions on the natural and built world" [2] (p. 240), has been regarded as one of the key challenges to achieving environmental sustainability [3]. Therefore, further insights on why individuals undertake pro-environmental behavior can provide a more in-depth understanding about how behavioral change can be made to address those environmental problems.

A review of the literature (e.g., [4, 5]) indicates that, while environmental education can assist in fostering pro-environmental behavior, this alone does not necessarily lead to a decision towards such a behavioral pattern change. Instead, some studies [4] argue that having relevant knowledge and information about an environmental issue can have little effect on decision-making but rather the understanding of subjective beliefs and attitudes that individuals hold towards the issue can reveal greater insights on how these beliefs and attitudes affect intention and pro-environmental behavior. It is through such an understanding that sustainable pro-environmental behavior can be facilitated and attained.

Researchers have typically used the theory of reasoned action (TRA) by Ajzen and Fishbein [6] and theory of planned behavior (TPB) by Ajzen [7, 8], to explain pro-environmental behaviors. These theories posit that engaging in pro-environmental behavior is mainly influenced by positive attitudes toward such behaviors and perceived social pressure to perform the behavior. The theory of planned behavior has been considered as a major theoretical framework used to explain the variables influencing pro-environmental behaviors in a range of settings [9–12]. For example, Rhodes et al. [12] indicate that intention, planning, and perceived behavioral control have independent effects on recycling behavior, and intention has a significant indirect effect on recycling through planning. In another study by Han et al. [9], the findings revealed that attitudes, subjective norms, and perceived behavioral control positively affected customers' intention to stay at a green hotel. Consistent with the theory, Lam [11] suggests that attitudes, perceived behavioral control, and normative belief are regarded as important factors affecting people's intentions to conserve water usage [13, 14]. Another study by Howell et al. [10] reveals that attitudes, perceived behavioral control and subjective norm can influence behavior and behavioral intention towards the anti-introduction of aquatic invasive species. However, it has been observed in these prior studies that subjective norms which is a key component of the theory of reasoned action and theory of planned behavior, has mainly been investigated as "the perceived social pressure to perform or not to perform the behavior" in question or in a specific situation, whereby the greater the pressure to support or oppose a norm, the stronger the effect of behaviors [7]. As a result, descriptive norms which represent individuals' perceived practices of most people with whom one has interacted in the same space, and its impact on human behaviors often seems unconscious [15], have not been adequately explored even though it is also important to the understanding of pro-environmental behaviors, particularly from the perspective of social psychology. Therefore, this study seeks to address this gap by adapting the theory of planned behavior model to investigate social norms (SNs), which will incorporate subjective norms, injunctive norms and descriptive norms. This contributes to the on-going refinement of the theory of planned behavior model.

On the other hand, the value–belief–norm theory (VBN) is useful for predicting pro-environmental behaviors through the examination of various psychological factors that are likely to encourage changes in individuals' behaviors. This theory holds that personal moral motives such as personal values, beliefs, and norms can encourage an individual to be more actively protecting the environment and therefore exhibit positive pro-environmental behaviors [1]. The value–belief–norm theory of environmentalism is a conceptual framework that has been developed and tested by several empirical studies [16, 17] to explain the relevant dimensions and variances of environmentally significant individual behavior. For example, Stern et al. [17] indicate that values, beliefs, attitudes, and norms have strong relations with the behavioral indicators that support environmental movement. Further in line with the value–belief–norm theory, Eriksson et al. [18] suggest that personal norms and beliefs are positively related to pro-environmental orientation and willingness to reduce car usage. Another study by Jansson et al. [19] also indicates that values, beliefs and personal norms are positively associated with the early adoption of a high involvement eco-innovation such as the alternative fuel vehicle. However, the value–belief–norm theory focuses mainly on values and personal (moral) norms with a prediction in all reasonably foreseeable circumstances on personal values, but lesser effects on social norms from an externalized sense of obligation. Thus, the propose inclusion of the social norms component in this study seeks to provide an extended understanding of the value–belief–norm theory from both external influences and internal effects.

Empirical analysis of the literature suggests that while studies [2, 16] have been conducted, mainly in Western countries (such as countries in Europe as well as the United States), to explore the impact of beliefs, attitudes and subjective norms on pro-environmental behavior, but there is limited evidence of such investigation from an Asia perspective. Such an understanding is important since there has been significant economic growth in Asia in recent years that resulted in increasing environmental problems (e.g., garbage issues) which prompted global concerns. This study provides further insights into the scarce understanding of pro-environmental behaviors in the little-known Asia context and seeks to offer evidence to complement previous studies [16, 20] that attempt to compare environmental related issues between Asian and Western countries. Some of these limited studies about pro-environmental behavior from an Asian perspective include, for example, Aoyagi-Usui et al. [20] indicating that environmental values are linked with both traditional and altruistic values in Japan, Bangkok, and Manila. In a study by Zheng [21], findings suggest that cultivation of environmental consciousness is an important factor in evoking people's pro-environmental behaviors in Beijing, Seoul, Taipei and Tokyo. Another study by Harris [22] reveals that people in China are unlikely to act to protect the environment unless doing so is necessary to protect their own interests or those of their loved ones. Hori et al. [23] also find that global warming consciousness, environmental behavior, and social interaction have significant effect on energy-saving behavior in Dalian, Chongqing, Fukuoka, Bangkok, and Ho Chi Minh.

Therefore, this study aims to fill the identified gaps in the extant literature by building on Ajzen's theory of planned behavior [7] and the value–belief–norm theory by Stern et al. [17], to explore specifically the influence of normative beliefs, attitudes, and social norms on pro-environmental behavioral intentions towards the usage of disposal tableware. This paper will begin by reviewing the relevant literature and then discusses the methodology including the data collection techniques used in this study. Next the analysis of data is described and the findings presented. These will be followed by the conclusions drawn from the research together with the implications and suggestions for future research on the role of waste reduction in the social construction.

1.1. Normative Beliefs

Normative beliefs refer to beliefs of an individual that are accepted by specific people or groups and dictate whether behaving in a particular fashion is appropriate [24]. Fishbein and Ajzen [6, 14] propose the theory of reasoned action and first used the term "normative belief" as antecedent variables of norms. Bicchieri [25] also states that "Only the joint presence of a conditional preference for conformity and the belief that other people will conform will produce an agreement between normative beliefs and behavior". Ajzen [24] discusses behavioral, normative, and control beliefs in the theory of planned behavior, which are antecedent variables of attitudes, subjective norm, and perceived behavioral control respectively. The value–belief–norm theory and other empirical studies [17, 26] have revealed that beliefs, and norms are related through a continuous process of causality. In addition, these empirical studies also indicate that normative beliefs affect subjective norms, attitudes, and behavioral intentions [6]. Ajzen and Fishbein [14] (p. 2) explain that "While a social norm is usually meant to refer to a rather broad range of permissible, but not necessarily required behaviors, normative belief refers to a specific behavioral act the performance of which is expected or desired under the given circumstances". In examining the relationships between social norms and behaviors, Lapinski and Rimal [27] conclude that findings of the effects of social norms (which include subjective norms, injunctive norms, and descriptive norms) on behavior are mixed in normative influences. They confirmed that norms are also "dynamic phenomena and individuals, acting on either self-interest or altruistic motives, continuously alter the normative contours" in beliefs [28]. Thus, the following hypotheses are formulated to be investigated in this study.

Hypothesis 1. *Visitors' normative beliefs on using reusable tableware have a positive influence on their social norms regarding the use of reusable tableware.*
Hypothesis 2. *Visitors' normative beliefs on using reusable tableware have a positive influence on their attitudes regarding the use of reusable tableware.*
Hypothesis 3. *Visitors' normative beliefs on using reusable tableware have a positive influence on their behavioral intention to use reusable tableware during their next visit.*

1.2. Attitudes

Blackwell et al. [29] define attitudes as favorable or unfavorable cognitive evaluations, emotional experiences, or behavioral tendencies that people constantly hold for certain situations or ideas. Modern psychologists believe that attitude is a type of inherent psychological property that comprises cognitive, affective, and conative tendencies that exhibits consistence and persistence actions [30]. In the field of social psychology, studies on the influence of attitudes on behavior have shown that educationalists believe that in a behavioral change system, if people obtain relevant knowledge that is helpful in improving the environment, they may gain further environmental awareness and a positive attitude towards the environment, and therefore exhibit a greater level of pro-environmental behavior [31]. Studies [32, 33] reveal a positive correlation between environmental attitudes and pro-environmental behavior. According to Gatersleben et al. [34], attitude has a significant influence on pro-environmental behavior, which subsequently help reduce energy consumption in Dutch households. However, some studies [35, 36] suggest that specific environmental behaviors could not be predicted on the basis of environmental attitudes. The findings of some studies [37, 38] on predictions of recycling behavior suggest that researchers should investigate peoples' attitudes towards recycling behavior rather than their attitudes towards the environment in general.

According to the theory of planned behavior model, attitudes are the judgments that people make about the behavior they like or dislike. The more positive a person's attitude towards a behavior, the greater the behavioral intention a person will have [8]. This suggests that positive attitudes towards a specific pro-environmental behavior could positively influence the intention to display pro-environmental behavior. Based on the above, a hypothesis on attitudes affect behavioral intentions is developed for this study as follow:

Hypothesis 4. *A positive attitude towards the use of reusable tableware significantly affects visitors' behavioral intention to use reusable tableware.*

1.3. Social Norms

From the perspective of social psychology, social influence refers to the influence of external social factors on individual behaviors, and could be categorized into: (1) normative social influence; and (2) informational social influence [39]. Rivis and Sheeran [40] have used the terms "injunctive norm" and "descriptive norm" to represent normative social influence, and informational social influence respectively. According to the definition of subjective norm (one of the social norms) in the theory of planned behavior, people engage in various types of specific behavior when they are under pressure to meet the requirement of normative social influence, which is also known as normative conformity. Thus, "subjective norm" in the theory of planned behavior and "normative social influence" in social psychology share a similar meaning.

Cialdini et al. [41] suggest that the concept of norms also include the constructs of "behavior supported by groups" and "how groups themselves behave". In other words, two concepts of "should do" and "how to do" could be considered, rather than simply "norms" as a single concept. In norms, descriptive norms are others' practices that people follow when they are not familiar with particular situations or ideas. Thus, normative values are shared understandings of actions that are obligatory, permitted, or forbidden [42]. As such, people make judgment on whether a behavior is typical, normal, effective, adaptive, or appropriate by observing the specific behavior displayed by people around them. Furthermore, Bicchieri [25] proposes that social norms should be understood as types of social interaction behaviors, and such behaviors enable people to understand what is acceptable in society or groups in a system of norms.

Although previous studies [43, 44] have widely discussed the positive correlation between norms and behavioral intentions in pro-environmental behavior contexts, a study by White et al. [45] reveal otherwise where the correlation between social norms and behavioral intentions is regarded as minor. Hence, the following hypothesis on social norms affect behavioral intentions is formulated for this study to further investigate the relationship.

Hypothesis 5. *Visitors' social norms positively and significantly affect their behavioral intention to use reusable tableware.*

1.4. Research Framework

Based on the literature reviewed, a research framework has been proposed for this study, as shown in Figure 1. Particularly, this research will seek to measure the following and in the context of using reusable tableware by theme park visitors: (1) the influence of normative belief on social norms (H1), attitudes (H2), and behavioral intentions (H3); (2) the impact of attitude towards behavioral intentions (H4); and (3) the effects of social norms on behavioral intentions (H5).

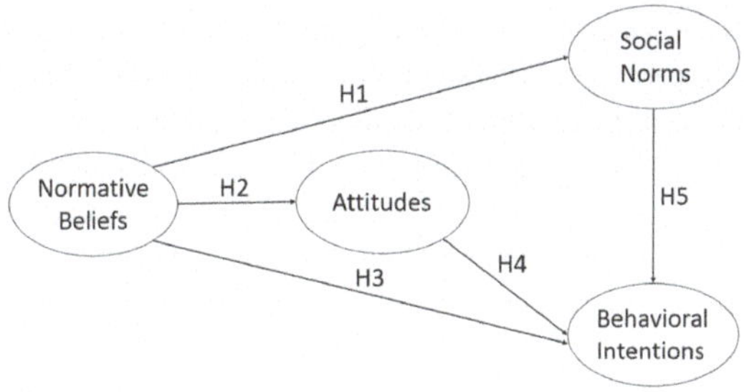

Figure 1. The research framework.

This framework intends to explain that the behavioral intentions of using reusable tableware are influenced by social norms (external factors), and attitudes (internal factors), whereas normative beliefs are regarded as the antecedents of social norms, attitudes, and behavioral intentions.

2. Materials and Methods

The photographs shown in Figure 2 have been presented in the questionnaires, so that there is a clear explanation about the domain ("tableware") of the study. The questionnaires have been referenced from the concepts of Ajzen's theory of planned behavior and the value–belief–norm theory by Stern [1]. We used the content validity

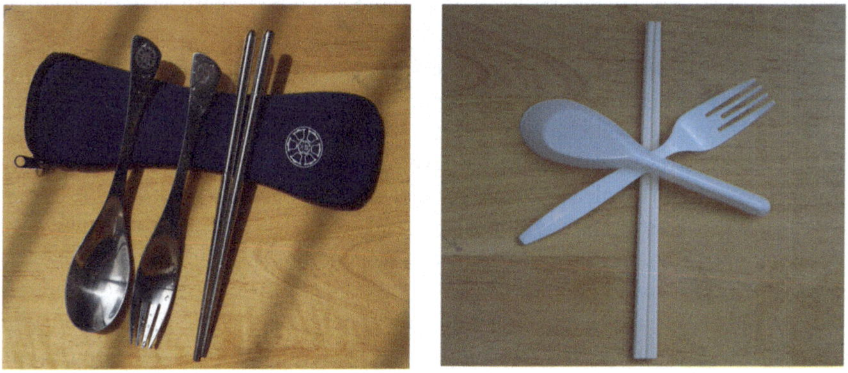

Figure 2. (**Left**) Consumers are encouraged to adopt reusable tableware, such as stainless-steel chopsticks, forks, and spoons used by the faculties and students at National Taiwan Normal University (NTNU) while going outing to eat. Copyright was authorized by National Taiwan Normal University (NTNU). (**Right**) Disposable tableware can be made of plastic, plastic-coated paper, and woody materials: such as chopsticks, forks, spoons, cups, plates, bowls, trays, and food containers. Photographed by Wei-Ta Fang.

of the questionnaire to compromise the degree that items reflecting any of the four facets of the domain (e.g., items on the magnitude of normative beliefs, social norms, attitudes, and behavioral intentions). We made reference to prior studies related to normative beliefs, social norms, attitudes, and behavioral intentions to explore the causal relationships [6, 46]. However, the questionnaire content (measures) has been rearranged so that they are well blended into the four key domain areas of normative beliefs, social norms, attitudes, and behavioral intentions (see Appendix A).

2.1. Survey Method

This study was carried out at Leofoo Village Theme Park in the Hsinchu County of Taiwan. Accordingly, Leofoo Village Theme Park is one of the most frequently visited leisure destinations by local Taiwanese and had consistently attracted a significant number of visitors over the years. This study had adopted the face-to-face questionnaire survey approach in which visitors to the Leofoo Village Theme Park were selected to complete the questionnaire. A non-probability convenience sampling technique was adopted at the theme park by interviewing visitors at three dining sites during 12:00–13:00 and 17:30–18:30 for seven consecutive days. To minimize potential selection biasness in the convenience sampling technique used in this study, all visitors who were at the three dining sites during the aforementioned periods were approached and requested to complete the questionnaire. To be eligible for the questionnaire study, screening questions were asked to determine participants were at least 16 years old and a local Taiwanese living in Taiwan. All valid participants were briefed on their questionnaires with self-reports without offering any incentives. Assistances were provided to participants who require any clarifications to the completion of the questionnaire.

A total of 410 questionnaires were distributed, of which 391 valid responses were obtained that represented a return rate of 95.4%. The questionnaire consisted of four key sections (including a demographic section) and several items aimed at investigating the research hypotheses developed for this study. The questionnaire used a 5-point Likert scale (1 = strongly disagree, 3 = neutral, 5 = strongly agree) to measure the items on the key dimensions (i.e., attitudes, social norms, behavioral intentions, and normative beliefs as shown in Appendix A) as outlined in the research framework. The National Taiwan Normal University Research Ethics Committee concluded that our study does not fall within the scope of the Human Subjects Research Act. They approved the study protocol (201707HS001) and agreed with active informed consent by the participants with the option to opt out of the study.

2.2. Data Analysis

In this study, SPSS 22 and LISREL 8.7 were used to analyze the data collected from the questionnaires and hypotheses were also tested on the dimensions as presented in the measurement hypothesis model. The conceptual foundation, development, construct validity and reliability, and psychometric qualities were examined through confirmatory factor analysis (CFA), which is a form of structural equation modeling (SEM) that used to verify the overall and inherent goodness of fit of the research framework. The method of maximum likelihood estimation was used to estimate model parameters. In addition, the overall goodness-of-fit test for the model was conducted using various fit indices as the bases for determination.

2.3. Structural Equation Modeling (SEM)

SEM, which comprises advanced statistical procedures that are increasingly applied for multivariate analysis, was employed to analyze and validate the scales used in this study. Its popularity is attributable to its confirmatory (as opposed to exploratory) nature and requires the researcher to possess an explicit model. SEM explicitly models observed as well as latent variables (i.e., the items and the concepts they are measuring) and can further model and correct measurement error, thus reducing inaccuracies. Finally, SEM enables analyzing a complete multivariate model including direct and indirect effects [47, 48].

2.4. Validity and Reliability Analysis

2.4.1. Content Validity

In our study, the questionnaire has been validated by a panel of judges who are professors in the field of environmental education and tourism education in Asian countries. Five professors, selected by their academic reputation, knowledge, as well as their familiarity with the fields, have been invited in the panel of judges who have been detected to be validated to examine the research questionnaire.

2.4.2. Composite Reliability (CR), and Average Variance Extracted (AVE)

Convergent validity is determined to be the degree of confidence we have measured by their indicators. According to Fornell and Larcker [49], convergent validity was assessed by standardized factor loading, Composite Reliability, and Average Variance Extracted [49]. The Composite Reliability value in our SEM Output is t-statistics value to be examined [49]. The Composite Reliability is there to check how well a construct factor is measured in the model. We detected the reports from convergent validity assessed from Composite Reliability, and Average Variance Extracted as follows: the Composite Reliability value of the normative beliefs, social norms, attitudes, and behavioral intentions are 0.58, 0.76, 0.74, and 0.85, respectively. Three of the Composite Reliabilities exceed the recommended threshold of 0.60 suggested by Bagozzi and Yi [50] (p. 82), and one of the Composite Reliabilities exceeds the

Table 1. Questionnaire questions and reliability analysis.

Model Dimension	Number of Questions	Cronbach's α
Normative beliefs	2	0.609
Social norms	5	0.756
Attitudes	2	0.731
Behavioral intentions	3	0.759

recommended threshold of 0.50 suggested by Raines-Eudy [51] (p. 126), which is the acceptable CR level. According to Raines-Eudy [51] (p. 128), an Average Variance Extracted value of greater than 0.50 indicates that the validity of both the construct and the individual variables is high. The Average Variance Extracted values derived from normative beliefs, social norms, attitudes, and behavioral intentions are 0.41, 0.40, 0.59, and 0.66, respectively. According to Hair et al. [52] (p. 808), all values of standardized factor loadings should exceed the values of 0.50, and the Average Variance Extracted values estimate should exceed $(0.5)^2 = 0.25$. Subsequently, we can confirm that the items measure in their construct and convergent validity of a model is satisfied.

2.4.3. Cronbach's α

We also conducted Cronbach's α reliability analysis to determine the level of consistency between each variable (i.e., item) and the measurement dimensions used in this study. Accordingly, Cronbach's α reliability analysis should be at least 0.50, preferably 0.70, to ensure the consistency and stability of the test results. When the α value is greater than 0.70, the reliability indicates sufficient reliability. In this study, SPSS 22 was used to conduct the reliability analysis of the questionnaire survey. The analytical results revealed high reliability for social norms, attitudes, and behavioral intentions, whereas normative beliefs attained a moderate level of reliability. A summary of the results of Cronbach's α reliability analysis is shown in Table 1.

2.5. Confirmatory Factor Analysis (CFA)

In structural equation modeling, Confirmatory Factor Analysis has been used to assess construct validity with clearly specified conceptual boundaries [53]. Ghadi et al. [54] suggest that "(t)he validation emphasizes a logical analysis and tests the relationships predicated based on theoretical considerations". Therefore, we focused on the selections of essential necessities in a combined model from Ajzen's theory of planned behavior and the value–belief–norm theory by Stern et al., to investigate the effects of normative beliefs, social norms, and attitudes on pro-environmental behavior in Figure 3. We used Confirmatory Factor Analysis to evaluate and refine

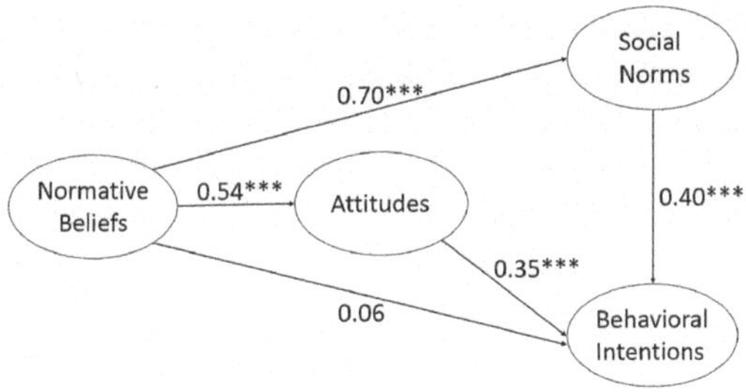

Figure 3. The path coefficients of normative beliefs (NBs), attitudes (ATs), social norms (SNs), and pro-environmental behavioral intentions (BIs) in model. *** $p < 0.001$.

the hypothesis model for the pro-environmental behavior of visitors to theme parks in Taiwan. The analysis explicitly tested a hypothesized measurement model (as opposed to an exploratory approach), accounts for sources of common measurement and method error that are inherent in survey research, and provides empirical justification. In other words, Confirmatory Factor Analysis was mainly applied to determine the goodness of fit between measurement items in the scale.

In this study, the model's overall goodness of fit to the data was good. The analytical results showed that the Goodness-of-fit index (GFI) and the Comparative index (CFI) were 0.924 and 0.951 respectively, which were above the acceptable value of 0.90. In addition, the Standard Root Mean Square (SRMR) and the Root Mean-Square Error of Approximation (RMSEA) were recorded at 0.064 and 0.086, respectively, demonstrating acceptable level on the goodness of fit of the model. Table 2 provides a brief summary of the results.

Table 2. The different indices and determination results of confirmatory factor analysis (CFA).

Index	Hypothesis Model Values	Standard and Determination Result	Citation
Goodness-of-fit index (GFI)	0.924	>0.90, acceptable	Hu and Bentler [55]
Comparative index (CFI)	0.951	>0.90, acceptable	Hu and Bentler [55] Byrne [56]
Non-normed fit index (NNFI)	0.934	>0.90, acceptable	Hu and Bentler [55]
Standardized root mean square residual (SRMR)	0.064	<0.08, acceptable	Hu and Bentler [55]
Root Mean-Square Error of Approximation (RMSEA)	0.086	Near 0.08, fair	McDonald and Ho [57] Browne and Cudeck [58]

In this study, SPSS 22 was used to conduct the reliability analysis of the questionnaire survey. The summary results of Cronbach's α reliability analysis, correlation coefficients, mean, and standard deviation are shown in Tables 1 and 5, respectively. The analytical results revealed high reliability for social norms, attitudes, and behavioral intentions, whereas normative beliefs attained a moderate level of reliability.

3. Results

3.1. Analysis of Respondent Backgrounds

We investigated visiting consumer behaviors by self-reports and by our personal observations. Only 20% (78 of the 391 respondents) of the visitors represent a significant contribution to reducing large amount of playground disposal tableware waste generated in our study. They have observed to be carried reusable tableware, such as chopsticks and/or forks and spoons while dinning in the theme park (the pro-environmental behaviors as we set to be acknowledged as a reference example in the left side in Figure 2, but it is not necessary to be set in the same style of tableware/dinnerware toolkits as we provided). Among the 391 respondents who completed the valid questionnaires, there were 263 (67.3%) female and the remaining 128 (32.7%) being male. In terms of marital status, majority (68%) of the respondents were single with the others being married with children (29.2%) and without children (2.8%). Respondents were mainly (53.5%) between 21 to 40 years old, and about more than half (55%) had attained an education qualification of a Bachelor's or higher degrees. Respondents were predominantly residing in northern Taiwan (40.2%) and this was followed by central Taiwan (32.7%), southern Taiwan (18.9%), eastern Taiwan (5.6%), and offshore island (2.6%).

We further examined the correlation through the independent-sample t test, based on gender, and one-way ANOVA between other demographic variables associated with normative belief, social norms, attitudes, and behavioral intentions. Results suggested no significant differences between females and males in terms of their normative beliefs ($df = 390$, two-tailed, $t = -0.56$, $p = 0.57$), social norms ($df = 390$, two-tailed, $t = -1.73$, $p = 0.09$), attitudes ($df = 390$, two-tailed, $t = -0.83$, $p = 0.41$), and behavioral intentions ($df = 390$, two-tailed, $t = 0.46$, $p = 0.65$) (see Table 3). However, there was a significant difference between visitors' marital status (i.e., single, married with children, married without children) in terms of their normative belief ($F_{0.05}(2, 388)$, $F = 4.615, p = 0.01$), social norms ($F_{0.05}(2, 388)$, $F = 5.2, p = 0.006$), and behavioral intentions ($F_{0.05}(2, 388)$, $F = 8.385$, $p = 0.000$).

With regards to age group, there was also a significant difference between visitors on their normative belief ($F_{0.05}(3, 387)$, $F = 4.198, p = 0.006$), social norms ($F_{0.05}(3, 387)$, $F = 3.236$, $p = 0.022$), and behavioral intentions ($F_{0.05}(3, 387)$, $F = 5.465$, $p = 0.001$). Visitors between 41 to 60 years old had a higher mean score than other age groups across the four items (normative belief, social norms, attitudes, and

Table 3. Descriptive statistics related to the gender question for normative beliefs, social norms, attitudes, and behavioral intentions items ($n = 391$).

Items	Categories	Frequency	Percent	Normative Beliefs				Social Norms				Attitudes				Behavioral Intentions			
				Mean	*SD*	*t*	*p*	*Mean*	*SD*	*t*	*p*	*Mean*	*SD*	*t*	*p*	*Mean*	*SD*	*t*	*p*
Gender	Male	128	32.7	4.30	0.67	−0.56	0.57	3.88	0.74	−1.73	0.09	3.58	0.91	−0.83	0.41	3.78	0.78	0.46	0.65
	Female	263	67.3	4.35	0.69			4.00	0.61			3.66	0.92			3.74	0.73		

behavioral intentions). While the results did not reveal any significant differences in the education qualification and household income categories, a significant difference existed between visitors' place of residence on their normative belief ($F_{0.05}(4, 386)$, $F = 6.128$, $p = 0.000$) and behavioral intentions ($F_{0.05}(4, 386)$, $F = 4.827$, $p = 0.001$). Table 4 presents a summary of the demographic results regarding normative belief, social norms, attitudes, and behavioral intentions.

3.2. Test and Analysis of PEB Intentions Using SEM

The SEM software LISREL8.7 was used to conduct SEM hypothesis testing and outcome analysis. Regarding measurement model of SEM, the results of the SEM analysis on correlation coefficients, means, and standard deviations are presented in Table 5.

If a variable's p value of a path coefficient, such as a t test result, passed the p value test (i.e., $p < 0.001$), the proposed hypothesis would be considered as accepted (Figure 3). In addition, the results of the SEM analysis on the effects of the latent variable in path analysis are presented in Table 6.

As shown in Figure 3, the results indicated that the path coefficient between normative beliefs and social norms is 0.70, reaching the significant level. Therefore, H1 is supported, whereby visitors' normative beliefs on using reusable tableware have a direct positive influence on their social norms regarding the use of reusable tableware. Regarding H2, the findings revealed that the path coefficient is significant between normative beliefs and attitudes at 0.54. Thus, visitors' normative beliefs on using reusable tableware have a direct positive influence on their attitudes towards the use of reusable tableware.

The path coefficient between normative beliefs and behavioral intentions is 0.06 and has not achieved the significance level. Hence, H3 is not supported with the indication that visitors' normative beliefs on the use of reusable tableware do not have a direct influence on their behavioral intention towards the use of reusable tableware. In relation to H4, the path coefficient between attitudes and behavioral intentions is loaded at a significant level of 0.35. Therefore, this hypothesized relationship is supported through which a positive attitude towards the use of reusable tableware can have a direct and significant influence on visitors' behavioral intention to use reusable tableware.

The results also highlighted a significant level of path coefficient between social norms and behavioral intentions at 0.40, thus revealing the support of H5 with visitors' social norms having a direct positive influence on their behavioral intention to use reusable tableware.

Table 4. Descriptive statistics related to the demographic questions for normative beliefs, social norms, attitudes, and behavioral intentions items ($n = 391$).

Items	Categories	Frequency	Percent	Normative Beliefs				Social Norms				Attitudes				Behavioral Intentions			
				Mean	*SD*	*F*	*p*	*Mean*	*SD*	*F*	*p*	*Mean*	*SD*	*F*	*p*	*Mean*	*SD*	*F*	*p*
Marital status	Single	266	68.0	4.28	0.68	4.615	0.01	3.89	0.66	5.2	0.006	3.56	0.89	2.983	0.052	3.66	0.73	8.385	0.000
	Married with children	114	29.2	4.48	0.62			4.09	0.63			3.79	0.97			3.99	0.71		
	Married without children	11	2.8	4.04	1.02			4.31	0.61			3.86	0.81			3.64	1.03		
Age	20 or younger	131	33.5	4.26	0.66	4.198	0.006	3.88	0.65	3.236	0.022	3.52	1.00	1.877	0.133	3.71	0.81	5.465	0.001
	21–40	209	53.5	4.31	0.72			3.96	0.64			3.64	0.82			3.69	0.69		
	41–60	40	10.2	4.68	0.45			4.24	0.75			3.90	0.96			4.19	0.69		
	61 or older	11	2.8	4.45	0.52			4.07	0.50			3.77	1.17			3.93	0.82		
Highest education qualification attained	Junior high school or lower	46	11.8	4.23	0.61	1.119	0.341	4.01	0.56	0.506	0.679	3.49	1.17	1.154	0.327	4.02	0.75	2.646	0.049
	High school	101	25.8	4.35	0.78			3.90	0.71			3.76	0.90			3.76	0.75		
	College	215	55.0	4.32	0.66			3.98	0.64			3.60	0.86			3.69	0.74		
	Master's degree or higher	29	7.4	4.52	0.56			4.01	0.80			3.66	0.91			3.82	0.69		
Residence	Eastern Taiwan	22	5.6	3.66	1.05	6.128	0.000	3.70	0.63	2.240	0.064	3.27	0.91	2.135	0.076	3.39	0.88	4.827	0.001
	Central Taiwan	128	32.7	4.38	0.64			4.06	0.61			3.78	0.99			3.94	0.73		
	Southern Taiwan	74	18.9	4.35	0.57			3.87	0.69			3.53	0.84			3.55	0.71		
	Northern Taiwan	157	40.2	4.39	0.66			3.98	0.68			3.60	0.87			3.76	0.73		
	Taiwan's offshore islands	10	2.6	4.25	0.54			3.74	0.47			3.75	0.79			3.83	0.62		
Annual household income (NT$) [1]	Less than 459,999	93	23.8	4.26	0.72	1.151	0.332	3.87	0.56	0.652	0.626	3.48	0.94	2.054	0.086	3.70	0.70	1.290	0.273

(continued)

Table 4. (continued)

Items	Categories	Frequency	Percent	Normative Beliefs				Social Norms				Attitudes				Behavioral Intentions			
				Mean	*SD*	*F*	*p*	*Mean*	*SD*	*F*	*p*	*Mean*	*SD*	*F*	*p*	*Mean*	*SD*	*F*	*p*
	460,000–759,999	106	27.1	4.30	0.70			3.95	0.65			3.54	0.84			3.66	0.74		
	760,000–1,009,999	86	22.0	4.44	0.60			3.98	0.77			3.83	0.87			3.79	0.75		
	1,010,000–1,309,999	58	14.8	4.41	0.70			4.00	0.62			3.74	0.92			3.86	0.76		
	Greater than 1,310,000	48	12.3	4.26	0.68			4.06	0.72			3.68	1.02			3.89	0.84		

1 The monetary unit is NT dollar. In 2017, one US dollar is equivalent to 30.63 NT dollars, 1 USD = NT 30.63, and admission tickets to Leofoo Village Theme Park were priced at NT $999 each, equivalent to $32.61.

Table 5. Correlation coefficient, mean, standard deviation of observed variables in the model.

Dimension	NBs	SNs	ATs	BIs	Mean	Standard Deviation
NBs	1				4.3325	0.68083
SNs	0.391 **	1			3.9637	0.65977
ATs	0.317 **	0.288 **	1		3.6330	0.91439
BIs	0.324 **	0.399 **	0.526 **	1	3.7564	0.74826

Note: NBs = normative beliefs; SNs = social norms; ATs = attitudes; BIs = behavioral intentions. ** When $p < 0.01$ (two-tailed), it represents the significance level.

Table 6. The effects of the latent variable in path analysis.

Independent Variables	Dependent Variables: Endogenous Variables					
	BIs		SNs		ATs	
	Effect	*t* value	Effect	*t* value	Effect	*t* value
Exogenous variable NBs						
Direct effects	–	–	0.7	9.55 ***	0.54	7.97 ***
Indirect effects	0.467	5.16 ***	–	–	–	–
Total effects	0.467	5.16 ***	0.7	9.55 ***	0.54	7.97 ***
Endogenous variables SNs						
Direct effects	0.40	4.18 ***				
Indirect effects	–	–				
Total effects	0.40	4.18 ***				
Endogenous variables ATs						
Direct effects	0.35	4.41 ***				
Indirect effects	–	–				
Total effects	0.35	4.41 ***				

Notes: t value > 1.96, * $p < 0.05$; t value > 2.58, ** $p < 0.01$; t value > 3.29, *** $p < 0.001$.

4. Discussion and Implication

In this research, we have investigated the influence of normative beliefs, social norms, and attitudes towards behavioral intentions, specifically in relation to the reusable tableware at a theme park in Taiwan. This study extended the work by Ajzen's theory of planned behavior and also on the value–belief–norm theory by Stern et al. [7, 17], from an Asian perspective (which is lacking in the literature) in which the region has sustained considerable economic growth in the past decade but at the same time this economic growth has also contributed to the increasing environmental problems that they are facing nowadays.

We have achieved consistent results and tested them for validity, to check that the questionnaire measures what it claims to measure. The results of the CFA revealed that the overall goodness of fit for the hypothesis model is excellent, indicating that the proposed framework (as shown in Figure 3) for theme park visitors using reusable tableware is acceptable. The findings from the SEM analysis showed strong support and had accepted four of the five hypotheses (i.e., H1, H2, H4 and H5), indicating significant positive direct relationship.

Results suggested that social norms had a stronger influence on behavioral intentions than attitudes on behavioral intentions with the path coefficient being 0.4 and 0.35 respectively. This implied that people were more likely to display and engage in the pro-environmental behavior of using reusable tableware when they felt under pressure to conform to the expectation and requirement of normative social influence. While this outcome aligns with some prior studies, it also offers a different view, and this might be explained by the differing cultural context (i.e., eastern versus western). Specifically, Confucianism, which is an East Asian ethical and philosophical system that emphasizes social values, has strong culture roots in countries such as China, Japan, Korea, Singapore and Taiwan (which is the key focus of this study). This strong emphasis on social values forms the basis for a sense of belonging to the group, which could have explained the results emphasizing a greater influence of social norms on behavioral intentions. This finding supported the model by Schütte and Ciarlante in Maslow's Hierarchy of Needs in Asia [59]. They developed an Asian equivalent model beyond the original top of the pyramid's shape in Maslow's, "personal self-actualization", and called it as "social contexts" (i.e., status, admiration, and affiliation) to regulate Asian behaviors in "Act" in their daily lives (p. 93). Therefore, we enforced one of the Asian empirical and normative expectations to be formed their motive actions in the model of "social norms" (see the definitions of Rimal and Lapinski, and see also Bicchieri, p. 41) [27, 28]. This diagnostic process of identifying collective model in Asians have been discovered from our observation in a collective pattern of their behaviors. Therefore, we refer to Maslow's Hierarchy of Needs in Asia to present a synthetized variable of "social norms" without separating variables reflecting different types of norms, i.e., subjective norm in theory of planned behavior, to re-analyze the data.

However, this "Act" of pro-environmental behavior might have only occurred because of the pressure to conform to this specific behavior displayed by other people around the theme park, but not necessarily a true reflection of the action to be taken when such a pressure is non-existent (e.g., in situations when no others are around). As such, it is important to have continual efforts towards environmental education in order to equip people with relevant knowledge that could help enhance their awareness and gaining a more positive attitude towards the environment. Therefore, social norms may be regarded as an external force influencing pro-environmental behavioral intentions whereas attitudes are needed as an internal source to shape pro-environmental behavioral intentions, and these two dimensions have a critical role to play in achieving sustainable pro-environmental behavior.

On the other hand, H3 is rejected since the causal relationship between normative beliefs and behavioral intentions is relatively weak with the coefficient being 0.06. This indicated that the visitors' normative beliefs on using reusable tableware had minimal influence on their behavioral intention to use reusable tableware during their next visit. As shown in Figure 3, although normative beliefs had no direct influence on behavioral intentions, but it had positive direct relationships with social norms (coefficient 0.70) and attitudes (coefficient 0.54) which in turn had an impact on behavioral intentions. Thus, social norms and attitudes could be considered as mediators between normative beliefs and pro-environmental behavioral intentions. The absence of this relationship between normative beliefs and behavioral intentions could be explained through the possible lack of one's desire to conform to the pro-environmental behavior and/or the lack of belief in other people who will also exhibit pro-environmental behavior. From this perspective, the emphasis on the development of psychological internal factors and external factors is essential and can potentially help enhance theme park visitors' behavioral intentions to use reusable tableware.

4.1. Study Implications

The findings from this study have several implications. Firstly, approximately 20% (78 of the 391 respondents) of the participants display pro-environmental behaviors when detected in our observations as well as examined in their self-report questionnaire. There is serious concern on the effects of social environmental education on people's social influences for the other 80% of the non-pro-environmental behaviors' participants. Therefore, we recommend that environmental education programs (formal and informal) at various levels (e.g., community, school) need to be targeting independent of knowledge so that people can better understand the relationship between their behavior and the environment. This helps towards attaining a more sustainable pro-environmental behavior of becoming a socially responsible citizen and not just conforming because of the social group pressure.

Next, while the Environmental Protection Administration in Taiwan has enforced a restriction on the use of disposable tableware at certain types of restaurant since 2002, it is necessary to review the effectiveness of this policy. The review of 80% of the non-pro-environmental behaviors' observation should not only emphasize on the aspect of enforcing relevant environmental protection policies but also the effects on pro-environmental behavior and the relevant initiatives.

Thirdly, promotional campaigns will be necessary to further increase people's awareness about the types of pro-environmental behavior, and information should also be included on how they can be implemented. Through these campaigns, people will be appropriately equipped with the knowledge that enables them to adequately implement their pro-environmental behavior, therefore contributing to achieving a greener environment.

4.2. Study Limitations

This study has four key limitations. The first is the research focused mainly on internal and external factors in the visitors' pro-environmental behavioral intentions towards using reusable tableware, including attitudes and social norms (i.e., injunctive, descriptive, and subjective norms). Furthermore, the questionnaire had adopted

Table 7. Types of norms.

Norms	Norm Levels	Types of Norms	Definition
Social	External	Injunctive Norms	Display behaviors others approve. Compliance is encouraged and violations will be punished [60–62].
	External	Subjective Norms	Behaviors expected or supported by people around you, such as family, peers, and colleagues [61, 62].
	External	Descriptive Norms	Respondents perceive the behaviors that most people are doing [60–62].
Personal	Internalized	Personal Norms	Self-expectations and feel obliged to do what morally right [60–62].

the social norm propositions to include injunctive norms, descriptive norms, and subjective norms, and there were no distinctions between them. Social cognition variables, such as social norms, in the theory of planned behavior predict behavior more strongly when the behavior is under volitional control. The different types of norms are summarized in Table 7.

Next, this study was conducted in Taiwan, specifically in a theme park context with a limited number of samples, investigating the use of reusable tableware as a form of pro-environmental behavior. Therefore, the findings cannot be generalized but instead provide further insights from an Asian perspective about the influence of normative beliefs, social norms and attitudes towards behavioral intentions and pro-environmental behavior. A more comprehensive sample is needed to for such a generalization.

Thirdly, due to the apparent and potential limitations of self-reporting data, evidence of validity on the findings can be limited. Although participation in this study is voluntary, self-report questionnaires rely on the honesty of the participants. In this study, participants might not have provided their honest answers to the questionnaire due to a variety of reasons (e.g., social pressure). Furthermore, participants might also have a varying degree of understanding or interpretation of particular questions, which could result in the provision of inaccurate responses. However, interviewers conducting the face-to-face questionnaire survey were available to participants who require any clarifications to completing the questionnaire.

Lastly, this study has inadequately operationalized the theory of planned behavior variables, perceived behavioral control. Only to some extent do our variables align with the principle of compatibility in theory of planned behavior and value–belief–norm theory. We assumed that, combining the two theories, especially selecting essential necessities from the theory of planned behavior and value–belief–norm theories, we would have successfully decomposed the antecedent variable and mediators in a refining model. However, perceived behavioral control (one of the key components of the theory of planned behavior) as indicated by Wubs et al. in South

Africa and Tanzania [63], as well as values (one of the key components of the value–belief–norm theory) represented a beyond meaning with a non-causal relationship compared to social norms in our pioneering studies in Asia.

4.3. Areas for Future Research

In terms of future research, several areas are proposed. As part of the limitations mentioned above, this study has not differentiated among the various types of norms (i.e., injunctive norms, descriptive norms, subjective norms, and personal norms), and therefore lacks the detailed investigation about these norms. However, prior studies investigated the predictions about pro-environmental behavior based on moral or ethical elements have found that personal norms played a crucial role. Likewise, injunctive norms, descriptive norms, and subjective norms should also be explored separately in more detail. Thus, it is suggested that separate detailed studies on personal norms, injunctive norm, descriptive norms, and subjective norms be conducted in the future to investigate the key variables influencing pro-environmental behavior.

Secondly, although environmentally friendly behavioral intentions were investigated in this study through a self-completed questionnaire survey, observation technique should be well used in similar pro-environmental behavior researches. The use of observation technique can further enhance the understanding of the pro-environmental behavior in which a more realistic and truthful behavior can be examined. Thirdly, other theories can be introduced for more advanced studies from "social contexts" (i.e., status, admiration, and affiliation) in Asian's specific value. Beyond social influences, the self-determination theory also suggested that internal motivation from one-self will have a greater impact on positive outcomes (i.e., exhibiting pro-environmental behavior) than from external motivation since it generates a higher level of self-satisfaction and happiness [64]. Therefore, the role of waste reduction in the social construction of the Eastern society could be re-built from an innovative social model related to real solutions.

Lastly, studies conducted in the future can also focus on other types of pro-environmental behavior that visitors might exhibit, such as purchasing green goods (green consumerism), recycling, and staying in green hotels and restaurants [65]. This enables other variables affecting pro-environmental behavior and causal relationships to be identified so that a more comprehensive framework can be developed.

Acknowledgments: We thank members of the Graduate Institute of Environmental Education, National Taiwan Normal University (NTNU) and the Department of Leisure and Recreation Management, Chung Hua University for their contributions to the manuscript. This work was supported by grants from the Ministry of Science and Technology (105-2511-S-003-021-MY3 to Wei-Ta Fang). This article was also subsidized by the National Taiwan Normal University (NTNU), Taiwan, ROC. The useful suggestions from anonymous reviewers were incorporated into the manuscript.
Author Contributions: W.-T.F. coordinated this research project as well as designed the questionnaire. He also carried out this survey. W.-T.F. and M.-L.H. analyzed the

data. C.-M.W. contributed in the discussion of results. W.-T.F., E.N., and M.-L.H. wrote the paper.

Conflicts of Interest: The authors declare no conflict of interest.

Appendix A

Questionnaire: Dimensions and items

Items on normative beliefs

1. I believe that the use of disposable tableware will affect the environment forever.
2. I believe that every visitor has a responsibility to use less disposable tableware.

Items on social norms

3. I am willing to follow the strategy that the government employed to limit the use of disposable tableware.
4. I would carry reusable tableware because those who travel with me already carry it.
5. I would decide to use reusable tableware because of the influence of my family.
6. I would decide to use reusable tableware because of other people's criticism.
7. I would be more willing to use reusable tableware if I were rewarded by the theme park.

Items on attitudes

8. It is inconvenient to carry reusable tableware. (Opposite)
9. As far as I am concerned, it is not fashionable to carry reusable tableware. (Opposite)

Items on behavioral intentions

10. I will carry my own reusable tableware when I visit theme parks in the future.
11. I will carry my own reusable cup when I visit theme parks in the future.
12. I will carry my own reusable chopsticks when I visit theme parks in the future.

References

1. Stern, P.C. New environmental theories: toward a coherent theory of environmentally significant behavior. *J. Soc. Issues* **2000**, *56*, 407–424.
2. Kollmuss, A.; Agyeman, J. Mind the gap: Why do people act environmentally and what are the barriers to pro-environmental behavior? *Environ. Educ. Res.* **2002**, *8*, 239–260.
3. Gifford, R.; Nilsson, A. Personal and social factors that influence pro-environmental concern and behaviour: A review. *Int. J. Psychol.* **2014**, *49*, 141–157.
4. Gifford, R.; Steg, L.; Reser, J.P. Environmental psychology. In *IAAP Handbook of Applied Psychology*; Martin, P.R., Cheung, F.M., Knowles, M.C., Kyrios, M., Overmier, J.B., Prieto, J.M. Eds.; Wiley Blackwell: Chichester, UK, 2011; pp. 440–470.

5. Jensen, B.B. Knowledge, action and pro-environmental behaviour. *Environ. Educ. Res.* **2002**, *8*, 325–334.
6. Ajzen, I.; Fishbein, M. Understanding Attitudes and Predicting Social Behavior. *J. Exp. Soc. Psychol.* **1980**, *22*, 453–474.
7. Ajzen, I. The theory of planned behavior. *Organ. Behav. Decis. Process.* **1991**, *50*, 179–211.
8. Ajzen, I. Attitude structure and behavior. In *Attitude Structure and Function*; Pratkanis, A.R., Breckler, S.J., Greenwald, A.G., Eds.; Erlbaum: Hillsdale, NJ, USA, 1989; pp. 241–274.
9. Han, H.; Hsu, L.-T.J.; Sheu, C. Application of the theory of planned behavior to green hotel choice: Testing the effect of environmental friendly activities. *Tour. Manag.* **2010**, *31*, 325–334.
10. Howell, A.P.; Shaw, B.R.; Alvarez, G. Bait shop owners as opinion leaders: A test of the Theory of Planned Behavior to predict pro-environmental outreach behaviors and intentions. *Environ. Behav.* **2015**, *47*, 1107–1126.
11. Lam, S.P. Predicting intentions to conserve water from the theory of planned behavior, perceived moral obligation, and perceived water right. *J. Appl. Soc. Psychol.* **1999**, *29*, 1058–1071.
12. Rhodes, R.E.; Beauchamp, M.R.; Conner, M.; de Bruijn, G.-J.; Kaushal, N.; Latimer-Cheung, A. Prediction of depot-based specialty recycling behavior using an extended theory of planned behavior. *Environ. Behav.* **2015**, *47*, 1001–1023.
13. Fishbein, M.; Ajzen, I. *Belief, Attitude, Intention, and Behavior: An Introduction to Theory and Research*; Addison-Wesley: Reading, MA, USA, 1975.
14. Ajzen, I.; Fishbein, M. Attitudes and normative beliefs as factors influencing behavioral intentions. *J. Personal. Soc. Psychol.* **1972**, *21*, 1–9.
15. Cialdini, R.B.; Demaine, L.J.; Sagarin, B.J.; Barrett, D.W.; Rhoads, K.; Winter, P.L. Managing social norms for persuasive impact. *Soc. Influ.* **2006**, *1*, 3–15.
16. Oreg, S.; Katz-Gerro, T. Predicting proenvironmental behavior cross-nationally values, the theory of planned behavior, and value-belief-norm theory. *Environ. Behav.* **2006**, *38*, 462–483.
17. Stern, P.C.; Dietz, T.; Abel, T.D.; Guagnano, G.A.; Kalof, L. A value-belief-norm theory of support for social movements: The case of environmentalism. *Hum. Ecol. Rev.* **1999**, *6*, 81–97.
18. Eriksson, L.; Garvill, J.; Nordlund, A.M. Acceptability of travel demand management measures: The importance of problem awareness, personal norm, freedom, and fairness. *J. Environ. Psychol.* **2006**, *26*, 15–26.
19. Jansson, J.; Marell, A.; Nordlund, A. Exploring consumer adoption of a high involvement eco-innovation using value-belief-norm theory. *J. Consum. Behav.* **2011**, *10*, 51–60.
20. Aoyagi-Usui, M.; Vinken, H.; Kuribayashi, A. Pro-environmental attitudes and behaviors: An international comparison. *Hum. Ecol. Rev.* **2003**, *10*, 23–31.
21. Zheng, Y. Association analysis on pro-environmental behaviors and environmental consciousness in main cities of East Asia. *Behaviormetrika* **2009**, *37*, 55–69.

22. Harris, P.G. Environmental perspectives and behavior in China: Synopsis and bibliography. *Environ. Behav.* **2006**, *38*, 5–21.
23. Hori, S.; Kondo, K.; Nogata, D.; Ben, H. The determinants of household energy-saving behavior: Survey and comparison in five major Asian cities. *Energy Policy* **2013**, *52*, 354–362.
24. Ajzen, I. Constructing a Theory of Planned Behavior Questionnaire. 2006. Available online: http://people.umass.edu/~aizen/pdf/tpb.measurement.pdf (accessed on 20 July 2017)
25. Bicchieri, C. *The Grammar of Society: The Nature and Dynamics of Social Norms*; Cambridge University Press: New York, NY, USA, 2006.
26. Nordlund, A.M.; Garvill, J. Effects of values, problem awareness, and personal norm on willingness to reduce personal car use. *J. Environ. Psychol.* **2003**, *23*, 339–347.
27. Lapinski, M.K.; Rimal, R.N. An explication of social norms. *Commun. Theory* **2005**, *15*, 127–147.
28. Rimal, R.N.; Lapinski, M.K. A re-explication of social norms, ten years later. *Commun. Theory* **2015**, *25*, 393–409.
29. Blackwell, R.D.; Miniard, P.W.; Engel, J.F. *Consumer Behavior*, 10th ed.; South-Western College Pub: Cincinnati, OH, USA, 2006.
30. Gifford, R. *Environmental Psychology: Principles and Practice*, 4th ed.; Optimal books: Colville, WA, USA, 2007.
31. Hungerford, H.R.; Volk, T.L. Changing learner behavior through environmental education. *J. Environ. Educ.* **1990**, *21*, 8–21.
32. Blake, D.E.; Guppy, N.; Urmetzer, P. Canadian public opinion and environmental action: Evidence from British Columbia. *Can. J. Political Sci.* **1997**, *30*, 451–472.
33. Oskamp, S.; Harrington, M.J.; Edwards, T.C.; Sherwood, D.L.; Okuda, S.M.; Swanson, D.C. Factors influencing household recycling behavior. *Environ. Behav.* **1991**, *23*, 494–519.
34. Gatersleben, B.; Steg, L.; Vlek, C. Measurement and determinants of environmentally significant consumer behavior. *Environ. Behav.* **2002**, *34*, 335–362.
35. De Oliver, M. Attitudes and inaction a case study of the manifest demographics of urban water conservation. *Environ. Behav.* **1999**, *31*, 372–394.
36. Do Valle, P.O.; Rebelo, E.; Reis, E.; Menezes, J. Combining behavioral theories to predict recycling involvement. *Environ. Behav.* **2005**, *37*, 364–396.
37. Taylor, S.; Todd, P. Decomposition and crossover effects in the theory of planned behavior: A study of consumer adoption intentions. *Int. J. Res. Market.* **1995**, *12*, 137–155.
38. Boldero, J. The prediction of household recycling of newspapers: The role of attitudes, intentions, and situational factors. *J. Appl. Soc. Psychol.***1995**, *25*, 440–462.
39. Deutsch, M.; Gerard, H.B. A study of normative and informational social influences upon individual judgment. *J. Abnorm. Soc. Psychol.* **1955**, *51*, 629–636.

40. Rivis, A.; Sheeran, P. Descriptive norms as an additional predictor in the theory of planned behaviour: A meta-analysis. *Curr. Psychol.* **2003**, *22*, 218–233.
41. Cialdini, R.B.; Kallgren, C.A.; Reno, R.R. A focus theory of normative conduct: A theoretical refinement and reevaluation of the role of norms in human behavior. *Adv. Exp. Soc. Psychol.* **1991**, *24*, 201–234.
42. Cummins, D.D. Evidence of deontic reasoning in 3-and 4-year-old children. *Mem. Cognit.* **1996**, *24*, 823–829.
43. Abrahamse, W.; Steg, L. How do socio-demographic and psychological factors relate to households' direct and indirect energy use and savings? *J. Econ. Psychol.* **2009**, *30*, 711–720.
44. Guagnano, G.A.; Stern, P.C.; Dietz, T. Influences on attitude-behavior relationships a natural experiment with curbside recycling. *Environ. Behav.* **1995**, *27*, 699–718.
45. White, K.M.; Smith, J.R.; Terry, D.J.; Greenslade, J.H.; McKimmie, B.M. Social influence in the theory of planned behaviour: The role of descriptive, injunctive, and in-group norms. *Br. J. Soc. Psychol.* **2009**, *48*, 135–158.
46. Haynes, S.N., Richard, D.C.S.; Kubany, E.S. Content validity in psychological assessment: A functional approach to concepts and methods. *Psychol. Assess.* **1995**, *7*, 238–247.
47. Byrne, B.M. *Structural Equation Modeling with AMOS: Basic Concepts, Applications, and Programming*, 3rd ed.; Routledge: New York, NY, USA, 2016.
48. Kline, R.B. *Principles and Practice of Structural Equation Modeling*; Guilford Publications: New York, NY, USA, 2015.
49. Fornell, C.; Larcker, D.F. Evaluating structural equation models with unobservable variables and measurement error. *J. Market. Res.* **1981**, *18*, 39–50.
50. Bagozzi, R.P.; Yi, Y. On the evaluation of structural equation models. *J. Acad. Market. Sci.* **1988**, *16*, 74–94.
51. Raines-Eudy, R. Using structural equation modeling to test for differential reliability and validity: An empirical demonstration. *Struct. Equ. Model.* **2000**, *7*, 124–141.
52. Hair, J.F.; Black, W.C.; Babin, B.J.; Anderson, R.E.; Tatham, R.L. *Multivariate Data Analysis*, 6th ed.; Prentice-Hall: Upper Saddle River, NJ, USA, 2006.
53. Jöreskog, K.G. A general approach to confirmatory maximum likelihood factor analysis. *ETS Res. Bull. Ser.* **1967**, 183–202.
54. Ghadi, I.; Alwi, N.H.; Bakar, K.A.; Talib, O. Construct validity examination of critical thinking dispositions for undergraduate students in University Putra Malaysia. *High. Educ. Stud.* **2012**, *2*, 138–145.
55. Hu, L.T.; Bentler, P.M. Cutoff criteria for fit indexes in covariance structure analysis: conventional criteria versus new alternatives. *Struct. Equ. Model.* **1999**, *6*, 1–55.
56. Byrne, B.M. *Structural Equation Modeling with EQS: Basic Concepts, Applications, and Programming*, 2nd ed.; Erlbaum: Mahwah, NJ, USA, 2006.
57. McDonald, R.P.; Ho, M.-H.R. Principles and practice in reporting statistical equation analyses. *Psychol. Methods* **2002**, *7*, 64–82.

58. Browne M.W.; Cudeck R. Alternative ways of assessing model fit. *Sociol. Methods Res.* **1992**, *21*, 230–258.
59. Schütte, H.; Ciarlante, D. *Customer Behaviour in Asia*; New York University: New York, NY, USA, 1998.
60. Heberlein, T.A. *Navigating Environmental Attitudes*; Oxford University Press: New York, NY, USA, 2012; pp. 53–68.
61. Hernández, B.; Martín, A.M.; Ruiz, C.; Hidalgo, M.D.C. The role of place identity and place attachment in breaking environmental protection laws. *J. Environ. Psychol.* **2010**, *30*, 281–288.
62. Thøgersen, J. Norms for environmentally responsible behaviour: An extended taxonomy. *J. Environ. Psychol.* **2006**, *26*, 247–261.
63. Wubs, A., Aarø, L.; Kaaya, S.; Onya, H.; Mathews, C. Social cognition variables and victimization as predictors of sexual debut among adolescents in South Africa and Tanzania: A multi-group SEM analysis. *AIDS Behav.* **2015**, *19*, 2141–2151.
64. Ryan, R.M.; Deci, E.L. Self-determination theory and the facilitation of intrinsic motivation, social development, and well-being. *Am. Psychol.* **2000**, *55*, 68–78.
65. Manuel, J.C.; Sunseri, M.A.; Olson, R.; Scolari, M.A diagnostic approach to increase reusable dinnerware selection in a cafeteria. *J. Appl. Behav. Anal.* **2007**, *40*, 301–310.

Appendix 2

Article

Environmental Literacy on Ecotourism: A Study on Student Knowledge, Attitude, and Behavioral Intentions in China and Taiwan

Wei-Ta Fang [1,*], Ching-Yu Lien [2], Yueh-Wen Huang [3], Guosheng Han [4], Guey-Shin Shyu [5,*], Jui-Yu Chou [6] and Eric Ng [7,*]

[1] Graduate Institute of Environmental Education, National Taiwan Normal University, Taipei 11677, Taiwan

[2] Department of Tourism Management, School of Geography and Tourism, Jiaying University, Meizhou 514015, China; cylien@jyu.edu.cn

[3] Department of Tourism, Shih Hsin University, Taipei 11604, Taiwan; hwhuang@mail.shu.edu.tw

[4] Department of Tourism Management, School of Business Administration, Shandong University at Weihai, Weihai 264209, China; hanguosheng@sdu.edu.cn

[5] Department of Tourism, Tungnan University, Shenkeng, New Taipei City 22202, Taiwan

[6] Department of Biology, National Changhua University of Education, Changhua 50007, Taiwan; jackyjau@cc.ncue.edu.tw

[7] School of Management and Enterprise, University of Southern Queensland, Toowoomba, QLD 4350, Australia

* Correspondence: wtfang@ntnu.edu.tw (W.-T.F.); gsshyu@mail.tnu.edu.tw (G.-S.S); eric.ng@usq.edu.au (E.N.); Tel.: +886-9-3985-9399 (W.-T.F.); Tel.: +886-2-8662-5958 (ext. 723) (G.-S.S); Tel.: +61-7-4631-1298 (E.N.)

Published: 5 June 2016 in open access, Sustainability

Abstract: This study aims to gain further insights to Chinese and Taiwanese university students' environmental literacy on ecotourism. A structural equation model

W.-T. Fang, *Tourism in Emerging Economies*,
https://doi.org/10.1007/978-981-15-2463-9

is developed and validated in an effort to explore the differences between Chinese and Taiwanese university students in terms of their environmental knowledge, environmental attitude, and behavioral intentions. The results showed that the ecotourism perception of Chinese and Taiwanese university students affect their behavioral intentions. Chinese university students exhibited a higher correlation between ecotourism knowledge and behavioral intentions than their Taiwanese counterparts. The findings also revealed differences between the Chinese and Taiwanese students in their perception of ecotourism, and this disparity was particularly evident with regards to how ecotourism should be governed. A moderate difference in ecotourism behavioral intentions was also identified, in which Taiwanese university students were less likely to engage in self-empowerment or private empowerment, to be more educated in the field of ecotourism than their Chinese counterparts.

Keywords: behavioral studies; ecotourism; education; environmental attitudes; environmentally responsible behavior

1. Introduction

The tourism industry is commonly accepted as an 'industry without chimney' as it does not discharge significant amounts of harmful substances or smokes like other industries (e.g., manufacturing, construction) that can have a devastating impact on the environment [1]. However, the argument about the extent to which the effects of the tourism industry have on the environment has been subject to controversial debates. In recent years, there has been an increasing focus on tourism activities, and this is accompanied by an increase in the tourism population. However, a rapid growth in tourism affects the natural environment and the socio-cultural structures of those places [2, 3]. Tourism growth relies on the environment of the tourism area being healthy. Since the beginning of 2002, which was the United Nation's International Year of Ecotourism, the global tourism industry has emphasized the growing importance of both sustainable and tourism development that aims at maximizing the economic, environmental, and social benefits from ecotourism [4, 5]. Furthermore, people have sought to strike a balance between economic growth and environmental protection in order to avoid the drawbacks associated with mass tourism. According to the Goal 17: Revitalize the global partnership for sustainable development, as outlined in the United Nations Sustainable Development Goals (SDG) in 2015, tourism activities are urged to protect the natural environment and emphasize respect for natural heterogeneity, and thereby regard nature as an independent form of life. The 2017 World Tourism Day theme was also declared as 'Sustainable Tourism—a Tool for Development'. Therefore, in addition to passively ensuring that ecotourism will not destroy the environment, tourists can actively understand and enjoy the natural environment, and thus promote the behavior of 'conservation, education, and restoration' [6, 7].

In East Asia, both China (People's Republic of China, PRC), which is home to one-fifth of the global population, and Taiwan, which is an island nation (Republic of

China, ROC), promote ecotourism. China has, in the past decade, seen a significant growth in its tourism market and this was evident by the very high domestic (4.44 billion) and international inbound (138 million) tourist trips in 2016, which accounted for a revenue worth of CNY$3.94 trillion and US$120 billion, respectively. While China has a large number of attractions, such as historic sites and relics, economic hot spots, and culturally diverse places of interest, more recently there has been a focus towards ecotourism whereby the government aims to improve China's reputation for sustainable tourism. Some of the ecotourism destinations include: Changqing National Nature Reserve, Jiuzhaigou National Park, Changbai Mountains, and Kanas Lake. As for Taiwan, the tourism industry has been growing gradually over the past decade and the number of international visitors had reached a record high of 10.69 million in 2016. Accordingly, the Taiwan tourism sector is expected to be worth NT$1.2 trillion by 2025 and is considered the backbone of the nation's economy. In recognition of the contributions made by this important sector, the government has undertaken significant amount of work towards quality tourism development and promotion, including ecotourism which was first embraced in 2002. Some of the ecotourism destinations include: Yangmingshan National Park, Sun Moon Lake, Taroko National Park, and Aboriginal Culture Village. While both countries share many similarities, particularly in traditional customs, cultures, and beliefs, because their political systems are at distinct stages of development, the environmental-governance behaviors, perceptions of ecotourism and the environmental education system at these two places are postulated to differ. Figure 1 shows the recent international inbound tourist trips to China and Taiwan.

A review of the literature revealed that few scholars have previously studied the relationship between ecotourism and the education environment, which is the focus of this study [8, 9]. This study seeks to investigate university students' environmental literacy on ecotourism in China and Taiwan, and aims to:

Figure 1. International inbound tourist trips.

(1) understand the impact of the concept of ecotourism on Chinese and Taiwanese university students' environmental knowledge, environmental attitude, and behavioral intentions;
(2) explore the ecotourism factors that affect Chinese and Taiwanese university students' environmental knowledge, environmental attitude, and behavioral intentions;
(3) analyze and discuss the factors that facilitate or hinder the impact of ecotourism on environmental knowledge, environmental attitude, and behavioral intentions; and
(4) propose recommendations on introducing environmental education in universities.

2. *Literature Review*

2.1. Ecotourism Courses

The history of ecotourism is short, spanning no more than 50 years. During 1960–1970, the immigrant tourist populations in Europe and USA grew rapidly, and this has affected the ecosystems in national parks and protected areas in numerous countries worldwide. With the development of the environmental movement, people begin to examine the environmental and social impacts of tourism and started exploring how to use the nature and seek to balance tourism, environmental protection, and sustainable development. Hetzer [10] indicated that as leisure travel becomes a trend, tourism activities in developing countries have affected the ecological environment. Hetzer [10] also proposed the concept of 'ecological tourism', which has the following connotations: (1) minimal environmental impact; (2) minimal impact on, and maximal respect for, host cultures; (3) maximal economic benefits to host country's people at the grassroots level; and (4) maximal recreational satisfaction derived from participating in the tour. In the early 1980s, the travel industry noted the potential benefits of ecotourism, and has since leased or purchased land in remote areas, and introduced ecological itineraries and tour-guide services. Numerous countries have identified ecotourism as a mean to not only achieve the goals of environmental protection and national development, but also to provide notable benefits to local communities through the protection of community resources that enable sustainable development [11, 12]. However, ecotourism cannot be sustainable if the tourism industry does not use the limited natural or cultural resources in specific locations wisely in the long term [12, 13]. In 1983, Ceballos-Lascuráin (a Mexican architect), who coined the word 'ecotourism', had in 1981 started using the Spanish phrase 'turisimo ecológico' in order to express the ecotourism approach [3, 14]. He considered ecotourism to be an alternative to mass tourism and is to be undertaken for the purpose of studying, enjoying, and experiencing the ecological environment in the absence of anthropogenic interference or contamination, with the goals being to promote local economic and ecological conservation [14]. Since the 1990s, ecotourism has emerged as a concept of environmental literacy and innovative business, and it has included adventure tourism, alternative tourism, environment-friendly travel, ethical tourism, green tourism, nature tourism, nature-oriented tourism, responsible tourism, soft tourism/tourisme doux, and sustainable tourism [15, 16]. Honey [11] defines ecotourism as "responsible travel to natural areas which conserves the environment and improves the welfare of local people." She has noted the promotion of ecotourism for economic development by non-industrialized countries in the early 1990s. For example, the former Soviet Union, Eastern Europe, China, and Vietnam, which once tightly controlled tourism, are now on the bandwagon (p. 18) [11]. Debates in previous studies on how ecotourism can be used to promote democratic reforms and human rights have resulted in an extended definition: "Ecotourism is travel to fragile, pristine, and usually protected areas that strives to be low impact and (usually) small scale. It helps educate the traveler; provides funds for conservation; directly benefits the economic development and political empowerment of local communities; and fosters respect for different cultures and for human rights" (p. 25) [11].

To date, only a few studies have focused on ecotourism that investigate tourists' political empowerment and human rights [17]. However, Hungerford and Volk [18] declared that empowerment variables are crucial in the training of environmental responsible citizens. Furthermore, few studies have also investigated the relationship between environmental literacy and ecotourism activities. Roth and Teksoz et al. considered environmental literacy to include ability, knowledge, skills, and actions that can be defined using behavioral intentions [19, 20]. Environmental literacy refers to people's knowledge associated with environmental matters, their positive environmental attitudes and values, and the skills, motivation, and commitment required to participate in solving environmental problems [21, 22]. This actively enables the environment movement in maintaining balance between the environment and the quality of life.

Ecotourism was first introduced in a few Chinese universities as a major study in the early 1990s. Since then, the number of Chinese universities (e.g., Peking University, Beijing, China, Shanghai Jiao Tong University, Shanghai, China) offering ecotourism as a specialization or an elective course (e.g., ecotourism principles) has increased, but it is still not a course that is commonly taught and undertaken. On the other hand, the concept of ecotourism was dated back to the late 1980s in Taiwan and its importance has grown considerably over the years. In recognizing the significant contribution of ecotourism to the tourism industry, many universities (e.g., TransWorld University, Douliu, Taiwan, National Taiwan Normal University, Taipei, Taiwan) have established disciplinary areas and departments of tourism, business, and environmental sciences that offer ecotourism as a major study, as well as an elective course (e.g., ecotourism and sustainable development). The increasing offerings of ecotourism courses at universities in China and Taiwan align closely to the respective governments' effort and acknowledgement towards the growing significance of ecotourism.

2.2. Ecotourism Knowledge

According to the definition that governmental bodies, scholars, experts, and private conservation groups have attempted to construct from 1960 until now, ecotourism refers not only to the enjoyment of the beautiful ecological landscape situated within tourism areas, it also emphasizes the evolution of environmental conservation behavior and thinking patterns [23, 24]. This mode of progression from thinking to action has been developed as an attempt to protect the quality of tourism resources through ecological protection. The goal is to ensure that tourism activities are long-lasting and can maintain the ecological environment, social culture, local economy, and a balanced development of sustainable tourism. Moreover, this process deepens our understanding of the local natural environment, people, and the social and traditional culture in natural areas. Thus, studies conducted by ecological and tourism scholars show that the conditions that facilitate ecotourism are based on the natural environment (including the aboriginal culture). The function of ecotourism is to protect the environment and enable sustainable local development, which can achieve common benefits for all stakeholders (e.g., regional ecological managers, community residents, local tourism operators, and tourists) [25].

To target the ecological traits of people who visit tourism destinations, tourists are provided with recreational experiences that guide their environmental awareness. However, this may not produce responsible tourism behavior [26]. Many studies [27, 28] showed that environmentally friendly attitudes do not necessarily lead to eco-friendly behaviors. These experiences from environmentally friendly behaviors should be regarded as 'necessary conditions' of ecotourism. In Asia, during the period when authoritarian states are being transformed into democratic states, few people engage in participating tourism in developing countries [29]. In this study, the governance of tourism was included, which is a form of tourism management that is based on the policies that Asian countries use in managing ecotourism areas. According to the norms and the legal system of ecotourism established by these nations, investment funds are input to strengthen the tourism governance manpower. In considering the aforementioned work, this study regards ecotourism to be a composition of the following:

(1) Responsible tourism: This type of tourism focuses on the responsibility of the people visiting natural areas to respect the local ecological environment, social culture, and traditional economy and to preserve local monuments [30, 31].
(2) Educational tourism: This is focused on providing education to enhance people's environmental knowledge, inducing people to develop favorable attitudes toward the environment, and making them take environmental actions to reduce the impact of tourism [32, 33].
(3) Participating tourism: This type of tourism focuses on participation, which allows tourists to give something back to the local economy, improves the well-being of local residents, and encourages all stakeholders to join in brainstorming [7, 25, 29].

(4) Governing tourism: This deals with management units that provide environmental governance strategies, programs, and plans, and invest funds and manpower in order to improve the local tourism environment [34, 35].

Given the above, this study seeks to investigate participants' understanding of ecotourism, level of awareness and perception, and judgment or evaluation, and the proposed 'knowledge' dimension of ecotourism, which is the first dimension of environment literacy. The definition of environmental knowledge adopted in this study is the mental process of acquiring understanding through thought, experience, and the senses [36, 37]. Thus, environmental knowledge contains the understanding of knowledge of ecotourism through thought, experience, and the senses from self-directed and school learning. Based on the literature reviewed, the knowledge dimension of ecotourism includes: responsible tourism, educational tourism, participating tourism, and governing tourism. Environmental-knowledge education is provided in order to develop environmental awareness through interactions between humans and the environment. This results in an appreciation, and exploration of the environment, which help people to be more aware of and acknowledge the existence of environmental problems.

2.3. Attitudes toward Ecotourism

Environmental literacy incorporates the dimensions of knowledge, attitude, and behavior. Diverse opinions exist regarding attitude, but it commonly includes the results of knowledge and learning processes, reactions to the feeling or perception of liking or disliking something, and judgment factors [38]. Hines et al. considered attitude to be a combination of people's views regarding the environment and matters that are part of a specific case related to the entire environment [39]. In this study, attitude is considered to be a dimension of the ecotourism. This study seeks to investigate participants' beliefs, values, and sentiments toward ecotourism, including their subjective judgment or evaluation.

2.4. Behavioral Intentions toward Ecotourism

Hines et al. identified certain environmental behavior patterns [39]. A major factor that elicits environmental behavior is people's intention to act, and therefore intention is the most favorable variable that influences action [30, 40]. Furthermore, studies have also demonstrated that intentions related to environmental protection can explain the observed behavior of people [41, 42]. Hungerford and Peyton proposed that environmental behavior can be classified into five categories: (1) persuasion; (2) consumer/economic action; (3) eco-management; (4) legal action; and (5) political action. Environmental behavior is measured here at the 'skill' level [43]. Behavioral practices that are observed in daily life can prevent the deterioration of the environment, and therefore environmental education, are critical to the development towards environmental behavior. In order to measure environmental behavior, behavioral intentions need to be measured. This study focuses on the students' behavioral intentions related to ecological management within the studied ecotourism area. Ecological management is presented as the third dimension of behavioral intentions, and based on the literature, it includes private empowerment (i.e., empowered personal capabilities), private action (i.e., personal action for privilege in private areas), and public service.

Previous studies related to this area have shown that environmental knowledge, environmental attitude, and environmental behavior are positively related to each other [44, 45]. In examining the structure of environmental literacy, the assumption in most studies has been that environmental knowledge affects environmental attitude, and the attitude affects environmentally friendly behavior. However, only few studies have assumed and directly confirmed that knowledge and attitude of the environment affect environmentally friendly behavior, which this study aims to further explore [46].

2.5. Relationship between Environmental Knowledge and Environmental Attitude

Pooley and O'Connor considered attitude domains to be affected by the knowledge domain [47]. Specifically, the likelihood of change is affected by a person's stance toward knowledge or what Kruglanski described as 'epistemic motives'. Based on the definition of McKechnie, attitude is the composition of beliefs, values, and sentiments [38, 48]. Here, the affective dimension of attitude deals with the emotional and non-emotional reactions toward the state of the environment (e.g., values, or likes/dislikes) [49]. Hence, the following hypothesis will be investigated in this study.

Hypothesis 1. *Environmental knowledge positively affects environmental attitude.*

2.6. Relationship between Environmental Knowledge and Environmental Behavioral Intentions

Hines et al. performed a meta-analysis on environmental behavior where the results indicate that people who understand environmental concerns and know how to act in response to such concerns tend to exhibit the attitude required to engage in environmentally responsible behaviors [39, 50]. Kaiser et al. [46] reported that environmental knowledge and environmental values explained 40% of the variance of ecological behavioral intentions. Bamberg [50] also suggested that situation-specific knowledge are direct determinants of specific behaviors. Based on these studies, the second hypothesis is proposed as follow:

Hypothesis 2. *Environmental knowledge positively affects environmental behavioral intentions.*

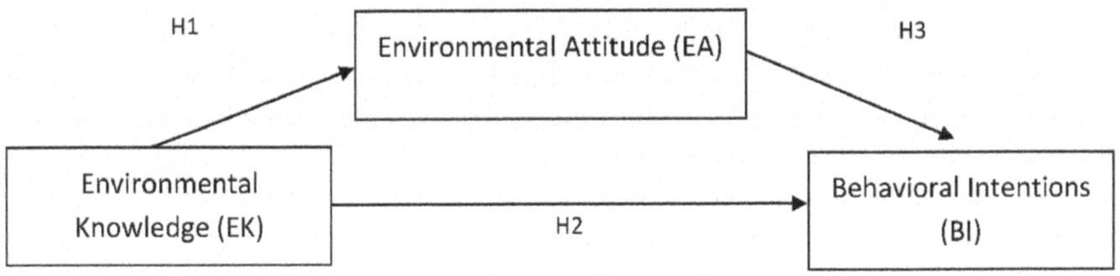

Figure 2. Hypothetical structure of environmental concern, H1: Environmental knowledge → Environmental attitude (Environmental knowledge has a positive impact on environmental attitude); H2: Environmental knowledge → Behavior intentions (Environmental knowledge has a positive impact on environmental behavioral intentions); H3: Environmental attitude → Behavior intentions (Environmental attitude has a positive impact on environmental behavioral intentions).

2.7. Relationship between Environmental Attitude and Environmental Behavioral Intentions

Grob [51] indicated that 39% of the variance in environmental behavior can be explained by attitudinal components. Pooley and O'Connor [47] also reported that values and attitudes affect environmentally conscious behaviors, and that changing environmental attitudes, emotions, and beliefs are more important than accumulating environmental related knowledge for influencing behaviors. Best and Mayerl [52] used traditional path analysis and latent structural equation modelling to test the hierarchical structure of environmental values, beliefs, attitudes, and their contribution to explaining recycling behavior, which conclude the direct effects of specific attitudes on behavior. This leads to the development of the third hypothesis as follows:

Hypothesis 3. *Environmental attitude positively affects environmental behavioral intentions.*

Using the above assumptions, a theoretical framework (Figure 2) is proposed for this study to investigate the relationships between environmental knowledge, environmental attitude, and behavioral intentions on ecotourism.

3. Materials and Methods

3.1. Research Areas and Participants

This study was conducted at two locations; People's Republic of China (P. R. China), and Republic of China (Taiwan, ROC). The participants of the study consisted of first-, second-, third-, and fourth-year undergraduate students in China and Taiwan, age between 18 to 23 years old. The students were from a diverse study background, including disciplinary areas in natural sciences and social sciences. Based on the geographic (i.e., north, east, south, and west regions) and demographic (size of universities) stratum, a stratified random sampling technique was used to select four universities each from China and Taiwan. The survey was conducted over an extended period from March 2011 to March 2014, and this was due to limited resources and the extensive geographical distances to be covered in China. The rationale for selecting the sample was primarily because they were considered being at a critical life stage to self-shaping their environmental knowledge, affective values, and environmental behavior from their environmental identity and decision-making process [53].

A total of 1,200 questionnaires were equally distributed and surveyed at the selected universities in China and Taiwan during 2011 to 2014. Of which, 835 (69.5%) valid questionnaires were returned, with China accounted for 420 and the remaining 415 were from Taiwan. Written consent forms were sent to class teachers and students to obtain their agreement to participate in this cross-national research study and inform them of their rights to withdraw at any time. The questionnaire survey was conducted in the classroom setting at the respective classes (50 in China and 48 in Taiwan) after permissions to participate had all been received. The participants had approximately 10 min to listen to relevant instructions and the opportunity to seek any clarifications (including Honey's definition of ecotourism) prior to completing the 30 min questionnaire in class.

3.2. Instrument and Instrumentation

When designing these questionnaire items, the following steps were undertaken to select ecotourism and environmental literacy instrument with consideration given to the alignment of the contextual issues. This study was based on a substantial consequence of literature reviews demonstrating the knowledge and attitudinal aspects of environmental behavioral research. Reference to previous studies related to knowledge, attitude, and behavioral intentions was attempted to explore the causal relationships.

The item analysis in terms of difficulty and discriminating the knowledge items for "yes/no" answers required the knowledge category to be changed to students' understanding through thought, experience, and the senses. The compiled pools of items were selected from the following steps [54]:

- Step 1: More than 25 research papers and documents related to ecotourism and environmental literacy in China, Taiwan, and abroad [55–59] were reviewed to establish these item pools.
- Step 2: The items in the pools were selected in accordance to the research objectives guided by the definition of each element and associated components. The initial compiled questions items were then prepared for panel review from seven experts.
- Step 3: The initially-compiled question items were given to a panel of seven experts (three from China, four from Taiwan) from different areas of specialization, such as tourism, psychology, geography, sociology, biology, ecology, and environmental education for their formal review and expert opinions. The experts were required

to assess the items on their appropriateness, relevance, and language used from which content validity had to be reached. Each question item had received at least 80% agreement by the experts.

- Step 4: The instrument was revised based on the experts' opinions and feedback, and subsequently pre-tested with 40 (20 each from China and Taiwan) randomly selected undergraduate students. The final questionnaire survey consisted of two main sections; (1) demographic items; and (2) environmental literacy and ecotourism assessment items. However, the questionnaire content (measures) had been rearranged so that they were well blended into the three key domain areas of knowledge, attitude, and behavioral intentions (see Appendix A). A five-point Likert scale (i.e., 1 = "Strongly disagree" to 5 = "Strongly agree") was adopted for the measurement. The following process was used to construct the factor dimensions and the measurement questions related to ecotourism environmental literacy. Therefore, this study specifically sought to measure two key dimensions: (1) knowledge; (2) attitude, that were considered influential to environmental behavioral intentions. Three factor dimensions (i.e., knowledge, attitude, and behavioral intentions) were identified with 38 measurement questions developed.

3.3. Measures

Next, Statistical Package for the Social Sciences (SPSS 22.0) statistical software was used to conduct the descriptive analysis and the reliability analysis of the samples before using structural equation modelling to perform confirmatory factor analysis. Frequency analysis was used to determine the total number of occurrences, the mean and standard deviation (SD) scores for the demographic questions and items in the key dimensions. The Pearson Correlation technique was used to measure the strength and direction of relationship that exists between these key dimensions. The multiple regression analysis was used to predict the influence of attitude on environmental behaviors.

4. Results

4.1. Descriptive Statistics

As shown in Table 1, majority (60.8%) of the participants were female with the remaining 39.2% being male. There were more male participants in China (40.7%) than in Taiwan (37.6%), whereas female participants in Taiwan (62.4%) accounted for a higher proportion than in China (59.3%). Most of the participants were fourth-year undergraduate students (31.6%), and this was followed by first-year (28.7%), second-year (21.1%), and third-year (18.6%) undergraduate students. In China, 35.5% were

Table 1. Gender and year level of participants.

	China		Taiwan		Total	
	Frequency	**Percentage**	**Frequency**	**Percentage**	**Frequency**	**Percentage**
Gender						
Male	171	40.7	156	37.6	327	39.2
Female	249	59.3	259	62.4	508	60.8
Year level						
First-year	95	22.6	145	34.9	240	28.7
Second-year	89	21.2	87	21.0	176	21.1
Third-year	87	20.7	68	16.4	155	18.6
Fourth-year	149	35.5	115	27.7	264	31.6

fourth-year undergraduate students, while majority of the participants in Taiwan were represented by first-year (34.9%) undergraduate students.

In relation to the question "How many travel days have you ever engaged in domestic tourism in the last year?", the findings (as shown in Figure 3) showed that a one-day trip was considered as most common in 'travel days' by 370 participants (44.3%), while a four-day trip (85 participants, 10.2%) was regarded as the least common. A comparison between Chinese and Taiwanese students revealed a notable difference in their travel days. While majority of the Taiwanese students (277 participants, 66.7%) engaged in a one-day trip, the travel days for Chinese students were more evenly spread across (i.e., ranging from a one-day trip to a five-day or more trip).

With regards to the question "Which travel period did you choose to engage in domestic tourism in the last year?", results (as shown in Figure 4) indicated that more than half of the participants (435 participants, 52.1%) would travel during Spring

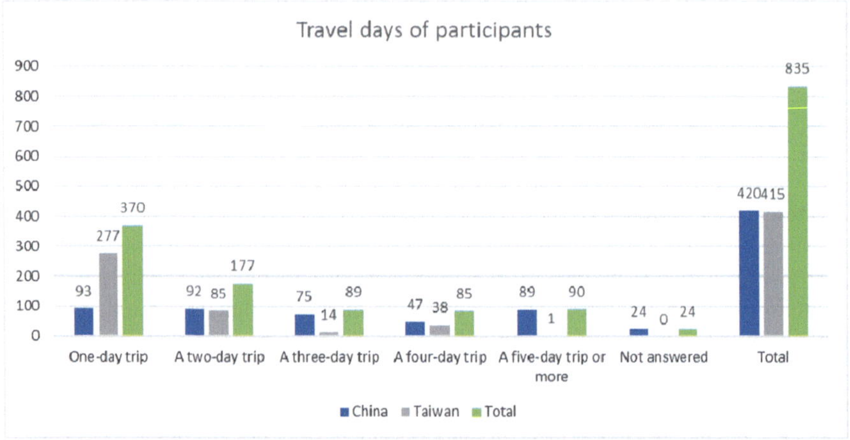

Figure 3. Travel days of participants.

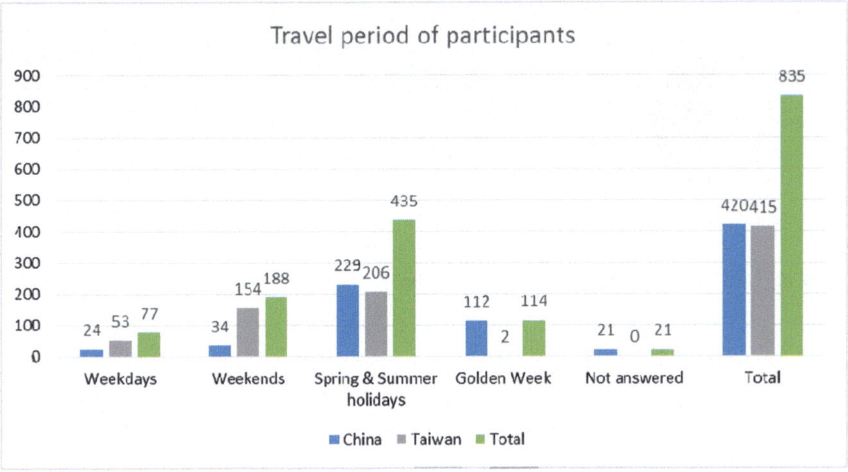

Figure 4. Travel period of participants.

and Summer holidays, and this was followed by weekends (188 participants, 22.5%), Golden Week (continuous holidays) (114 participants, 13.7%), and weekdays (77 participants, 9.2%). Although both the Chinese and Taiwanese students agreed that they were most likely to travel during the period of Spring and Summer holidays, they differed in their responses about travelling during weekends and Golden Week (continuous holidays). Approximately 26.8% of the Chinese students (112 responses) favored travelling during the Golden Week (continuous holidays), while Taiwanese students (154 participants, 37.1%) preferred travelling during weekends.

In terms of the question about ecotourism participation "Have you ever participated in ecotourism activities?", the findings (as shown in Figure 5) suggested that majority (64.9%) of the participants (542 responses) had not participated in any ecotourism activities, while the remaining 35.1% (293 participants) had some level of participation. The results also revealed a high proportion (374 participants, 89%) of the Chinese students who had not participated in any ecotourism activities, with only a marginal 11% (46 responses) indicated their involvement in such activities. In

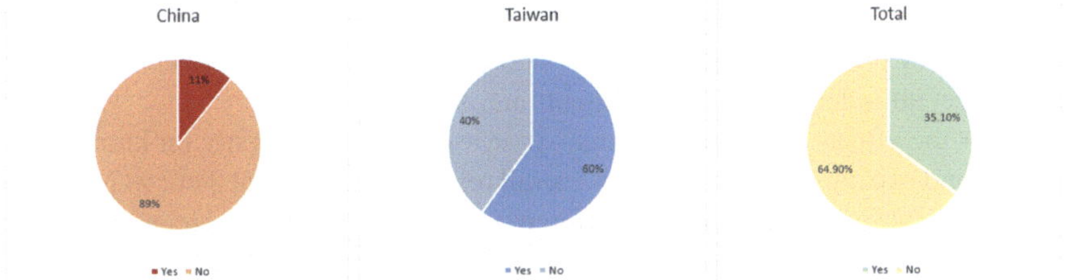

Figure 5. Ecotourism participation.

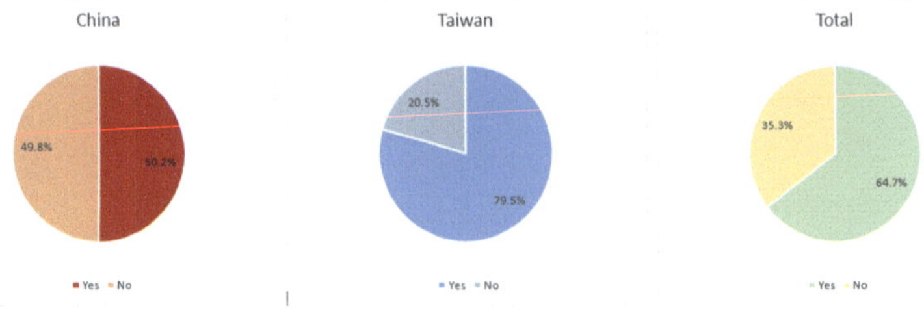

Figure 6. Learn the concepts of ecotourism at school.

contrast, 247 (60%) Taiwanese students had responded to having participated in ecotourism activities, and 40% (168 participants) responded with no participation. This suggested that Taiwanese students had a greater level of participation in ecotourism activities than their Chinese counterparts.

In response to the question "Have you ever learn the concepts of ecotourism at school?" about learning the concepts of ecotourism as a part of their university curricula, the results (as shown in Figure 6) showed that 540 (64.7%) participants agreed that they had been taught the concepts of ecotourism, with the remaining 35.3% (295 participants) suggested that such knowledge had not been taught at their university. The findings revealed a higher proportion of Taiwanese students (329 responses, 79.5%) had been taught the concept of ecotourism at university than the Chinese students (211 responses, 50.2%). This could possibly indicate a greater level of emphasis on teaching the concepts of ecotourism as part of the university curricula in Taiwan than those in China.

In terms of the effects of the gender variable on knowledge, attitude, and behavioral intentions, the overall findings revealed significant differences between females and males in their knowledge (df = 833, two-tailed, $t = 3.801 > 1.967$, $p = 0.000$), attitude (df = 833, two-tailed, $t = 4.476 > 1.967$, $p = 0.000$) and behavioral intentions (df = 833, two-tailed, $t = 2.745 > 1.967$, $p = 0.006$). Results also suggested that there were significant differences between females and males in China about their knowledge (df = 418, two-tailed, $t = 3.070 > 1.967$, $p = 0.002$), attitude (df = 418, two-tailed, $t = 2.238 > 1.967$, $p = 0.026$) and behavioral intentions (df = 418, two-tailed, $t = 2.966 > 1.967$, $p = 0.018$). Whereas in Taiwan, significant differences between females and males existed in their knowledge (df = 413, two-tailed, $t = 2.593 > 1.967$, $p = 0.010$), attitude (df = 413, two-tailed, $t = 4.044 > 1.967$, $p = 0.000$). Table 2 below provides a brief summary about the effects of the gender variable on knowledge, attitude, and behaviors intentions.

As for the effects of the participants' year level at the university on knowledge, attitude, and behavioral intentions, the overall results indicated that there were no significant differences between all the different year levels on knowledge, attitude, and behavioral intentions. However, the findings for the Chinese participants suggested that there existed significant differences between different year level of undergraduate students on behavioral intentions (df = 418, two-tailed, $F = 3.691 > 1.967$, $p = 0.012$). With regards to the Taiwanese participants, the findings indicated significant

Table 2. The effects of the gender variable on knowledge, attitude, and behavioral intentions.

Gender	Knowledge				Attitude				Behavioral Intentions			
	Mean	SD	*t*	*p*	Mean	SD	*t*	*p*	Mean	SD	*t*	*p*
Total												
Female	3.83	0.45	3.801	0.000	4.03	0.39	4.476	0.000	3.21	0.47	2.745	0.006
Male	3.72	0.43			3.90	0.46			3.12	0.47		
China												
Female	3.91	0.47	3.070	0.002	4.04	0.33	2.238	0.026	3.37	0.46	2.966	0.003
Male	3.78	0.38			3.95	0.41			3.23	0.44		
Taiwan												
Female	3.76	0.43	2.593	0.010	4.03	0.43	4.044	0.000	3.05	0.42	1.496	0.136
Male	3.64	0.47			3.84	0.51			2.98	0.46		

differences between different year level of undergraduate students on knowledge (df $= 413$, two-tailed, $F = 2.674 > 1.967$, $p = 0.047$) and behavioral intentions (df $= 413$, two-tailed, $F = 2.639 > 1.967$, $p = 0.049$). A brief summary about the effects of the year level variable on knowledge, attitude, and behaviors intentions is provided in Table 3 below.

4.2. Data Analysis of Study Variables

Structural equation modelling was used to validate the measurement model and the structural model. Prior to conducting the modelling measurement, the data on the research variables were analyzed. The structural equation estimation is affected by the distribution properties of variables, and thus the distribution of variables must be tested first. According to the argument of Kline [60], when the absolute value of the skewness of a variable is >3, the value can be regarded as being an extreme value; in the case of kurtosis, absolute values >10 are considered extreme values [40, 56]. If these problems are created by the skewness and kurtosis values, bias will be produced when the maximum-likelihood method is used for estimating correlation coefficients. If the coefficient lies within the aforementioned ranges, the bias of the estimated coefficient is considered acceptable. As such, each dimension's skewness and kurtosis values was calculated, and the results are presented in Table 4. In the three dimensions of the research variables, the absolute values of maximal skewness and maximal kurtosis were 1.084 and 1.521, respectively; thus, in terms of the argument of Kline [60], extreme values were not measured.

When examining the Chinese and Taiwanese students' environmental knowledge, environmental attitude, and behavioral intentions, the findings (as shown in Table 4) showed that Chinese students had exceeded the mean values of 3.50 for all three dimensions, i.e., environmental knowledge (3.966 ± 0.393), environmental attitude (4.229 ± 0.486), behavioral intentions (3.713 ± 0.564). Conversely, the Taiwanese students only had the environmental attitude dimension (3.834 ± 0.478) exceed the mean value 3.50. In the three dimensions of environmental knowledge, environmental attitude, and behavioral intentions, the mean values measured in the case of Chinese students were higher than those measured in the case of Taiwanese students. Refer to Appendix A for the list of questions asked in the survey for these three dimensions.

4.3. Confirmatory Factor Analysis

Confirmatory factor analysis was performed mainly to understand the state of the fit between each item of a scale and the measurement questions. The following conventional goodness-of-fit criteria was recorded in order to evaluate the model [47–49]. Chinese students: Goodness of Fit Index (GFI) $= 0.96$, Comparative Fit Index (CFI) $= 0.94$, Non-Normed Fit Index (NNFI) $= 0.92$, the standardized root mean square residual (SRMR) $= 0.056$, the root mean square error of approximation (RMSEA) $= 0.075$; and Taiwanese students: GFI $= 0.96$, CFI $= 0.89$, NNFI $= 0.84$,

Table 3. The effects of the year level variable on knowledge, attitude, and behavioral intentions.

Year Level	Knowledge				Attitude				Behavioral Intentions			
	Mean	SD	*F*	*p*	Mean	SD	*F*	*p*	Mean	SD	*F*	*p*
Total												
First-year	3.76	0.55	1.654	0.176	3.97	0.44	0.628	0.597	3.16	0.50	1.253	0.289
Second-year	3.75	0.42			3.96	0.42			3.12	0.42		
Third-year	3.80	0.36			3.96	0.42			3.19	0.45		
Fourth-year	3.83	0.41			4.00	0.41			3.20	0.49		
China												
First-year	3.92	0.64	1.011	0.388	3.40	0.40	0.843	0.471	3.43	0.46	3.691	0.012
Second-year	3.82	0.36			3.40	0.37			3.26	0.41		
Third-year	3.83	0.35			3.96	0.40			3.22	0.44		
Fourth-year	3.86	0.36			4.04	0.32			3.33	0.48		
Taiwan												
First-year	3.65	0.44	2.674	0.047	3.96	0.46	0.142	0.935	2.98	0.44	2.639	0.049
Second-year	3.68	0.48			3.93	0.47			2.98	0.40		
Third-year	3.76	0.36			3.96	0.46			3.15	0.46		
Fourth-year	3.79	0.46			3.97	0.50			3.05	0.45		

Table 4. Test value of the data of the study variables.

	China				Taiwan			
	Mean	**SD**	**Skewness**	**Kurtosis**	**Mean**	**SD**	**Skewness**	**Kurtosis**
Environmental knowledge	3.966	0.393	−0.265	0.532	2.620	0.253	−0.124	0.362
Environmental attitude	4.2294	0.486	−1.084	1.521	3.834	0.479	−0.564	−0.382
Behavioral intentions	3.713	0.564	−0.458	0.208	2.924	0.456	−0.069	−0.008

SRMR = 0.061, RMSEA = 0.068. The measured criteria reveal that the test results are in line with the criteria used for these indicators [60–62].

4.4. Testing the Research Hypotheses

The maximum-likelihood estimation method was used to estimate the path coefficients of the theoretical model, and also to test each of the hypotheses proposed. Accordingly, a sample size of 100 to 150 is considered most suitable for use in the maximum-likelihood estimation method when used in structural modelling [63]. As the number of valid samples collected was 420 in the case of Chinese students and 415 in the case of Taiwanese students, therefore the sample size is considered appropriate for the use of the estimation methods.

According to the path analysis, results (as shown in Table 5) revealed that Hypotheses 1 and 2 were supported in the case for the Chinese students, which indicated that environmental knowledge had positive impact on environmental attitude, and environmental knowledge also had a positive impact on environmental behavioral intentions. On the other hand, findings for the Taiwanese students showed that environmental knowledge had a positive impact on environmental behavioral intentions, and environmental attitude also had a positive impact on environmental behavioral intentions. Therefore, Hypotheses 2 and 3 were both accepted.

Table 5. The result of test for path analysis.

Path	**China**				**Taiwan**			
	Standard Error	**t-Value**	**Significant**	**Result**	**Standard Error**	**t-Value**	**Significant**	**Result**
H1: EK→EA	0.75	5.50	***	Support	−0.62	6.85	***	Non-support
H2: EK→BI	0.76	2.31	***	Support	0.16	1.26	***	Support
H3: EA→BI	−0.42	−1.10	***	Non-support	0.21	1.36	***	Support

4.5. Relationships between Environmental Knowledge, Environmental Attitude, and Behavioral Intentions

Table 6 lists the total and indirect effects obtained as the output measured using LISREL 8.80, and Figures 7 and 8 show the results of the structural modelling path analysis.

Results indicated that, in the case of Chinese students, there was a significant positive-impact-relationship existed between environmental knowledge and environmental attitude and between environmental knowledge and behavioral intentions. On the other hand, in the case of Taiwanese students, there existed a significant positive-impact-relationship between environmental knowledge and behavioral intentions and between environmental attitude and behavioral intentions.

4.6. Findings

The results shown in Figure 7 reflect the relative levels of conceptual emphasis on the causal relationships among these three dimensions in the ecotourism environmental literacy of Chinese university students. In the case of Chinese university students, the correlation coefficients (β, γ) calculated between the three factor dimensions of environmental knowledge, environmental attitude, and behavioral intentions in ecotourism environmental literacy were as follow: 0.76, calculated for 'environmental knowledge affects behavioral intentions' (Hypothesis 2: supported; $\gamma = 0.76$, t-value $= 2.31$); 0.75, calculated for 'environmental knowledge affects environmental attitude' (Hypothesis 1: supported; $\gamma = 0.75$, t-value $= 5.50$); and -0.42, calculated for 'environmental attitude affects behavioral intentions' (Hypothesis 3: not supported; $\beta = -0.42$, t-value $= -1.10$). The values of the correlation coefficients obtained in this analysis indicated that among the three factor dimensions of ecotourism environmental literacy, the explanatory power was highest in the case of the factor dimension 'environmental knowledge affects behavioral intentions', and it was followed by 'environmental knowledge affects environmental attitude'.

In the case of Taiwanese university students (see Figure 8), the correlation coefficients (β, γ) calculated between the three factor dimensions of environmental attitude, environmental knowledge, and behavioral intentions were as follows: 0.21, calculated for 'environmental attitude affects behavioral intentions' (Hypothesis 3: supported; $\beta = 0.21$, t-value $= 1.36$); 0.16, calculated for 'environmental knowledge affects behavioral intentions' (Hypothesis 2: supported; $\gamma = 0.16$, t-value $= 1.26$); and -0.62, calculated for 'environment knowledge affects environmental attitude' (Hypothesis 1: not supported; $\gamma = -0.62$, t-value $= 6.85$). The size of the λ coefficient in the analysis indicated that among the three factor dimensions of ecotourism environmental literacy, explanatory powers were exhibited by 'environmental knowledge affects behavioral intentions' and 'environmental knowledge affects environmental attitude', but the correlation was extremely weak. Therefore, the results confirmed the findings of Ajzen and Fishbein, and Best and Mayerl [52, 64]. Weak bivariate correlations existed between environmental attitude (i.e., attitudes and values) and

Table 6. Total and indirect effects between environmental knowledge, environmental attitude, and behavioral intentions.

	Effect	**China**				**Taiwan**			
		Environmental Attitude		**Behavioral Intentions**		**Environmental Attitude**		**Behavioral Intentions**	
		Effect	***t*-Value**	**Effect**	***t*-Value**	**Effect**	***t*-Value**	**Effect**	***t*-Value**
Environmental knowledge	Direct effect	0.75	5.50	0.76	2.31	−0.62	6.85	0.16	1.26
	Indirect effect			−0.32	−1.05				
	Total effect	0.75	5.50	0.44	1.26	−0.58	1.26	0.16	5.09
Environmental attitude	Direct effect			−0.42	−1.10			0.21	1.36
	Indirect effect								
	Total effect	—	—	−0.42	−1.10	—	—	0.21	1.36

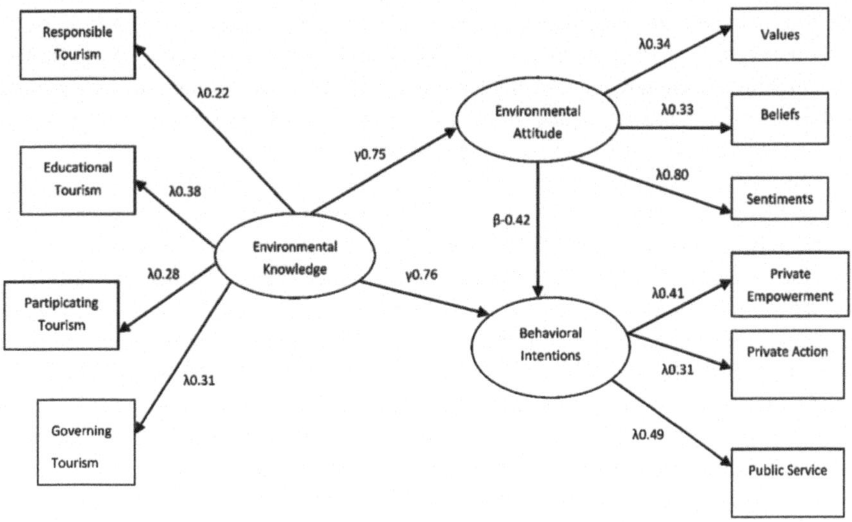

Figure 7. Empirical structure of environmental concern in Chinese students ($n = 420$).

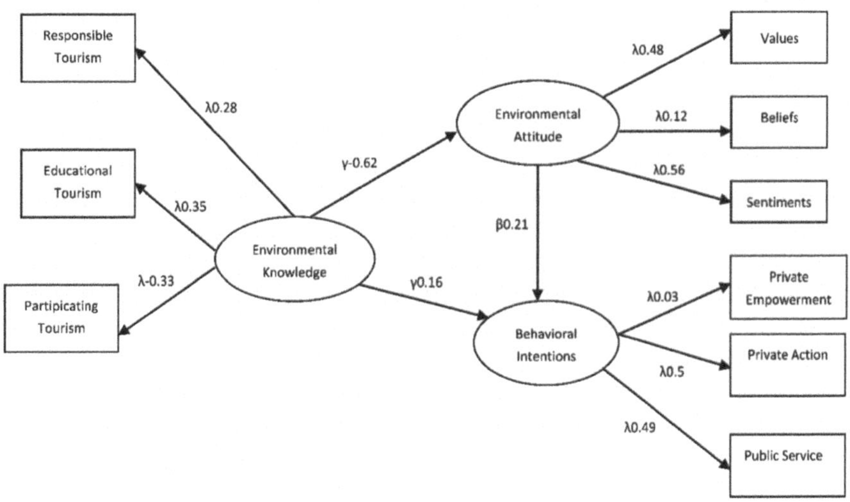

Figure 8. Empirical structure of environmental concern in Taiwanese students ($n = 415$).

environmental behavior because of the absence of direct effects in both Chinese and Taiwanese students. The results also showed that the environmental knowledge of ecotourism affect behavioral intentions in both Chinese and Taiwanese university students. In terms of correlation, the environmental ecotourism behavioral intentions of Chinese students were stronger than those of Taiwanese students and exhibited a positive relationship ($\gamma = 0.76$). Based on the λ coefficient, the behavioral intentions of Chinese university students were ranked in the following sequence: public service ($\lambda = 0.49$), private empowerment ($\lambda = 0.41$), and private action ($\lambda = 0.31$). In the aforementioned knowledge of Chinese university students, ecotourism was ranked in the following sequence: educational tourism ($\lambda = 0.38$), governing tourism ($\lambda = 0.31$), participating tourism ($\lambda = 0.28$), and responsible tourism ($\lambda = 0.22$).

Among the Taiwanese university students, those who exhibited high environmental knowledge in ecotourism also exhibited behavioral intentions in ecotourism, but the correlation was weak ($\gamma = 0.16$). The behavioral intentions of Taiwanese university students were ranked in the following sequence: private action ($\lambda = 0.52$), public service ($\lambda = 0.49$), and private empowerment ($\lambda = 0.03$). The Taiwanese university students' knowledge of the definition of ecotourism included educational tourism ($\lambda = 0.35$) and responsible tourism ($\lambda = 0.28$), but not participating tourism ($\lambda = -0.33$). Furthermore, in the confirmatory factor analysis, the measurement results of governing tourism deviated considerably from the Taiwanese university students' knowledge of ecotourism and did not constitute the sub-dimensions, and thus these measurement results were omitted.

5. Discussion

In East Asia, both China and Taiwan are increasingly promoting ecotourism, but ecotourism has been developed to distinct levels because of the use of dissimilar political systems in these places. Both regions had been subjected to some controversial debates for true ecotourism. In Taiwan, for example, six Indigenous Seidok tribes had jointly developed ecotourism in Nantou County. This ecotourism program was supported by the government, with the intention to combine factual history associated with Seidok culture and local tourism. The successful stories were detected in Taiwan because this island nation features a prosperous economy, universal education, and a solid middle class, and Taiwan has successfully transformed from an authoritarian political system to a Western-style democracy.

In China, ecotourism was also an emerging sector which had gained substantial attention [65]. For example, new ecotourism resorts (e.g., Naked Stables Private Reserve, Red Capital Ranch) were built to promote ecotourism and sustainability as China sought to improve its global environmentalism reputation. Although the market is gradually opening up, and with people's lives improving, the voice for freedom and democracy is getting increasingly loud; however, China still adheres to a mixed socialism and state-capitalism system [66].

5.1. Ecotourism Knowledge Affects Behavioral Intentions

Results revealed that ecotourism knowledge affects the behavioral intentions of both Chinese and Taiwanese university students, and that this correlation was stronger in the case of Chinese university students than to their Taiwanese counterparts. There is inadequate environmental education intervention provided at the university level in China and Taiwan, and thus it is challenging to construct an environmental teaching module featuring a theoretical background that has been demonstrated to be effective. Therefore, in the planning of environmental education curriculum in China and Taiwan, the 'knowledge–attitude–behavior' theory is still accepted: the education system accepts the notion that increasing environmental knowledge can change a person's attitude toward the environment and produce responsible environmental behavior [67]. Thus, considerable argument and controversy exists regarding whether or not environmental attitude is the intermediary variable of environmental knowledge and environmental behavior. Because the concept of the 'knowledge–attitude–behavior' theory is used, the environmental education curriculum in China and Taiwan leans toward a one-way transfer of knowledge. However, a transfer of knowledge is not equal to the construction of environmental values and judgment. Grob [51] reported that no effects on environmental behavior stemming from factual knowledge could be identified, thus the relationship between knowledge and behavior could not be established. Moreover, several studies have shown that a high degree of correlation does not necessarily exist among knowledge, sentiment, and behavior, and Fishbein reported that sentiment, knowledge, and behavior are not always strongly correlated [68]. However, the findings in this study suggested that an accurate understanding of ecotourism by Chinese and Taiwanese university students will elicit environmentally friendly behavioral intentions, which in turn lead to having comprehensive knowledge of ecotourism that will produce a self-driven ability to engage in environmentally friendly behaviors.

5.2. Disparities in Chinese and Taiwanese University Students' Ecotourism Behavioral Intentions

With regards to self-empowerment for personal improvement, the Taiwanese university students responded that they did not commonly read explanatory signs and billboards posted by the government in ecotourism areas, and the students displayed a high level of personal awareness of ecotourism and many expressed distrusts in the government. In terms of self-learning behaviors in ecotourism, Taiwanese university students stressed autonomy, and from the perspective of receiving multiple messages from communication media, the students pursued diverse values, expressed autonomous awareness, and opposed the self-centered concept of a centralized consciousness than the Chinese university students did.

The curriculum structures used in university education in both China and Taiwan emphasize the civil rights and obligations stipulated by the Constitution and the protection of the law. The university education in Taiwan introduced 'beneficial interest', university students have exhibited increased self-interest and have received increased personal rights protections. In contrast, China's university education introduced 'supervisory powers', and the university students have enjoyed limited freedom of speech and increased powers of self-advocacy compared to Post-Mao era (1976–1989).

In promoting the concept of ecotourism, Chinese and Taiwanese university students might differ in terms of what influences the formation of the environmental protection concept or in terms of the factors of educational environment and intention of participation. Nevertheless, the results indicated that the cross-strait university students who participated in this study displayed certain similarities in environmental literacy related to ecotourism. Although students in both China and Taiwan fully affirm the prospects of ecotourism, but Chinese university students expressed a greater desire for ecotourism than did the Taiwanese university students, who have been living in a free society.

6. Conclusions, Recommendations, Limitations and Future Research

6.1. Conclusions

In conclusion, this study has compared university students in China and Taiwan in terms of their ecotourism environmental literacy. The results showed that the knowledge of ecotourism affects behavioral intentions. Specifically, the Chinese university students demonstrated a higher correlation between knowledge of ecotourism and behavioral intentions than the Taiwanese students. Furthermore, Chinese university students considered ecotourism to be a business that should be under governmental control. In terms of behavior, the desire for gaining knowledge for the purpose of self-empowerment was weaker in the Taiwanese university students than in the Chinese students.

6.2. Recommendations

Based on the aforementioned findings, it is recommended that the level of importance attached to 'environmental knowledge' in Taiwan ecotourism education be slightly reduced because knowledge alone cannot always generate positive value judgment. Therefore, environmental education in Taiwan should focus instead on nurturing the training provided for environmental attitude and environment-friendly behaviors, to develop environmentally conscious citizens raised with high-quality environmental literacy. In addition, the participation rates of ecotourism are low, and learning opportunities of ecotourism concepts through relevant courses at the university are limited. We recommend that both regions increase the number of courses that offer greater insights about environmental education on ecotourism. A closer exchange between Chinese and Taiwanese students will allow the mutual communication to alter the knowledge of university students in both places and generate a fusion of ideas and action. As Taiwan gradually receives increased numbers of tourists/sightseers from China, the frequency of exchanges between the two sides will also increase

[69]. Cross-strait exchanges in ecotourism will be promoted, and the political gap between the two sides will be eliminated. This study is intended to provide a new discourse and new ideas for higher education and research on ecological tourism in China and Taiwan and thus improves tourism education and environmental education in both places.

6.3. Limitations and Future Research

This study is only a preliminary exploration of university students' ecotourism and environmental literacy with selected universities in China and Taiwan, and therefore cannot be generalized to other contexts. Another limitation in this study is the absence of interviews with academics who can assist in determining if the course syllabi cover the relevant tourism contents, which may help enhance the research methodology and instrument design. Although the purpose for the provision of a clear definition of the term "ecotourism" to the participants prior to the survey is to ensure a consistent understanding of what "ecotourism" refers to in the questionnaire, but this may potentially affect the results of some questions related to ecotourism.

The relation between behavioral intentions and the observed behavior of people is needed for further studies, to determine if intentions are reflected as actual actions. Further studies from personal observation associated self-reports studies are required to solve this research gap toward sustainable tourism. In addition, future research can explore possible correlation between political systems and university student knowledge and attitudes toward ecotourism, and also the student's area of study to determine if their attitudes towards ecotourism may be affected.

Appendix A

Questions about environmental knowledge, environmental attitude, and behavioral intentions

Reverse scoring using (R) (x, '1 = 5; 2 = 4; 3 = 3; 4 = 2; 5 = 1')

1. **Environmental Knowledge (16 Items)**
    - 1.1. *Responsible Tourism*
        - 1.1.1. Ecotourism should consider wildlife habitats
        - 1.1.2. Ecotourism should pay attention to stakeholders' rights
        - 1.1.3. Ecotourism should conserve natural and human resources toward sustainability
        - 1.1.4. Ecotourism should pay attention to protect local historical images
    - 1.2. *Educational Tourism*
        - 1.2.1. Ecotourism should persuade tourists to actively participate in environmental protection through environmental education
        - 1.2.2. Ecotourism should cooperate with full and complete interpretation content
        - 1.2.3. Ecotourism should be a small-team learning tour and not a bunch of people's consuming tour

1.3. *Participating Tourism*
   1.3.1. Environmental groups should assist in ecological interpretation and environmental protection in ecotourism areas
   1.3.2. Residents of ecotourism areas should have the right to participate in tourism development, decision making, and interpretation design
   1.3.3. The operating income of the ecotourism areas should be returned to the local community

1.4. *Governing Tourism*
   1.4.1. Government units should formulate regulations on ecological protection in ecotourism areas
   1.4.2. Environmentalists should strictly play the role of supervisor and provide advice in ecotourism areas
   1.4.3. Government units should ensure that resources are used correctly in ecotourism areas
   1.4.4. Ecotourism operators should receive professional certification
   1.4.5. Ecotourism should utilize legislation to regulate the use of natural resources
   1.4.6. Ecotourism should delineate more animal protection areas

2. **Environmental Attitude (14 Items)**

2.1. *Values*
   2.1.1. Natural resources are inexhaustible. (R)
   2.1.2. It is feasible to convert all agroforestry to other uses. (R)
   2.1.3. Humans can control other organisms in order to survive. (R)
   2.1.4. Interpretation is very important for ecotourism
   2.1.5. Compared with environmental protection, economic development should be emphasized. (R)
   2.1.6. Ecotourism should be paid more than other travel types

2.2 *Beliefs*
   2.2.1. Ecotourism should emphasize the fairness of animals and plants
   2.2.2. Ecotourism should emphasize the importance and rarity of the environment
   2.2.3. Visitors should learn to respect nature and humanistic environment in ecotourism
   2.2.4. Ecotourism should respect and understood local customs and lifestyles

2.3. *Sentiments*
   2.3.1. China's environmental protection work has done very well so far/Taiwan's environmental protection work has been done very well so far. (R)
   2.3.2. What I thought that the process of ecotourism can have a serious impact on wildlife and their habitats. (R)
   2.3.3. It is an immoral act to destroy or contaminate natural resources in protected areas

2.3.4 Once the natural environment is destroyed or contaminated, it is difficult to recover

3. **Behavioral Intentions (8 Items)**

3.1. *Private Empowerment*
   3.1.1. During travelling, I will carefully read the guidance to the wildlife and plants in details
   3.1.2. During travelling, I will join in listening to natural interpretation from tour guides
   3.1.3. During travelling, I will communicate with local residents to understand the local culture

3.2. *Private Action*
   3.2.1. During travelling, I will endure less convenient public flitiesaci
   3.2.2. During travelling, I would refuse to buy if the stores' salesmen are not eco-friendly (i.e., if they provide disposable tableware in restaurants)
   3.2.3. During travelling, I will bring my own tableware and water bottles

3.3. *Public Service*
   3.3.1. Seeing other tourists littering, I will persuade him
   3.3.2. I will take the initiative to report to the relevant authorities when I detect the environment to be damaged

Author Contributions: W.-T.F. conceived and designed the experiments; W.-T.F., C.-Y.L., and G.H performed the experiments; W.-T.F. analyzed the data; S.-W.L., W.-T.F., Y.-W.H, G.H., G.-S.S, J.-Y.C., E.N. contributed reagents/materials/analysis tools; W.-T.F., C.-Y.L., and E.N. wrote the paper.

Acknowledgments: We thank members of the Graduate Institute of Environmental Education, National Taiwan Normal University (NTNU, Taipei, Taiwan) and the Department of Tourism and Leisure, Chung Hua University for their contributions to the manuscript. This work was supported by grants from the Ministry of Science and Technology under Grant numbers MOST 103-2511-S-237-003 to Wei-Ta Fang & Ching-Yu Lien; MOST-105-2511-S-003-021-MY3 to Wei-Ta Fang. This article was also subsidized by the National Taiwan Normal University (NTNU) and Taiwan Nature Trail Society, Taiwan, ROC. The useful suggestions from anonymous reviewers were incorporated into the manuscript.

Conflicts of Interest: The authors declare no conflicts of interest.

References

1. Wu, K.-S.; Teng, Y.-M. Applying the extended theory of planned behavior to predict the intention of visiting a green hotel. *Afr. J. Bus. Manag.* **2011**, *5*, 7579–7587.
2. Gössling, S.; Peeters, P. Assessing tourism's global environmental impact 1900–2050. *J. Sustain. Tour.* **2015**, *23*, 639–659.
3. Wearing, S.; Neil, J. *Ecotourism: Impacts, Potentials and Possibilities*, 2nd ed.; Butterworth-Heinemann: Oxford, UK, 2009.

4. Butcher, J. The United Nations International Year of Ecotourism: A critical analysis of development implications. *Prog. Dev. Stud.* **2006**, *6*, 146–156.
5. Weaver, D.B.; Lawton, L.J. Twenty years on: The state of contemporary ecotourism research. *Tour. Manag.* **2007**, *28*, 1168–1179.
6. Blangy, S.; Mehta, H. Ecotourism and ecological restoration. *J. Nat. Conserv.* **2006**, *14*, 233–236.
7. Stem, C.J.; Lassoie, J.P.; Lee, D.R.; Deshler, D.D.; Schelhas, J.W. Community participation in ecotourism benefits: The link to conservation practices and perspectives. *Soc. Nat. Resour. Int. J.* **2003**, *16*, 387–413.
8. Russell, C.L. Ecotourism as experiential environmental education? *J. Exp. Educ.* **1994**, *17*, 16–22.
9. Powell, R.B.; Ham, S.H. Can ecotourism interpretation really lead to pro-conservation knowledge, attitudes and behaviour? Evidence from the Galapagos Islands. *J. Sustain. Tour.* **2008**, *16*, 467–489.
10. Hetzer, N.D. Environment, tourism, culture. *LINKS* **1965**. Reprinted in *Ecosphere* **1970**, *1*, 1–3.
11. Honey, M. *Ecotourism and Sustainable Development, Who Owns Paradise?* Island Press: Washington, DC, USA, 2008.
12. Saarinen, J. Traditions of sustainability in tourism studies. *Ann. Tour. Res.* **2006**, *33*, 1121–1140.
13. Bianchi, R. Tourism restructuring and the politics of sustainability: A critical view from the European Periphery (The Canary Islands). *J. Sustain. Tour.* **2004**, *12*, 495–529.
14. Ceballos-Lascurain, H. The future of ecotourism. *Mex. J.* **1987**, *1*, 13–19.
15. Dolnicar, S.; Crouch, G.I.; Long, P. Environment-friendly tourists: What do we really know about them? *J. Sustain. Tour.* **2008**, *16*, 197–210.
16. Xu, F.; Fox, D. Modelling attitudes to nature, tourism and sustainable development in national parks: A survey of visitors in China and the UK. *Tour. Manag.* **2014**, *45*, 142–158.
17. Scheyvens, R. Ecotourism and the empowerment of local communities. *Tour. Manag.* **1999**, *20*, 245–249.
18. Hungerford, H.R.; Volk, T.L. Changing learner behavior through environmental education. *J. Environ. Educ.* **1990**, *21*, 8–21.
19. Roth, C.E. *Environmental Literacy: Its Roots, Evolution and Directions in the 1990s*; ERIC/CSMEE: Columbus, OH, USA, 1992.
20. Teksoz, G.; Sahin, E.; Tekkaya-Oztekin, C. Modeling environmental literacy of university students. *J. Sci. Educ. Technol.* **2012**, *21*, 157–166.
21. Goldman, D.; Pe'er, S.; Yavetz, B. Environmental literacy of youth movement members—Is environmentalism a component of their social activism? *Environ. Educ. Res.* **2017**, *23*, 486–514.
22. McBride, B.B.; Brewer, C.A.; Berkowitz, A.R.; Borrie, W.T. Environmental literacy, ecological literacy, ecoliteracy: What do we mean and how did we get here? *Ecosphere* **2013**, *4*, 1–20.

23. Gupta, G.; Chopra, P. Eco-tourists and environment protection: A pro-environment behavioural segmentation approach. *Amity Glob. Bus. Rev.* **2014**, *9*, 90–95.
24. Wurzinger, S.; Johansson, M. Environmental concern and knowledge of eco-tourism among three groups of Swedish tourists. *J. Travel Res.* **2006**, *45*, 217–226.
25. Tsaur, S.H.; Lin, Y.C.; Lin, J.H. Evaluating ecotourism sustainability from the integrated perspective of resource, community and tourism. *Tour. Manag.* **2006**, *27*, 640–653.
26. Juvan, E.; Dolnicar, S. The attitude-behaviour gap in sustainable tourism. *Ann. Tour. Res.* **2014**, *48*, 76–95.
27. Bergin-Seers, S.; Mair, J. Emerging green tourists in Australia: Their behaviours and attitudes. *Tour. Hosp. Res.* **2009**, *9*, 109–119.
28. Jackson, T. Motivating sustainable consumption. *Sustain. Dev. Res. Netw.* **2005**, *29*, 30.
29. Tosun, C. Limits to community participation in the tourism development process in developing countries. *Tour. Manag.* **2000**, *21*, 613–633.
30. Lai, P.H.; Nepal, S.K. Local perspectives of ecotourism development in Tawushan Nature Reserve, Taiwan. *Tour. Manag.* **2005**, *27*, 1117–1129.
31. Medina, L.K. Ecotourism and certification: Confronting the principles and pragmatics of socially responsible tourism. *J. Sustain. Tour.* **2005**, *13*, 281–295.
32. Ohe, Y. Evaluating operators' attitudes to educational tourism in dairy farms: The case of Japan. *Tour. Econ.* **2012**, *18*, 577–595.
33. Samah, A.A.; Ahmadian, M. Educational tourism in Malaysia: Implications for community development practice. *Asian Soc. Sci.* **2013**, *9*, 17.
34. Maikhuri, R.K.; Rana, U.; Rao, K.S.; Nautiyal, S.; Saxena, K.G. Promoting ecotourism in the buffer zone areas of Nanda Devi Biosphere Reserve: An option to resolve people—Policy conflict. *Int. J. Sustain. Dev. World Ecol.* **2000**, *7*, 333–342.
35. Sofield, T.; Li, S. Tourism governance and sustainable national development in China: A macro-level synthesis. *J. Sustain. Tour.* **2011**, *19*, 501–534.
36. Bamberg, S. How does environmental concern influence specific environmentally related behaviours? A new answer to an old question. *J. Environ. Psychol.* **2003**, *23*, 21–32.
37. Grob, A. A structural model of environmental attitudes and behaviour. *J. Environ. Psychol.* **1995**, *15*, 209–220.
38. McKechnie, G.E. Measuring environmental dispositions with the environmental response inventory. In Proceedings of the 2nd Annual Environmental Design Research Association Conference, Pittsburgh, PA, USA, October 1970; pp. 320–326.
39. Hines, J.M.; Hungerford, H.R.; Tomera, A.N. Analysis and synthesis of research on responsible environmental behaviour: A meta-analysis. *J. Environ. Educ.* **1987**, *18*, 1–8.
40. Ajzen, I. The theory of planned behaviour. *Organ. Behav. Hum. Decis. Process.* **1991**, *50*, 179–211.

41. Chen, M.F.; Tung, P.J. Developing an extended theory of planned behavior model to predict consumers' intention to visit green hotels. *Int. J. Hosp. Manag.* **2014**, *36*, 221–230.
42. Cheng, T.M.; Wu, H.C. How do environmental knowledge, environmental sensitivity, and place attachment affect environmentally responsible behavior? An integrated approach for sustainable island tourism. *J. Sustain. Tour.* **2015**, *23*, 557–576.
43. Hungerford, H.R.; Peyton, R.B. A Paradigm of Environmental Action. ERIC Documentation Service ED137116. 1977. Available online: https://eric.ed.gov/?id=ED137116 (accessed on April 10, 2018).
44. Lee, K. The role of media exposure, social exposure and biospheric value orientation in the environmental attitude-intention-behavior model in adolescents. *J. Environ. Psychol.* **2011**, *31*, 301–308.
45. Zsóka, Á.; Szerényi, Z.M.; Széchy, A.; Kocsis, T. Greening due to environmental education? Environmental knowledge, attitudes, consumer behavior and everyday pro-environmental activities of Hungarian high school and university students. *J. Clean. Prod.* **2013**, *48*, 126–138.
46. Kaiser, F.G.; Wölfing, S.; Fuhrer, U. Environmental attitude and ecological behaviour. *J. Environ. Psychol.* **1999**, *19*, 1–19.
47. Pooley, J.A.; O'Connor, M. Environmental education and attitudes emotions and beliefs are what is needed. *Environ. Behav.* **2000**, *32*, 711–723.
48. Kruglanski, A.W. *Lay Epistemics and Human Knowledge: Cognitive and Motivational Bases*; Plenum: Newark, UK, 1989.
49. De Rosis, F.; Castelfranchi, C.; Goldie, P.; Carofiglio, V. Cognitive Evaluations and Intuitive Appraisals: Can Emotion Models Handle Them Both? In *Emotion-Oriented Systems*; Springer: Berlin/Heidelberg, Germany, 2011; pp. 459–481.
50. Best, H.; Mayerl, J. Values, beliefs, attitudes: An empirical study on the structure of environmental concern and recycling participation. *Soc. Sci. Q.* **2013**, *94*, 691–714.
51. Freed, A. The relationship between university students' environmental identity, decision-making process, and behavior. *Environ. Educ. Res.* **2018**, *24*, 474–475.
52. Neisser, U. *Cognitive Psychology*; Prentice-Hall: Englewood Cliffs, NJ, USA, 1967.
53. Revlin, R. *Cognition: Theory and Practice*; Worth Publishers: New York, NY, USA, 2013.
54. Hollweg, K.S.; Taylor, J.R.; Bybee, R.W.; Marcinkowski, T.J.; McBeth, W.C.; Zoido, P. *Developing a Framework for Assessing Environmental Literacy*; North American Association for Environmental Education: Washington, DC, USA, 2011. Available online: http://www.naaee.net (accessed on April 10, 2018).
55. Tuncer, G.; Tekkaya, C.; Sungur, S.; Cakiroglu, J.; Ertepinar, H.; Kaplowitz, M. Assessing pre-service teachers' environmental literacy in Turkey as a mean to develop teacher education programs. *Int. J. Educ. Dev.* **2009**, *29*, 426–436.
56. Erdogan, M.; Ok, A. An assessment of Turkish young pupils' environmental literacy: A nationwide survey. *Int. J. Sci. Educ.* **2011**, *33*, 2375–2406.

57. McBeth, W.; Volk, T.L. The national environmental literacy project: A baseline study of middle grade students in the United States. *J. Environ. Educ.* **2009**, *41*, 55–67.
58. McKeown-Ice, R. Environmental education in the United States: A survey of preservice teacher education programs. *J. Environ. Educ.* **2000**, *32*, 4–11.
59. Walter, P.G. Theorising visitor learning in ecotourism. *J. Ecotour.* **2013**, *12*, 15–32.
60. Kline, R.B. *Principles and Practice of Structural Equation Modelling*, 4th ed.; Guilford: New York, NY, USA, 2016.
61. Bentler, P.M. *EQS 6 Structural Equations Program Manual*; Multivariate Software: Encino, CA, USA, 2006.
62. Hu, L.; Bentler, P.M. Cutoff criteria for fit indexes in covariance structure analysis: Conventional criteria versus new alternatives. *Struct. Equ. Model.* **1999**, *6*, 1–55.
63. McDonald, R.P.; Ho, M.H.R. Principles and practice in reporting statistical equation analyses. *Psychol. Methods* **2002**, *7*, 64–82.
64. Ding, L.; Velicer, W.F.; Harlow, L.L. Effects of estimation methods, number of indicators per factor, and improper solutions on structural equation modeling fit indices. *Struct. Equ. Model. Multidiscip. J.* **1995**, *2*, 119–143.
65. Qian, J.; Shen, H.; Law, R. Research in sustainable Tourism: A longitudinal study of articles between 2008 and 2017. *Sustainability* **2018**, *10*, 590, https://doi.org/10.3390/su1003059060.
66. Bremmer, I. State capitalism comes of age: The end of the free market? *Foreign Aff.* **2009**, *88*, 40–55.
67. Shi, L.; Wang, G. A survey on environmental awareness and behaviours of college-student tourists in China. *Environ. Sci. Technol.* **2012**, *35*, 193–196.
68. Fishbein, M. Attitude and the prediction of behaviour. In *Readings in Attitude Theory and Measurement*; Fishbein, M., Ed.; John Wiley & Sons: New York, NY, USA, 1967; pp. 477–492.
69. Mainland China, Visitors Arrivals 2012–2017. Available online: http://stat.taiwan.net.tw/system/country_years_arrival_result.asp?101&106&1&2&%E4%B8%AD%E5%9C%8B%E5%A4%A7%E9%99%B8%3Cbr%3E(Mainland%20China) (accessed on 31 May 2018).

Index

W.-T. Fang, *Tourism in Emerging Economies*,
https://doi.org/10.1007/978-981-15-2463-9

T

U

V

The manufacturer's authorised representative in the EU is Springer Nature Customer Service Centre GmbH, Europaplatz 3, 69115 Heidelberg, Germany. If you have any concerns regarding our products, please contact ProductSafety@springernature.com

Printed and bound by CPI Group (UK) Ltd, Croydon, CR0 4YY
30/09/2026
02225556-0010